ELEMENTARY APPLIED STATISTICS

Marvin Lentner
New Mexico State University

Bogden & Quigley, Inc., Publishers
Tarrytown-on-Hudson, N.Y. / Belmont, California

Text and cover design by William Bennett and Maura Fitzgerald

Library of Congress Catalog Card No.: 78-182360

ISBN: 0-8005-0022-9

Printed in the United States of America

1 2 3 4 5 6 7 8 9 10—76 75 74 73 72

ELEMENTARY APPLIED STATISTICS

GLOSSARY OF SYMBOLS

Symbol	Page of First Occurrence	Meaning
$n!$	30	n factorial
p	123	proportion
A	18	an event (B, C, D, E, etc. are also used)
$\bar{A}$	20	complement of event A
E	83	expected value operator
CI	157	confidence interval
H_0, H_1	172	hypotheses
L, U	157	endpoints of confidence interval
N	6	number of elements in a finite set
R	100	sample range
S^2	102	sample variance
S_e^2	225	regression variance
S	103	sample standard deviation
$S_{\bar{X}}$	106	sample standard error of the mean
SS_X	101	sum of squares for X
T_v	139	Student T variable
T^*	163	pseudo T variable
X	76	random variable
x	76	value of random variable X
$x_{(i)}$	90	i-th order statistic
$x.$	99	total of X values
$\bar{X}$	95	sample mean of X
$\tilde{X}$	96	sample median of X
$\hat{Y}$	222	least-squares equation
Z	135	standard normal variable
F_{v_1, v_2}	141	F variable
$f(x)$	7, 82	function of x; equation describing the distribution of random variable X
$f(x, y)$	85	equation describing the bivariate distribution of random variables X and Y
$F(x)$	82	cumulative distribution of random variable X
${}_nP_r$	31	permutations of n elements taken r at a time
$\binom{n}{r}$	32	combinations of n elements taken r at a time
$P(A)$	42	probability of event A
$P(B\|A)$	46	conditional probability of event B given event A
$\mathcal{S}$	18	sample space
$\mathcal{E}$	16	experiment
$\mathcal{P}$	9	population

To Helen, Scott, Christine, and Monte

PREFACE

This book was written for a first course in statistical methods for persons with minimum mathematical backgrounds. Familiarity with algebra, either a course in college algebra or mastery of high school algebra, is adequate preparation. Calculus is neither required nor used; only a few references to the calculus occur in the entire book.

The material is not slanted toward any discipline. Problems are chosen from a wide range of applications and many are based on actual research projects.

Statisticians have differing views concerning the amount of probability which should be taught in a first course in statistical methods: they vary from those who believe that students need only to understand the fundamentals of probability to those who believe that students should encounter a rigorous treatment of the subject. The part of this book dealing with probability was developed with this range of views in mind. Accordingly, the first part of Chapter 4 has been written for those who wish a brief study; the second part is for those who wish a more thorough coverage, including the calculation of probability.

Tables and graphs are used extensively to solve problems, in particular for setting confidence intervals and power calculations. All statistical tables that appear in the appendices are given in cumulative form to facilitate reading.

Because extra material has been included to enhance the book's value as a reference, instructors will find that *all* the material probably cannot be covered in a 3-hour quarter or semester course. The book is adaptable, however, to one-quarter, one-semester, or two-quarter courses, depending upon deletions or the use of supplements. Starred sections and problems, dealing with specialized or advanced topics, may be omitted without disrupting the continuity of the remaining material.

For a 4-hour quarter or a 3-hour semester course covering a minimum amount of probability, the following outline is suggested:

Chapters 1, 2, 3
Chapter 4, omitting Section 4.2
Chapters 5, 6
Chapter 7, omitting Sections 7.8.7, 7.8.8, 7.10, 7.11
Chapter 8, omitting Section 8.5
Chapters 9 through 13, omitting the starred sections

Sections 14.2.1, 14.2.2, 14.2.3, 14.3.1, 14.4
Chapter 15, omitting the starred sections

For a course covering a maximum amount of probability and less statistics, the following outline is suggested:

Chapters 1, 2, 3, 4, 6, 7, 8, 9, 10, in their entirety

For a course covering a moderate amount of probability, the following outline is suggested:

Chapters 1, 2, 3, 4, 6
Chapter 7, omitting Sections 7.8.7, 7.8.8, 7.10, 7.11
Chapter 8, omitting Section 8.5
Chapters 9 through 13, omitting the starred sections
Sections 14.2.1, 14.2.2, 14.2.3, 14.3.1, 14.4

Numerous persons contributed directly or indirectly to the development of this book. I would like to express my appreciation for the many helpful suggestions and comments from students, colleagues, and friends. Particular thanks are due to Professors Thomas J. Boardman, Colorado State University, Mary Lou Harkness, Pennsylvania State University, and S. K. Katti, University of Missouri, for their detailed reviews of an earlier version; to the staff of Bogden & Quigley for their editorial and other assistance; to Richard Glaze and William Bovermann for their computational assistance; to Professor Morris Finkner for his comments and encouragement and for providing a setting conducive to writing; and to Mrs. Dorothy A. Bassett for her typing of the final draft. Special thanks are due to Professor N. Scott Urquhart for his painstaking review of the final draft, for his constructive criticism, and for his willingness to listen when I imposed upon his time.

For permissions to reproduce material, I am indebted to Professor F. N. David, to Professor E. S. Pearson for the Biometrika Trustees, to the Addison-Wesley Publishing Company, to the McGraw-Hill Book Company, and to the RAND Corporation.

Finally, I would like to express special thanks to my family for their patience and understanding and for the sacrifices they made during this writing.

Las Cruces, New Mexico *Marvin Lentner*

TO THE READER

Each section is identified by numbers separated from the chapter number by a period. Thus, Section 3.6 means the sixth section of Chapter 3. Subdivisions of a section (subsections) are identified by the inclusion of a third number; thus, the number 14.5.4 means the fourth subdivision of the general Section 14.5.

Equations, formulas, and other displayed material are identified by two numbers appearing at the left margin, in parentheses, separated by a period. Thus, (11.24) refers to the twenty-fourth identified item of Chapter 11.

Each example begins and ends with a wedge, ▶.

References are given in alphabetical order by author at the end of each chapter and numbered in square brackets. Thus, [2] appearing in the text refers to the second reference at the end of the chapter in which it appears.

Problems appear at the end of each chapter, except the first. At the end of many sections, a few related problems are specified in parentheses; the reader may want to attempt some of these before continuing on. Answers to odd-numbered problems are given at the back of the book.

CONTENTS

ELEMENTARY APPLIED STATISTICS

Chapter 1

INTRODUCTION

Statistical procedures serve as an effective vehicle for interpreting and communicating research findings. Properly applied, statistics allow the researcher to make inferences in the face of uncertainty, an element present in virtually all research. For this reason, statistics are gradually becoming a universal language of research.

Statistical procedures might be used in several phases of a project: (1) to implement a research program, (2) for systematic evaluation of research results, or (3) as a means of deciding on a particular avenue of further work. Although each of these phases, and possibly others, could be pursued in great detail, this book will place primary emphasis on the second—analysis. Even restricting attention to this phase, we cannot begin to provide an extensive coverage of statistical procedures. Our aim is to give an exposition of the language and elementary procedures used by statisticians.

1.1 DIFFERENT MEANINGS OF STATISTICS

The term *statistics* is frequently applied to the reporting of recorded data, either in original or summary form. We hear of half-time statistics of a football game, baseball statistics, agricultural statistics, enrollment statistics, and so on. In other words, any set of reported data might be referred to as statistics.

In other contexts, *statistics* denotes a discipline, the art and the science of making inference from the particular to the general. It is an art to design an effective research project and to decide appropriate assumptions upon which a proposed analysis rests; it is a science to draw sound conclusions from results obtained within the framework of proper assumptions. Because we are unable to observe most phenomena in their entirety, our alternative is to observe a portion of the general situation and, from this, to make inferences about the entire phenomenon. This book is concerned with the scientific approach to this inductive process.

Finally, *statistics* is used to denote quantities calculated from gathered data. This can be viewed as a special summary of observed information. An entire chapter, Chapter 7, is devoted to this topic.

1.2 A BRIEF HISTORY OF STATISTICS

Statistics started thousands of years ago when man began counting. Collection and the recording of data have been common for centuries. In 2000 B.C. the population of Judea was estimated to be 3,800,000. Military leaders, such as Caesar Augustus, ordered censuses of the world. For military and taxation purposes, William the Conqueror of England ordered a census, and the results were published in the *Domesday Book* of 1086 A.D. In the development of statistics, the period before 1500 A.D. was primarily devoted to simple data recording.

From 1500 A.D. until 1800 A.D. more sophisticated types of analysis emerged. When gambling among the nobility became one of the pastimes of the day, mathematicians and logicians were called upon to work out the odds for card games, dice games, and the like. Such men as DeMere, Fermat, and Pascal were great contributors to this probability. During the same era the analysis of measurement errors and the study of distributions was begun by Laplace (1749–1827), DeMoivre (1667–1754), and Gauss (1777–1855).

Wider emphasis on applications and statistical education became rather common in the 1800's and early 1900's. The number of statisticians grew, and Quetelet (1796–1874), a Belgian astronomer and statistician, set up the first statistical congress in 1853. Some of the first applications of statistics were made by astronomers in their attempt to fix planetal orbits. Galton (1822–1911) postulated a "law of universal regression" in the study of inheritance (1888); this theory was subjected to a statistical challenge by Karl Pearson (1857–1936), who is now known as the "founder of the science of statistics."

Statistics as we know the subject today is considered to have begun in the work of Sir Ronald A. Fisher (1890–1962), a mathematical geneticist at the Rothamsted Experimental Station in England. Known as the "father of modern statistics," Fisher published numerous articles and was actively engaged in statistical research until his death. Other great contributors to modern statistics have been W. S. Gossett (1876–1937), a student of Karl Pearson, Abraham Wald (1902–1950), Jerzey Neyman (1894–   ), and E. S. Pearson (1895–   ), the son of Karl Pearson.

A detailed history of statistics and probability for the period 1500 A.D.

until 1800 A.D. is given in the excellent book by Todhunter [6].† Other aspects of the history are covered in [4] and [7].

Since 1925, statistics have been widely applied in virtually every area of scientific endeavor. Agriculture was one of the first areas in which statistical methods were extensively used. As is frequently the case, known methods are modified to handle new situations. Today researchers use techniques of probability and statistics to solve problems in agriculture, biology, social sciences, engineering, transportation, wildlife preservation, economics, history, and various environmental issues—indeed every subject you can name. Perhaps the most familiar use of probability and statistics is in determining the figures you hear reported as "precipitation probabilities," which utilize rainfall data compiled for many years. Among the extensive rainfall studies are projects [1], [2], and [3]. Statistical techniques have also been applied recently to a classical problem in history by Mosteller and Wallace [5], who discuss the various analyses used in trying to ascertain whether Hamilton or Madison wrote the twelve Federalist papers whose authorship is uncertain.

1.3 THE ROLE OF COMPUTERS IN STATISTICAL ANALYSIS

Without today's modern, high-speed computers, many projects simply would not be feasible. Consider the computations involved in summarizing rainfall data recorded daily for 30 years in 100 cities, or during the launching and tracking of a satellite, where speed is also essential. Because computers can perform elaborate analyses, they are used for summarizing and plotting data, for tabulating complex functions, for graphing functions, for simulation studies, and so on. Most of the tables and graphs found in Appendixes A and B are products of computers.

But, for economic reasons as well as input–output problems, computers are best utilized for large-scale problems. To square and sum ten thousand numbers takes a computer but a few milliseconds longer than the time required by the same computer to square and sum one hundred numbers. This is one reason why computer usage is not emphasized in this book; all the problems and examples can be readily solved with a desk calculator, the forerunner of the computer.

Statistics and computing are basically different disciplines, each of which may be studied without the other. But since computers are so useful in statistics for performing difficult and/or extensive calculations, we shall occasionally mention problems where computers have been used, or might be used, in statistical analyses.

† This book is now out of print, but most university libraries have a copy or can get one on interlibrary loan.

REFERENCES

[1] Feyerherm, A. M., and L. D. Bark, "Probabilities of Sequences of Wet and Dry Days in Kansas," Technical Bulletin 139a, Kansas State University, Manhattan, Kansas, 1965.

[2] Gifford, R. O., G. L. Ashcroft, and M. D. Magnuson, "Probability of Selected Precipitation Amounts in the Western Region of the United States," Western Regional Research Publication, Bulletin T-8, Agricultural Experiment Station, University of Nevada, Reno, Nevada, 1967.

[3] Heerman, D. F., M. D. Finkner, and E. A. Hiler, "Probability of Sequences of Wet and Dry Days for Eleven Western States and Texas," Western Regional Research Publication, Bulletin TB 117, Colorado State University Experiment Station, Fort Collins, Colo., 1971.

[4] Kendall, M. G., "Where Shall the History of Statistics Begin?" *Biometrika*, 1960. (This is Paper X in a series entitled *Studies in the History of Probability and Statistics* published since 1955 in *Biometrika.*)

[5] Mosteller, F., and D. L. Wallace, *Inference and Disputed Authorship: The Federalist*, Addison-Wesley, Reading, Mass., 1964.

[6] Todhunter, I., *A History of the Mathematical Theory of Probability*, Chelsea Publishing Company, New York, 1949 (first published Cambridge, Eng., 1865).

[7] Walker, H. M., *Studies in the History of Statistical Method*, Williams and Wilkins, Baltimore, Md., 1929.

Chapter 2

FUNDAMENTALS OF THE STATISTICAL LANGUAGE

In statistics, as in most disciplines, a special language is created to effect precise communication among its users. Undefined terms—or at least terms so "primitive" that they defy definition—are used in all specialized disciplines. Statistics is no exception, and such terms form the basis of its particular language.

Statistics depends heavily on mathematics, and therefore a portion of the language used by mathematicians is also used by statisticians. Some mathematical language useful in the study of statistics is presented in the next section, and basic principles of the statistical language are discussed in the remainder of the chapter. It is important to realize that most of the terms in our language have precise meanings, although these same terms may have quite different meanings in everyday usage or in other disciplines.

Additional terminology is introduced and discussed as needed in later chapters.

2.1 SOME MATHEMATICAL CONCEPTS

Two primitive terms in mathematics are *point* and *line*. A portion of a line is an *interval*. An *open interval* is the portion of a line between two distinct points; all points between 1 and 3 constitute the open interval (1, 3). If the endpoints of an interval are included, the result is a *closed interval*; including the points 1 and 3 gives the closed interval [1, 3].

2.1.1 Sets

The notion of a *set* is basic to many discussions in statistics. Synonyms for set are "group" or "collection": an aggregate of entities identified by some rule. Each entity belonging to a particular set is said to be an element of that set.

A "set of tableware for eight persons" is a collection of 8 dinner knives, 8 dinner forks, 8 soup spoons, 8 salad forks, 16 teaspoons (each person gets 2), 1 butter knife, and 1 sugar shell. A second collection of 8 dinner knives, 8 dinner forks, and so on, would be a "second set of tableware for eight persons." We have painstakingly listed all elements to avoid ambiguity. And this detailed description makes it easy to decide whether any item does or does not belong to a given set of tableware. An apple could not be an element of a tableware set but a soup spoon could be, provided the set does not already contain 8.

Two sets are *equal* if they contain identical elements. Thus, two sets of tableware are *not* equal if the elements of the two sets differ in number, in identification (such as pattern, metallic structure, or type of items), or in both number and identification.

A portion of a set is called a *subset of that set*. Therefore, "the spoons" is a subset of a set of tableware. Likewise, "the teaspoons" is a subset of "the spoons," as well as being a subset of the full set of tableware.

Sets and subsets are frequently represented by enclosing either their elements or a word description in curly braces. The "set of digits" might be represented in either of the following two ways:

$$\{\text{the digits}\} = \{0, 1, 2, 3, 4, 5, 6, 7, 8, 9\}$$

One subset of {the digits} would be

$$\{\text{factors of } 9\} = \{1, 3, 9\}$$

Sets may be classified according to the number of elements they contain. This set classification has three categories: (1) finite, (2) countably infinite, and (3) uncountably infinite.

A *finite* set contains a fixed number of elements, say N. Even though N may be very large, it must be a positive integer. A "set of tableware for eight persons" is a finite set containing between 24 and 70 items (usually 50). And, although quite large, the set of all cars registered in the United States is finite.

Whereas all elements of a finite set could eventually be recorded or labeled, those of an *infinite* set cannot. An infinite set contains more elements than every positive integer, N. Some infinite sets are

S_1 = {all integers},
S_2 = {all real numbers},
S_3 = {all possible temperatures that could be recorded at O'Hare International Airport in Chicago}, and
S_4 = {different angles through which a telescope might be elevated}.

Any fixed set whose elements can be placed in one-to-one correspondence with the positive integers is said to be *countable* and to have a countable infinity of elements. Sets S_1 and S_3 are countably infinite. On the other hand, any infinite set whose elements cannot be placed in a one-to-one correspondence with the positive integers is said to be *uncountable* and to have an uncountable infinity of elements. Sets S_2 and S_4 each have an uncountable infinity of elements. The number of elements in an uncountable set is at a *level of magnitude* higher than that of a countable set. In fact, uncountable sets have subsets that are countable; for example, S_1 is a subset of S_2.

(Problems 1, 3)

2.1.2 Functions

Function is another concept of considerable importance. Three items—two sets and a special rule relating the sets—are necessary for a function.

Definition 2-1

A function is a rule that associates the elements of two sets—*domain* and *range*—such that every element in the domain corresponds to a unique element in the range.

Note that any element in the range may correspond to more than one element in the domain; a duplicate correspondence like this is given in Example 2-1.

All functions used in this book are "real-valued functions"—that is, the range is always a set of real numbers. It is not necessary for the domain to be a set of real numbers; a nonnumerical domain is given in Example 2-2.

► ***Example 2-1***

The elements of the sets

$$D = \{-1, 0, 1, 2\} \text{ and } R_1 = \{0, 1, 4\}$$

together with the equation

$$f(x) = x^2, \text{ any } x \text{ from } D$$

define a function having domain D and range R_1. Under the correspondence denoted by $f(x)$, each x in D is associated with a single element in R_1. However, the element 1 in R_1 is associated with two elements in D, namely, -1 and 1. ◄

The quantity x in Example 2-1 is called a *variable*; it varies or takes on different values (namely, the elements of D). In the same example, $y = x^2$ is a variable that takes on the different values of R.

► ***Example 2-2***

Consider the set of eight elements suggested by the tossing of three coins (H = heads, T = tails):

$$S = \{\text{HHH, HHT, HTH, THH, HTT, THT, TTH, TTT}\}$$

The rule

$$X = \text{number of H's in an element of } S$$

corresponds the set S to the set

$$R_2 = \{0, 1, 2, 3\}$$

Consequently, the three items—X, S, and R_2—form a function. ◄

(Problems 5, 9)

2.2 *STATISTICAL PRELIMINARIES*

The three main phases of any statistical analysis are: (1) collecting data, (2) extracting information from these data, and (3) making inferences, interpreting the results. *Inference* is a collective term which includes statements, explanations, comparisons, projections, and the like.

The source of data has a crucial influence on the type and quality of inference. If data are collected from only one state of the fifty forming the United States, it makes little sense to use these data to predict national sentiment. Inferences should therefore be restricted to the set that provides the data. In most cases, the observed data form a subset of a larger set of data, a *population*.

Definition 2-2

A *population* is a set of elements, some or all of which can be observed.

A population may be: (1) a set of objects, such as all photographs of Mars; (2) a set of classification values, such as auto names; (3) a set of real numbers, such as weights of people; or (4) a set of envisioned elements, such as all problems that could be solved by a newly installed computer.

► ***Example 2-3***

Some populations are

$\mathscr{P}_1$ = {the credit hours, say x, taken by each student currently enrolled at your school},
$\mathscr{P}_2$ = {the age, say y, of each U.S. citizen},
$\mathscr{P}_3$ = {the number of heads, say z, showing when ten pennies are tossed indefinitely on a table},
$\mathscr{P}_4$ = {the brand name of each car registered at 8 A.M. today in Iowa}, and
$\mathscr{P}_5$ = {the nationality of each person employed by the U.S. Forest Service}.

The first two populations consist of real numbers, $\mathscr{P}_3$ is an envisioned population, and the last two populations consist of classification values. Population $\mathscr{P}_3$ is countably infinite in size; the rest are finite. ◀

Definition 2-3

Any subset of a population is called a *sample* of that population.

The credit hours of each sophomore currently enrolled at your school would be one sample from $\mathscr{P}_1$. The ages of the fifty state governors of the United States constitute one sample from $\mathscr{P}_2$; so do the ages of any fifty persons who were born in Denver, Colorado.

Sampling may be done (1) *with replacement*, where each element is returned to the population before the next element is selected, or (2) *without replacement*, where an element is *not* returned to the population before the next is selected. Sampling is more often done without replacement, since there is usually no desire to observe the same elements over and over again.

The sample contains pertinent information about the population. Our intuition warns us not to expect the same information about the population from every sample. Of all the possible samples imaginable, a few might perfectly reflect all features of a population; however, most samples miss the target by varying degrees. Therefore, inference based on any one sample may be either erroneous or misleading; in any event, it is subject to uncertainty.

(Problem 13)

2.2.1 Why Sample?

Since any sample might give erroneous or misleading inferences, why not observe the entire population? Sampling is actually necessary for various reasons. Following are some of the more important and obvious ones.

First of all, the limitations of time and money make it unrealistic to observe all the elements of many populations. To obtain one measurement from a complex chemical process might cost thousands of dollars and take many weeks. Likewise, it would take considerable time and money to contact all people living in the United States.

For some populations, it is simply absurd to consider observing all of the elements. This would be the case in the manufacture of photo flash bulbs; if all were tested, there would be none left to sell. Not only is sampling necessary in cases such as this, but it must be done without replacement.

Finally, and obviously, infinite populations cannot be observed in their entirety. The population of "all official high temperatures at Albuquerque, New Mexico" is envisioned as well as infinite, and hence cannot be completely observed.

2.2.2 Statistical Inferences

A primary objective of research is making inferences about populations. An economist ascertains the income for a sample of sugar beet growers and uses this information to make inferences about the average income of all sugar beet growers. A biologist observing the growth rate of bacteria cultured at different humidities hopes to find out whether different humidities give different growth rates.

Inferences like these, made on the basis of sample information, are statistical in nature. They rest on statistical approaches.

Definition 2-4

Conclusions about populations—when drawn from sample information—are called *statistical inferences.*

2.2.3 Randomness

What does an economist mean if he says, "The sugar beet growers that I contacted were randomly selected from all sugar beet growers"? Or, what does any research worker mean by saying, "I have obtained a random sample from the population"?

Much can be learned about randomness from the simplest of all sampling—selecting a single element from a finite population. After we discuss this simple case, we will consider more general sampling.

Definition 2-5

A single element is *randomly* drawn from a finite population if the element is selected by any procedure that gives every population element the same chance of being selected.

No population element is favored by a random-selection procedure. A randomly selected element is not influenced by personal preferences, biases, systematic variations, and so on.

A simple example of random sampling is the following. If a prize is to be given away by a raffle, the personnel conducting the raffle should take certain precautions to guard against awarding the prize unfairly. All eligible tickets should be placed in a mixing receptacle. The person (frequently a small child) who is to draw the winning ticket should be blindfolded and, after the tickets have been thoroughly mixed, one ticket is selected. A winning ticket selected in this manner is randomly drawn from the set of eligible tickets.

When sampling from finite population, Definition 2-5 is easily modified to cover samples of a general size, say n. The sampling may be done with or without replacement.

Definition 2-6

A sample of n elements is *randomly* drawn from a finite population if the sample is selected by any procedure that gives every sample of n population elements the same chance of being selected.

Ten sugar beet growers, selected from a population of 200 sugar beet growers, represent a random sample if every sample of 10 growers was given an equal opportunity of being selected.

When sampling from an infinite population, a sample is *randomly* drawn if every population element has an *equal* and *independent* chance of being selected. "Independence" is used here to mean that the selection of any elements has no influence on the chances of selecting any other elements.

2.2.4 The Role of Randomness

As we have indicated, there is always a possibility of drawing erroneous inferences from any sample. Only with random sampling can we *validly* apply the rules of probability to calculate the chances of such errors. Knowing the error rates is essential if a meaningful "degree of confidence" is to be attached to statistical inference.

If a lettuce grower weighs twenty heads of lettuce and finds their average weight is 33 ounces, is this value a "good" estimate of the average weight of all heads of lettuce? Or can he be more confident that the average weight of all heads of lettuce is between 28 and 38 ounces? How confident? Answers to such questions are given later as part of our study of statistical inference.

A lettuce grower hopes that the twenty selected heads of lettuce are perfectly "representative" of his entire crop. (Achieving this ideal situation would allow him to make error-free statements about his crop.) How might he obtain twenty representative heads of lettuce? He shouldn't select the twenty heads he "thinks" are representative: if he knows enough about his crop to say which twenty heads are representative, he certainly doesn't need sampling and statistics.

This is true in general: we would like to obtain the most representative sample of the population. Not knowing which population elements might be representative forces us to rely on randomization. Remember that randomization does not guarantee a representative sample. Instead, randomization gives the same chance of selection to every sample—the nonrepresentative ones as well as the representative ones.

2.2.5 Achieving Randomness

In the raffle mentioned earlier, randomness is achieved by mixing the tickets and blindfolding the drawer. These actions are done to eliminate circumstances that might otherwise favor the selection of certain tickets—for example, the last ones placed in the ticket container.

Techniques used to achieve randomness include mixing, shuffling, shaking, blindfolding, using random-numbers tables, using computers to generate random identification numbers, and so on. Several of these methods are discussed in the next section.

If research consists of many operations, it may be necessary to apply randomization techniques at many stages. The same technique may be applied over and over—mixing after each element is drawn, for example. Or, different techniques may be used—mixing and blindfolding, for example.

Applying randomization to every research operation may be time-consuming. But the consequences of omitting any randomization should be carefully considered. Randomization is like insurance: you should have it even though you're not certain you need it. When you need it, you'd better have it; if you don't need it, it doesn't matter that you have it. The rule is, When in doubt, obtain it—the insurance is worth the cost.

2.2.6 Drawing Random Samples

The first step in drawing a random sample from any population is to give each element a unique identification. This does not mean that all elements and their identifying labels must be listed, but the correspondence between any element and its identification must be known. Then the population elements designated for observation by the randomization scheme can be readily withdrawn and observed.

The elements of a small population may be numbered, say, on slips of paper. The numbered slips can be placed in a hat or some other receptacle, thoroughly mixed, and without looking, someone can withdraw a set of

n slips whose numbers tell which population elements comprise the sample. This is perhaps the simplest way of drawing a random sample.

Another method is to generate a set of random numbers that will identify the population elements to be observed. Such a set of random numbers may be generated in a variety of ways. For example, a spinner might be divided into ten equal sectors upon which the digits 0 through 9 are placed. Each spin gives one digit of a random number. Suppose a random sample of size 50 is to be obtained from a population of 9500 elements. Fifty random numbers of four digits each are needed; four spins are needed for each four-digit number. Numbers greater than 9500 must be eliminated, and for sampling without replacement, duplications also must be eliminated.

Several extensive tables of random numbers are the ones listed in [1], [2], and [3]. A portion of the RAND Tables appears in Appendix A. To obtain a random starting point in the tables, drop a pencil or other pointed object on any page. Then read off the digits nearest the pencil point to determine page and column numbers† for your starting point. Any path or sequence of numbers emanating from the starting point gives a set of random numbers. As before, duplications and numbers too large must be eliminated.

To illustrate the use of Table A.1, let us discuss drawing a random sample of size 15 from a population of 450 elements. The rows and columns of Table A.1 are numbered from 00 through 99, so we need four digits to designate the starting point. If a pencil is dropped on page 3 (an arbitrary choice), so that the four digits nearest its point are 5561, our starting point is row 55 and column 61 where we can read off three-digit numbers going down columns 61, 62, 63. This procedure gives us:

430	242	257	097	113
376	304	120	380	401
048	437	286	122	272

Note that many other paths could have been taken from the starting point; for the path we took, twelve numbers greater than 450 were eliminated. The fifteen numbers listed are *not* the sample observations; population elements corresponding to the numbers 430, 376, 048, and so on must be withdrawn, the sample data would be obtained from those elements.

(Problem 15)

PROBLEMS

1. List the members of each of the following sets:

(a) the telephone area codes of California and Texas.

(b) the chapter numbers of this book.

(c) the four major U.S. auto makers.

(d) when a dozen trees are planted, the number that might survive.

† If rows and columns are numbered anew on each page, digits for a page number will also be needed.

2. List the members of each of the following sets:
(a) your brothers and sisters.
(b) the books you own.
(c) the automobiles you own or have owned.
(d) the addresses where you have lived.

3. Specify the number of elements in each finite set of Problem 1.

4. Specify the number of elements in each finite set of Problem 2.

5. A function is defined as follows:

$$y = f(x) = 2x - 3; \qquad x = 1, 2, 4, 7$$
$$= 0; \text{ otherwise}$$

Determine

(a) $f(1)$ (b) $f(4)$ (c) $f(1) + f(2)$
(d) $f(0)$ (e) $5f(7)$

6. The sets $X = \{1, 3, 4\}$ and $Y = \{0, 8, 15\}$ together with the correspondence $y = x^2 - 1$ form a function.
(a) Show that the same correspondence holds if X is replaced by $\{-1, -3, -4\}$.
(b) Show that X and Y are also related by the correspondence $y = x^3 - 7x^2 + 19x - 13$.

7. A function is defined by the equation:

$$y = f(x) = 3A^x - 2; \qquad \begin{cases} x = 0, 1, 2, 3 \\ A = \text{nonzero constant} \end{cases}$$
$$= 0; \text{ otherwise}$$

Evaluate
(a) $f(0)$ (b) $f(1.5)$ (c) $f(0) + f(1)$
(d) $f(4)$ (e) $f(2)$ (f) $4f(0)$

8. Plot each of the following functions for values of X between 0 and 6:
(a) $y = x$ (b) $y = 2x$ (c) $y = x + 2$
(d) $y = x - 2$ (e) $y = -x$ (f) $y = 3 - x$
(g) $y = 4 - 2x$ (h) $y = 5$

9. Let $f(x) = x^2 - 2x$ for $x = 0, 1, 2, 3$ denote a function.
(a) What is the range of this function?
(b) What is the domain of this function?
(c) Evaluate $f(1), f(2) + f(3), 3f(1) - 2f(3)$, and $f(y/2)$.

10. Define a function $g(x) = 3x - x^2 + 4$ for $x = -1, 0, 1, 2$.
(a) What is the range of this function?
(b) What is the domain of this function?
(c) Evaluate $g(1), 2g(0) - 3g(-1), g(-1) + g(0) + g(1) + g(2)$, and $g(y^2)$.

11. (a) For Problem 1(a), define a correspondence, X, to be the sum of the three digits in an area code. Give the domain and range to complete this function.
(b) Is there any duplicate correspondence with X? For what area codes?

12. Take the set $\{0, 1, \ldots, 9\}$ to be a population of ten elements.
 (a) List all different samples of size 9 that might be drawn without replacement from this population.
 (b) Would the number of different samples of size 9 be the same for sampling with replacement? Discuss.

13. Take the set in Problem 1(c) to be a population. List all samples
 (a) of size 1 that might be drawn from this population.
 (b) of size 2 that might be drawn without replacement from this population.
 (c) of size 2 that might be drawn with replacement from this population.

14. Use Table A.1 to obtain identification numbers for drawing a random sample of size 10
 (a) with replacement from a population of 100 elements.
 (b) without replacement from a population of 100 elements.
 (c) without replacement from a population of 1,000 elements.

15. Use Table A.1 to obtain identification numbers for drawing a random sample of size 15
 (a) with replacement from a population of 100 elements.
 (b) without replacement from a population of 100 elements.
 (c) without replacement from a population of 200 elements.
 (d) without replacement from a population of 1,000 elements.

16. On ten identical objects (slips of paper, cardboard, plastic chips, etc.) place the ten digits $0, 1, \ldots, 9$, one on each object. Then, place these objects in a box, hat, or other receptacle.
 (a) After thoroughly mixing the objects in the receptacle, randomly draw three objects without replacement and record the numbers obtained. Replace the three objects drawn and repeat this process until you obtain twenty random samples of size 3.
 (b) Change the make-up of the objects in the receptacle by adding two more objects each having the digit 2 on it. There will now be twelve objects to draw from, three of which will have the digit 2 on them. Now obtain twenty random samples of size 3 as above.
 (c) Is there any difference in the frequency with which the digit 2 appears in your samples of (a) and (b)? Did you expect any difference?

REFERENCES

[1] Interstate Commerce Commission, Bureau of Transport Economics and Statistics, *Tables of 105,000 Random Decimal Digits*, U.S. Government Printing Office, Washington, D.C., 1949.

[2] Kendall, M. G., and B. B. Smith, *Tables of Random Numbers, Tracts for Computers No. XXIV*, Cambridge University Press, Cambridge, Eng., 1939.

[3] RAND Corporation, *A Million Random Digits with 100,000 Normal Deviates*, Free Press of Glencoe, Inc., New York, 1954.

Chapter 3

EXPERIMENTS

We will use *experiment* as a collective term to refer to any research activity where the data are randomly collected. A home economist performs an experiment when she prepares 24 cakes, randomly assigns four to each baking temperature, and measures cake moistness. An educator performs an experiment when he randomly selects student teachers and interviews them about the teacher education program. An agronomist performs an experiment when he randomly assigns five varieties of sweet corn to field plots and measures yields.

In conducting an experiment, the researcher might perform a single operation, many different operations, or the same operation repeatedly. For experiments where the same operation is repeated, each repetition is frequently called a *trial*. In a simple experiment of tossing a die 10 times, each toss of the die represents one trial. For an experiment where 20 trees are planted, the planting of one tree represents a trial.

While many variables might be observed in any experiment, the experimenter usually restricts his attention to a few pertinent ones and ignores all others which have little or no bearing on the information sought. Thus, when tossing several coins, we are usually interested in the number of heads showing on the upper faces of the coins. Such factors as (1) the distances of the coins from the tosser, (2) the distances of the coins from one another, and (3) the orientations of the coins may all be ignored, since they provide no information about the number of heads showing.

► ***Example 3-1***

Consider the following five experiments, designated by $\mathscr{E}_1$ through $\mathscr{E}_5$ for future reference.

$\mathscr{E}_1$: Toss a regular six-sided die.

The die will land with one of the six faces up. Customarily, these six results are the only ones considered relevant for this experiment.

$\mathscr{E}_2$: Bend a four-foot section of copper tubing until it cracks.

Since different four-foot sections (all made by the same process) will require different amounts of bending to crack them, the experimental results will be all possible real values in some interval.

$\mathscr{E}_3$: Plant a certain variety of strawberries on 20 plots.

The experimenter usually will be interested in the yield per plot. If this is the case, his experimental result will be a set of 20 yields, each yield being a real number in some interval.

$\mathscr{E}_4$: Deal a person 13 cards from a bridge deck.

An experimental result will be any specific set of 13 cards; the order of receiving the cards may or may not be considered.

$\mathscr{E}_5$: Select a pair of digits at random from a random-numbers table.

Here there are 100 experimental results—the pairs 00, 01, . . . , 99. ◀

3.1 OUTCOMES AND EVENTS

At each performance of an experiment, a particular result is obtained. The result need not be a single observation; a sample of n observations might be obtained. In experiment $\mathscr{E}_3$ every sample of 20 yields represents one experimental result. In experiment $\mathscr{E}_4$ every sequence of 13 cards is a single experimental result.

Definition 3-1

Each of the distinct results of an experiment is called an *outcome*. A set of outcomes is called an *event*. An event is said to occur if any one of its outcomes occur.

Note: Every individual outcome is an event, every pair of outcomes is an event, every three outcomes is an event—any set of outcomes is an event. Some events defy concise description. For example, when drawing one card from a bridge deck, the event

$$\{3 \text{ of spades, } 10 \text{ of clubs, queen of hearts, ace of diamonds}\}$$

is best described by listing the outcomes, as we have done.

▶ ***Example 3-2***

Barring such oddities as the die landing on one corner, mysteriously disintegrating, or being swooped out of the air by a hungry bird, each toss of the die in experiment $\mathscr{E}_1$ will produce one of six outcomes:

O_1: Die shows 1 spot.
O_2: Die shows 2 spots.

.
.
.

O_6: Die shows 6 spots.

Thus some events that can be defined relative to this experiment are as follows:

E_1: Die shows even spots (O_2 or O_4 or O_6).
E_2: Die shows odd spots (O_1 or O_3 or O_5).
E_3: Die shows at least 3 spots (O_3 or O_4 or O_5 or O_6).
E_4: Die shows 3 or 4 spots (O_3 or O_4).
E_5: Die shows 5 or 6 spots (O_5 or O_6).
E_6: Die shows 3 or 5 spots (O_3 or O_5).
E_7: Die shows 2 spots (O_2). ◀

▶ ***Example 3-3***

Experiment $\mathscr{E}_5$ has 100 outcomes—the pairs 00 through 99. There are millions of events for this experiment. A few of them are listed here:

A: The first digit is a 2.
B: The second digit exceeds the first.
C: Both digits are the same.
D: The two digits differ.
E: The sum of the two digits is 4.
F: The two digits differ by 3. ◀

(Problems 1, 3)

3.2 SAMPLE SPACE

We now give a name to the collection of distinct results of an experiment.

> ***Definition 3-2***
>
> The set of all possible outcomes of an experiment is its *sample space*, denoted by $\mathcal{S}$.

In general, the sample space for any experiment may be given in a number of ways. Words, symbols, pictures, or a combination thereof may be used in giving $\mathcal{S}$. This is illustrated in the next example.

▶ ***Example 3-4***

The sample space for $\mathscr{E}_1$ could be given in any of the following ways:

$\mathcal{S}$ = {upper face shows one spot; upper face shows two spots; $\cdots$; upper face shows six spots}

$\mathcal{S} = \{1, 2, \cdots, 6$, where the integer denotes the number of spots showing on the upper face}

$\mathcal{S}$ = {⚀, ⚁, ⚂, ⚃, ⚄, ⚅} ◀

▶ ***Example 3-5***

The sample space for $\mathscr{E}_5$ might be given as

$$\mathcal{S} = \{00, 01, \cdots, 99\}$$

or, in slightly more general terms, as

$$\mathcal{S} = \{(x, y): x = 0, 1, \cdots, 9; y = 0, 1, \cdots, 9\}. ◀$$

An event is merely a subset of the sample space. The terms "event" and "subset" will be used interchangeably. The symbolism

$$A \subset \mathcal{S} \tag{3.1}$$

is to be read as "A is a subset of $\mathcal{S}$" or "A is an event of $\mathcal{S}$." It is customary to consider the *null set* (the empty set, the set having no outcomes) as being a subset of every set, including itself and the sample space. The null set is denoted by $\varnothing$. As we will see later, certain combinations of events contain no outcomes; these are termed "null events."

3.3 GRAPHIC REPRESENTATION OF EVENTS

To aid in examining the structure of various events, we often use a graphic representation called a *Venn diagram* (after the English logician John Venn, 1834–1923). In general, each of the experimental outcomes is represented by a point within a rectangle. For convenience, the sides of the rectangle serve as boundaries of the sample space $\mathcal{S}$. A subset of the sample space, an event, is represented by a circle containing the relevant outcomes (sometimes squares, triangles, or other closed curves are used).

Figure 3-1 gives the Venn diagram representation of a sample space consisting of five outcomes and one event consisting of two outcomes.

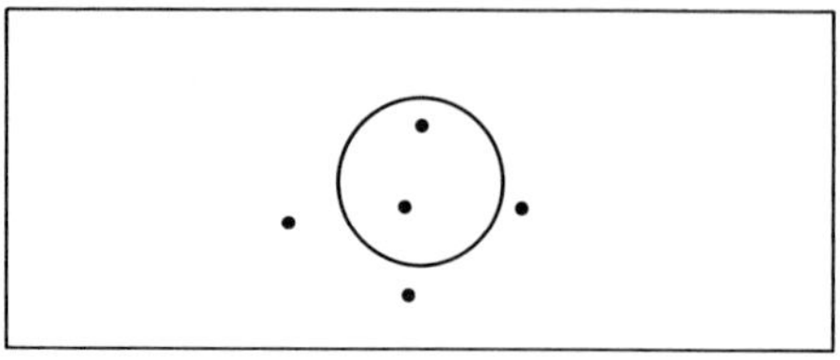

Figure 3-1 Venn Diagram: Sample Space and One Event

(Problem 7)

3.4 COMBINING AND SUBDIVIDING EVENTS

The solutions to many problems associated with events can be simplified by considering special related events, by subdividing events into component parts (subevents), or by combining events in special ways. We consider some special events and event modifications in the remainder of this section.

3.4.1 Complement of an Event

If A is an event defined for an experiment having sample space $\mathcal{S}$, specific outcomes of $\mathcal{S}$ belong to A. The outcomes of $\mathcal{S}$ not belonging to A constitute an event called the *complement* of A, or the event "not A."

Definition 3-3

If A is an event of a sample space, $\mathcal{S}$, the *complement* of A is $\bar{A}$, the set of all outcomes of $\mathcal{S}$ that are not in A.

Figure 3-2 gives a Venn diagram representation of a sample space and events A and $\bar{A}$.

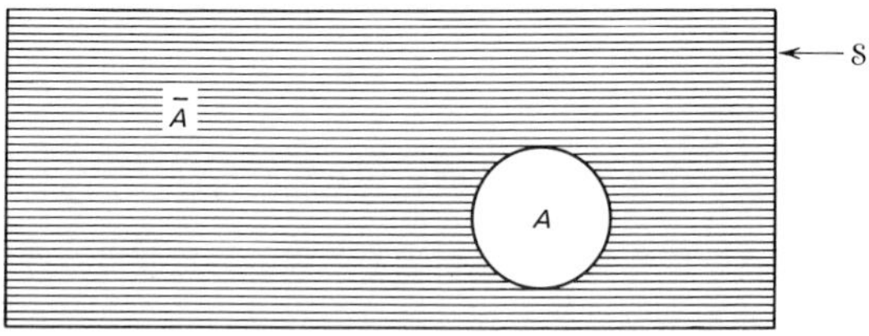

Figure 3-2 Venn Diagram: Events A and $\bar{A}$

▶ ***Example 3-6***

If a sample space is defined by

$$\mathcal{S} = \{0, 1, 2, 4, 5, 7, 8\}$$

event $A = \{0, 1, 7\}$ has complement $\bar{A} = \{2, 4, 5, 8\}$,
event $B = \{1, 2, 4, 5\}$ has complement $\bar{B} = \{0, 7, 8\}$. ◀

(Problems 5(d), (n), (p))

3.4.2 Intersection of Events

Some pairs of events have common outcomes. The "overlap" or common portion of two events is another event called their *intersection*.

> ***Definition 3-4***
>
> If A and B are events of a sample space, $\mathcal{S}$, the *intersection* of A and B, denoted by $A \cap B$, is the event consisting of all outcomes of $\mathcal{S}$ that are in *both* A and B.

Thus, in Example 3-6 $A \cap B = \{1\}$ and $A \cap \bar{B} = \{7\}$.

The crosshatched area of the Venn diagram in Figure 3-3 represents the intersection of A and B. Repeated application of Definition 3-4 shows that the intersection of more than two events gives the set of outcomes that are common to all events involved. The intersection is a subset of each event involved in the intersection; thus in Figure 3-3 we see that

$$(A \cap B) \subset A \quad \text{and} \quad (A \cap B) \subset B \tag{3.2}$$

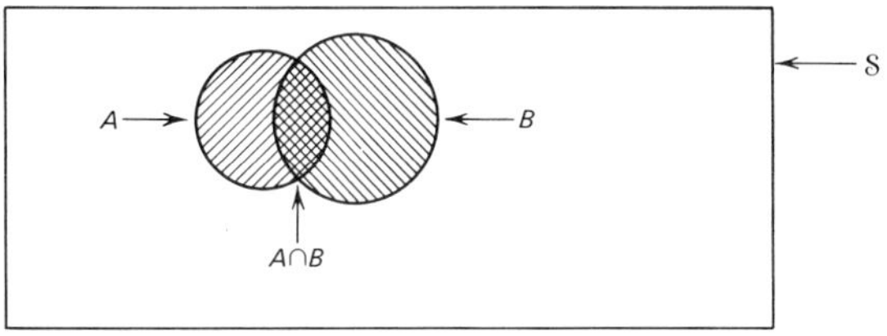

Figure 3-3 Venn Diagram: Events A, B, and $A \cap B$

▶ ***Example 3-7***

Referring to Example 3-3, where some events of $\mathscr{E}_5$ are listed, it is seen that

$A \cap B = \{23, 24, 25, \cdots, 29\}$
$A \cap C = \{22\}$
$A \cap D = \{20, 21, 23, 24, \cdots, 29\}$
$A \cap E = \{22\}$
$A \cap F = \{25\}$
$C \cap D = \varnothing$

$A \cap B \cap F = \{25\}$
$E \cap F = \emptyset$
$B \cap D \cap E = \{04, 13\}$. ◀

(Problems 5(e), (f), (h), (k), (m), 15(d))

3.4.3 Union of Events

In some situations there is interest in the set of outcomes belonging to one or both events A and B. This set represents a special combination of events A and B, and is now defined.

Definition 3-5

If A and B are events of a sample space, $\mathcal{S}$, the *union of A and B*, denoted by $A \cup B$, is the event consisting of all outcomes of $\mathcal{S}$ that are in A, or in B, or in both A and B. (The symbol for union is easy to remember, note that it resembles the first letter of "Union.")

If $A = \{1, 3, 9\}$ and $B = \{2, 4, 6\}$, then $A \cup B = \{1, 2, 3, 4, 6, 9\}$. Notice that in forming the union of two events, their elements are not duplicated. Thus $\{1, 2\} \cup \{2, 3\} = \{1, 2, 3\}$.

Referring to the Venn diagram in Figure 3-3, the union of A and B is represented by the set of points within the two circles. A union of events never leads to a reduction in outcomes (as does intersection). The union of A and B will always be larger than either event unless one event happens to be a subset of the other, in which case there is no gain in outcomes. For example,

$$A \cup B = B \quad \text{if and only if} \quad A \subset B \tag{3.3}$$

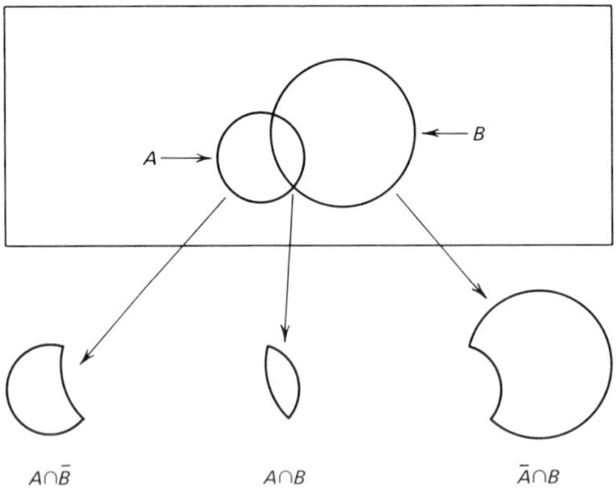

Figure 3-4 A Subdivision of $A \cup B$

In Figure 3-4 three subsets of $A \cup B$ are displayed—one is $A \cap B$. It is always true that the intersection is a subset of the union, or

$$(A \cap B) \subset (A \cup B) \tag{3.4}$$

The union of more than two events of a sample space, $\mathcal{S}$, can be obtained by repeated application of Definition 3-5. The union of several events is the set of outcomes of $\mathcal{S}$ that belong to one or more of the component events. Thus

$$\{1, 3, 4, 5\} \cup \{2, 4, 7\} \cup \{3, 4, 5, 7\} = \{1, 2, 3, 4, 5, 7\}$$

▶ ***Example 3-8***

The following are some unions of those events for the die experiment, $\mathscr{E}_1$, stated in Example 3-2:

$$E_4 \cup E_5 = E_3$$
$$E_2 \cup E_6 = E_2$$
$$E_1 \cup E_2 = \mathcal{S}$$
$$E_5 \cup E_3 = E_3$$
$$E_5 \cup E_6 = \{O_3, O_5, O_6\}$$
$$E_2 \cup E_4 \cup E_5 = \{O_1, O_3, O_4, O_5, O_6\}$$
$$E_1 \cup E_4 \cup E_6 = \{O_2, O_3, O_4, O_5, O_6\}.$$ ◀

▶ ***Example 3-9***

For the events of $\mathscr{E}_5$ listed in Example 3-3, some of their unions are:

$B \cup C$: second digit is greater than or equal to the first

$C \cup D = \mathcal{S}$

$C \cup E = \{00, 11, 22, ---, 99, 04, 13, 31, 40\}$

$E \cup F = \{04, 13, 22, 31, 40, 03, 14, 25, 36, 47, 58, 69, 30, 41, 52, 63, 74, 85, 96\}$

$A \cup C \cup E = \{20, 21, 23, 24, ---, 29\} \cup (C \cup E)$. ◀

▶ ***Example 3-10***

If an experiment $\mathscr{E}$ has 20 outcomes, the sample space, $\mathcal{S}$, can be represented by the Venn diagram below.

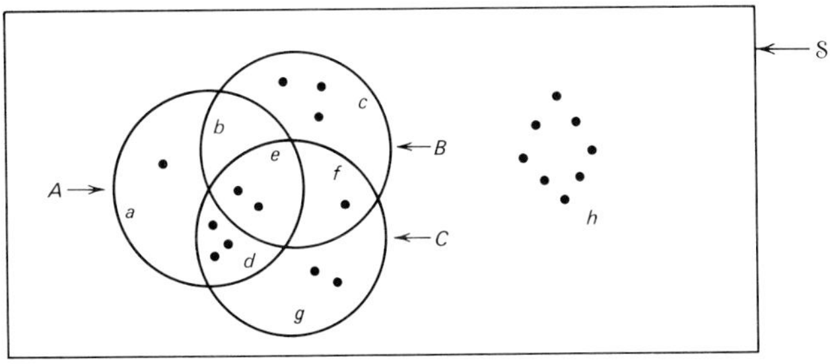

Event A has 6 outcomes and consists of components *a*, *b*, *d*, and *e*. Event $A \cap B$ has components *b* and *e*, with 2 outcomes. Components *e* and *f* with their 3 outcomes comprise $B \cap C$. Component *b*, a null set, represents $A \cap B \cap \bar{C}$, the outcomes of A and B but not C. Component *d* represents a similar event, $A \cap \bar{B} \cap C$. Component *e* is $A \cap B \cap C$, which in this case, is also $A \cap B$. Component *h* represents $\overline{(A \cup B \cup C)}$, all outcomes outside the union of A, B, and C. ◀

(Problems 5(a), (b), (c), (g), (i), (j), 15(c))

3.5 *MUTUALLY EXCLUSIVE EVENTS*

The union of two events is never the null event unless both component events are null. On the other hand, the intersection of many pairs of events is the null event. In Example 3-7 we saw that $C \cap D = \varnothing$ and $E \cap F = \varnothing$. Two events having this property have no points in common; if either event occurs, the other does not. Such events have simple mathematical properties (as we will see shortly). Hence, we explicitly define this concept.

Definition 3-6

Two events, A and B, are *mutually exclusive* if $A \cap B = \varnothing$.

The events $A = \{\text{temperature below freezing}\}$ and $B = \{\text{temperature above } 80°\}$ are mutually exclusive, since a temperature cannot belong to both A and B.

▶ ***Example 3-11***

For experiment $\mathscr{E}_1$, events E_1 and E_2 are mutually exclusive, since $E_1 \cap E_2 = \varnothing$. Also, the events E_4 and E_5 are mutually exclusive. ◀

(Problem 15(a))

3.6 *EXHAUSTIVE EVENTS*

The intersection operator and the null event were used in defining the mutually exclusive concept. In a somewhat opposite sense, the union operator and the sample space, $\mathbb{S}$, are used in defining another concept.

Definition 3-7

If a group of events is such that one of them must occur, the group of events is said to be *exhaustive*, and their union is the sample space.

When a regular die is tossed (experiment $\mathscr{E}_1$), events E_1 = {even spots} and E_2 = {odd spots} are exhaustive, since $E_1 \cup E_2$ = {any spots} = $\mathcal{S}$.

▶ ***Example 3-12***

Some events of die tossing are:

E_1 = {even spots}
E_2 = {odd spots}
E_3 = {at least 3 spots}
E_4 = {3 or 4 spots}
E_5 = {5 or 6 spots}
E_6 = {no more than 4 spots}

A pair of events is given for each of the four combinations listed; other possibilities exist.

	Exhaustive	Not exhaustive
Mutually exhaustive	E_1 and E_2	E_4 and E_5
Not mutually exhaustive	E_3 and E_6	E_2 and E_3

◀

(Problem 15(b))

3.7 *PARTITIONS*

Exhaustiveness of a group of events insures that each outcome of the sample space is "present and accounted for." Mutual exclusiveness means that the events have no overlapping. To have both of these properties means that the group of events represents a nonoverlapping subdivision of the sample space. In later work, especially probability, these special subdivisions produce numerous simplifications. This important concept is now defined and discussed.

Definition 3-8

A set of mutually exclusive and exhaustive events is said to form a *partition* of $\mathcal{S}$.

It is always true that an event and its complement form a partition of $\mathcal{S}$; that is, $A \cup \bar{A} = \mathcal{S}$ and $A \cap \bar{A} = \emptyset$ for every subset A of $\mathcal{S}$.

The idea of partitioning is not restricted to sample spaces; it applies to events and populations as well. Basically, partitioning allows us to look at a sample space, an event, or a population "one segment at a time." One component of a partitioning may be a miscellaneous, catchall category.

► ***Example 3-13***

(1) The cards of a bridge deck may be partitioned into 4 subsets: spades, clubs, hearts, and diamonds. They may also be partitioned into 2 subsets: black and red.
(2) The students at your school may be partitioned into 5 subsets: freshmen, sophomores, juniors, seniors, and graduate students; or into 2 subsets: undergraduates and graduate students; or into another 2 subsets: men and women; or into 3 subsets: freshmen, sophomores, and "all others."
(3) The days of a calendar year may be partitioned into 12 subsets according to months or into 7 subsets according to days of the week.
(4) This book may be partitioned into 16 subsets according to the chapters plus the miscellaneous material (table of contents, index, and so on). ◄

► ***Example 3-14***

In Figure 3-4 the three subdivisions of $A \cup B$ represent a partitioning of the union of two events.

And in Example 3-10 the eight components labeled *a* through *h* represent a partitioning of the sample space. ◄

(Problems 9, 10)

3.8 COUNTING PROCEDURES

In the preliminary stages of an experiment a researcher may need to know how many outcomes belong to certain events. Other researchers may need to know the number of different ways of performing a series of operations. Suppose a testing bureau has six different tests (labeled *A*, *B*, *C*, *D*, *E*, *F*) to administer. In how many different orders may these tests be given? That is, how many different randomizations are there for six different tests? How many if *C* must follow *A*? Answers to these questions require counting. We could list all different orders of giving the tests and do the necessary counting, but this would be a tedious task, to say the least—there are 720 different orders. With a simple counting rule, we can obtain the value 720 without any listing.

Counting problems are also encountered in the calculation of probability. There the calculations involve the number of outcomes in the sample space and in the events of interest.

Some basic principles of counting are discussed below; only the finite case is treated. These concepts are part of a larger field of study called *combinatorial analysis*. For a treatment of advanced topics in combinatorics, the interested reader should refer to [1], [2], [3], and [4].

3.8.1 Unions of Events

Let n_A be the number of outcomes in event A, n_B the number of outcomes in event B, and n_{AB} the number of outcomes in the intersection of A and B.

Mutually exclusive events have no outcomes in common, in which case $n_{AB} = 0$. Figure 3-5 represents this situation. The number of outcomes in the union of A and B equals the number of outcomes within the two circles. This number is clearly the sum of n_A and n_B.

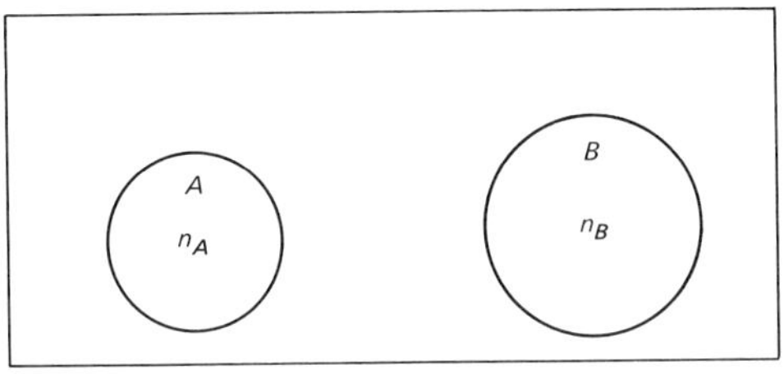

Figure 3-5 Venn Diagram: Two Mutually Exclusive Events

•• *Rule C-1:* If A and B are mutually exclusive events, the number of outcomes in $A \cup B$ is $n_A + n_B$.

If a car dealer sold $n_A = 14$ new cars and $n_B = 6$ new trucks during March, he had $n_A + n_B = 20$ sales during March that were either new cars or new trucks.

▶ ***Example 3-15***

A two-digit random number is selected, and we define events A and B as follows:

$$A = \{\text{first digit of the number is 0 or 1}\}$$
$$B = \{\text{number is greater than 50}\},$$

these events are mutually exclusive, with $n_A = 20$ and $n_B = 49$. Hence, $A \cup B$ has 69 outcomes. ◀

When events A and B are not mutually exclusive, their overlap contains n_{AB} outcomes. Consider Figure 3-6. Again, the number of outcomes in the union of A and B equals the number of outcomes within the two circles. Adding the three numbers displayed in Figure 3-6 gives $n_A + n_B - n_{AB}$ outcomes in the union.

•• *Rule C-2:* The number of outcomes in $A \cup B$ is $n_A + n_B - n_{AB}$.

(We point out that Rule C-1 is a special case of Rule C-2 with $n_{AB} = 0$.) If one card is drawn from a bridge deck, event $A = \{\text{red card}\}$ has $n_A = 26$ outcomes and event $B = \{\text{king, queen, or jack}\}$ has $n_B = 12$ outcomes and

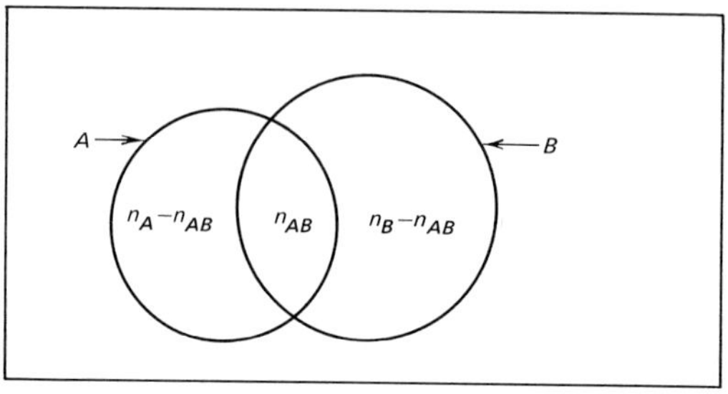

Figure 3-6 Venn Diagram: Two Events Not Mutually Exclusive

$A \cap B = \{$red king, red queen, red jack$\}$ has $n_{AB} = 6$ outcomes. Hence the event $\{$king, queen, jack, or red card$\}$ has $n_A + n_B - n_{AB} = 32$ outcomes.

► ***Example 3-16***

An experiment consists of tossing a regular die once, and the following events are defined:

A: at least four spots on uppermost face
B: even number of spots on uppermost face

Here $n_A = 3$, $n_B = 3$, $n_{AB} = 2$, and the number of outcomes in $A \cup B$ is $n_A + n_B - n_{AB} = 4$. ◄

► ***Example 3-17***

If a stockbroker considers the change (recorded only as $-$, 0, or $+$) in weekly price of each stock traded on the New York Stock Exchange, he may be interested in the number of stocks that had an increase during either of the first two weeks of January. He can easily determine n_A and n_B, where

$A = \{$stocks that increase during the first week of January$\}$
$B = \{$stocks that increase during the second week of January$\}$

but he must also determine n_{AB}, the number that showed an increase *both* weeks. ◄

(Problem 11)

3.8.2 Composite Operations

A Boston business man has appointments on April 10 at 1:00 P.M. in Baltimore, and at 8:00 P.M. in Dallas. He can take any one of 4 airline flights from Boston to Baltimore and any one of 3 flights from Baltimore to Dallas. He really can choose from 12 composite flights, since any one of the flights from Baltimore to Dallas can be taken with any one of the 4 flights from Boston to Baltimore.

The two operations, (1) flying from Boston to Baltimore and (2) flying from Baltimore to Dallas, could be performed in 4 and 3 ways, respectively. More generally, the two operations might be performed in m_1 ways and m_2 ways. Figure 3-7 shows the composite of these two operations. The second operation can be performed m_2 ways with each way of performing the first operation; thus, there is a total of $m_1 m_2$ ways of performing the two operations. Earlier, we had $m_1 = 4$ and $m_2 = 3$ giving a total of $m_1 m_2 = 12$ ways of flying from Boston to Dallas (subject to the conditions stated).

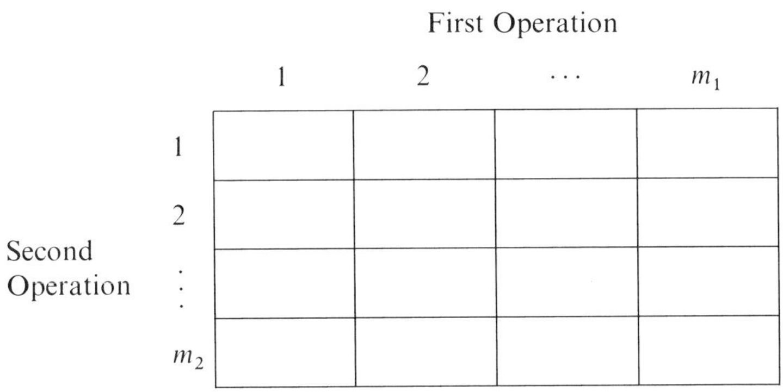

Figure 3-7 Ways of Performing Two Operations

•• *Rule C-3:* If an operation can be performed in m_1 ways, and if after performing this operation, a second operation can be performed in m_2 ways, then together the two operations can be performed in $m_1 m_2$ ways.

Rule C-3 easily extends to more than two operations. With a third operation that can be performed in m_3 ways, the three operations can be performed together in $m_1 m_2 m_3$ ways.

▶ *Example 3-18*

Suppose two elements are to be drawn from the set of ten digits. Let the first operation be the drawing of the first digit; then $m_1 = 10$. If the first digit is *not* replaced in the set, the second operation of drawing another digit has $m_2 = 9$, and there are $10(9) = 90$ ways of drawing two different digits from the set. However, if the first digit is replaced before the second is drawn, $m_2 = 10$, and there are then $(10)^2 = 100$ ways of drawing two digits from the set.

If a third digit is to be drawn from this set without replacement of the first two digits, $m_3 = 8$, and the number of ways of drawing three different digits is $10(9)(8) = 720$. With replacement of each digit, $m_3 = 10$, and the number of ways of drawing three digits from the set is $(10)^3 = 1{,}000$. ◀

Rule C-3 and extensions of it involve products of integers. A particular product of integers which occurs in many counting rules is defined.

Definition 3-9

The product of all integers from 1 through n is called *n factorial.* In symbols

(3.5) $$n! = n(n-1)\cdots(3)(2)(1)$$

► ***Example 3-19***

$3! = 3(2)(1) = 6$
$7! = 7(6)(5)(4)(3)(2)(1) = 5{,}040$ ◀

By grouping the factors in (3.5), we see that $n!$ satisfies the following relationship:

(3.6) $$n! = n[(n-1)!]$$

So that this relationship is meaningful when $n = 1$, it is customary to define $0! = 1$.

(Problems 17, 25, 31)

3.8.3 Permutations

At the beginning of Section 3.8 we discussed a task of administering six different tests. If all six tests are to be used, one student might receive them in the order *ABCDEF*, another might receive them in the order *ABCDFE*, and so on. Each different order is called a *permutation* of the six tests.

If only three tests are given each student, *ABC*, *ACB*, and *AEF* would be some permutations of three tests.

Definition 3-10

Each ordered arrangement of r elements is called a *permutation of r elements.*

Counting the number of different permutations is accomplished by repeated application of Rule C-3. Thus, in the case of giving the six tests to a student, there are $m_1 = 6$ ways of giving the first one (all six are available at the outset), $m_2 = 5$ ways of giving the second one (those remaining after the first one has been given), $m_3 = 4$ ways for the third test, $m_4 = 3$ ways for the fourth test, $m_5 = 2$ ways for the fifth test, and $m_6 = 1$ way for the sixth test. Therefore, the number of different orders of giving six different tests is the number of permutations of six elements, or $6(5)(4)(3)(2)(1) = 6! = 720$.

If only three tests are to be given to a student, instead of all six, there are still $m_1 = 6$ choices for the first test, $m_2 = 5$ for the second test, and $m_3 = 4$ for the third test. Now, there are $6(5)(4) = 120$ different orders; this is the number of permutations of three elements taken from a set of six distinct elements.

•• *Rule C-4:* The number of different permutations (ordered arrangements) of r elements taken from a set of n *distinct* elements is

$$ {}_nP_r = n(n-1)(n-2)\cdots(n-r+1); \qquad r = 0, 1, \ldots, n \tag{3.7} $$

If the elements of a permutation are considered to be occupants of r positions on a line, then by repeated application of Rule C-3, the number of ways of filling these positions is clearly the right-hand side of (3.7). The form of ${}_nP_r$ given in (3.7) is appropriate for calculating its numerical value. A more condensed form results when the right-hand side of (3.7) is multiplied by $(n-r)!/(n-r)! = 1$ and is

$$ {}_nP_r = \frac{n!}{(n-r)!}; \qquad r = 0, 1, \ldots, n \tag{3.8} $$

When $r = n$, all elements are taken from the set to form an arrangement; in this case, there are ${}_nP_n = n!$ different permutations.

► ***Example 3-20***

If a fraternity of 40 men is to select 4 of its members for the offices of president, vice president, secretary, and treasurer, the respective set of officers Roy, Bill, Carl, and Allan is different from the respective set Roy, Bill, Allan, and Carl. Since the order of selecting the men dictates the offices they hold, each permutation of 4 from 40 is a different way of selecting officers. The total is

$$ {}_{40}P_4 = 40(39)(38)(37) = 2{,}193{,}360. $$ ◄

► ***Example 3-21***

If an agronomist investigating eight strains of wheat has eight plots of land available and allocates one strain to each plot, he has ${}_8P_8 = 8! = 40{,}320$ possible allocations. ◄

(Problems 14, 21(b), 23, 25)

3.8.4 Combinations

When the order of drawing elements from a set or the order of assigning elements to certain positions is immaterial, a second class of arrangements is encountered. Thus, if a student takes three tests, and if we are interested only in which three tests he took, the specific order (permutation) is not important. Actually, the same three tests may be given to six students in different orders; for example: *ACD*, *ADC*, *CAD*, *CDA*, *DAC*, *DCA*.

> ***Definition 3-11***
>
> Any set of *r* elements, ordered in any manner, is called a *combination of r elements.*

Combinations are distinguished solely on the basis of composition. In contrast, permutations are distinguished by both composition and order of arrangement. Some examples of combinations are groups of tests, committees to be formed, teams of volunteers, and teams of players—in short, any subset of *r* chosen from *n* distinct elements.

The next rule covers the counting of combinations.

•• *Rule C-5:* The number of different *combinations* (unordered arrangements) of *r* elements taken from a set of *n* distinct elements is

(3.9)
$$\binom{n}{r} = \frac{n!}{r!(n-r)!}; \qquad r = 0, 1, \ldots, n$$

▶ ***Example 3-22***

If the fraternity of 40 men is to select a committee of 4 (rather than the four officers mentioned earlier), the order in which each person is placed on the committee is of no consequence. The number of different committees of 4 members selected from the 40 men is

$$\binom{40}{4} = \frac{40(39)(38)(37)}{4(3)(2)(1)} = 91{,}390. ◀$$

▶ ***Example 3-23***

A person selecting a new car finds that 6 minor options are available, each costing about \$30. He feels, however, that he should spend no more than \$100 on options and must limit his purchase to 3 options. There are $\binom{6}{3} = 20$ different sets of 3 options that he could consider. ◀

Except in those trivial cases where $r = 0$ or $r = 1$, the number of permutations exceeds the number of combinations (because each combination can be rearranged giving different permutations). In particular, when $r = n$, all elements are taken from the set; there is only one combination but $n!$ permutations.

The quantity $\binom{n}{r}$ appears as the coefficient of the term involving $x^r y^{n-r}$ in the expansion of $(x + y)^n$. For this reason, it is called a *binomial coefficient* and has the property

$$\binom{n}{r} = \binom{n}{n-r} \tag{3.10}$$

Table A.2 gives binomial coefficients for $n = 1(1)25$.[†]

(Problems 13, 19, 21)

3.8.5 Permutations of Nondistinct Elements

Another common counting problem is determining the number of different permutations formed by a set of elements that are not all distinct. Consider the four letters of the word OTTO. These letters can be rearranged to form only six different permutations, not $4! = 24$. They are:

(3.11)

OTTO	TOTO
OTOT	TTOO
OOTT	TOOT

If the four letters of OTTO were made distinguishable by attaching subscripts, the four new elements $\{O_1, T_1, T_2, O_2\}$ could be rearranged to give 24 different permutations. But these 24 can be obtained from the 6 "patterns" listed in (3.11) by simply adding subscripts to the letters. For example, the pattern OTOT gives the following four permutations:

$$O_1T_1O_2T_2 \qquad O_1T_2O_2T_1$$
$$O_2T_1O_1T_2 \qquad O_2T_2O_1T_1$$

An identical assignment of subscripts to each of the other patterns listed in (3.11) would give the 24 different permutations. Note that the subscripts are attached to the O's in $2!$ ways, and to the T's in $2!$ ways. Thus, the $4! = 24$ different permutations of $\{O_1, T_1, T_2, O_2\}$ would be reduced to only $4!/2!2! = 6$ different permutations *when the subscripts are removed.*

The following general counting rule can be established by applying the same argument as above.

[†] The quantity in parentheses denotes the increment of values between the numbers adjacent to the parentheses.

•• *Rule C-6:* Suppose a set of n elements can be partitioned into $k \geqslant 2$ distinguishable subsets such that the n_i elements of the ith subset are indistinguishable. Then, the number of different permutations of *all* n elements is

$$\frac{n!}{n_1!n_2!\cdots n_k!}; \qquad k = 2, 3, \ldots, n. \tag{3.12}$$

► ***Example 3-24***

If a well-balanced coin is tossed ten times, in how many different ways can four heads and six tails occur? One possibility is HHHHTTTTTT. But the 4 H's can occupy any four of the ten positions; hence, by Rule C-6, the number of different sequences containing 4 H's is $10!/4!6! = 210$. ◄

► ***Example 3-25***

A gift shop owner wishes to display a certain type of flower vase on a shelf in the store's window. If the shelf holds 6 vases and he uses 2 blue, 3 red, and 1 white, how many color patterns are possible? One possibility is BBRRRW. By Rule C-6, the number of different color patterns of 2 B's, 3 R's, and 1 W is $6!/2!3!1! = 60$. ◄

The special case of Rule C-6 where $k = n$, meaning all elements are distinguishable, forces each $n_i = 1$. Then, the right-hand side of (3.12) reduces to $n!$ and coincides with ${}_nP_n$.

If the multinomial function $(x_1 + x_2 + \cdots + x_k)^n$ is expanded, it can be shown that the coefficient of $x_1^{n_1}x_2^{n_2}\cdots x_k^{n_k}$ is precisely $n!/(n_1!n_2!\cdots n_k!)$. Appropriately, then, the quantity in (3.12) is called a *multinomial coefficient.* Few, if any, multinomial coefficients are ever present in tables because of the many possible values of the n_i for a given n and k. If an extensive table of binomial coefficients is available, multinomial coefficients can be calculated from the relationship

$$\frac{n!}{n_1!n_2!\cdots n_k!} = \binom{n}{n_1}\binom{n-n_1}{n_2}\binom{n-n_1-n_2}{n_3}\cdots\binom{n-n_1-\cdots-n_{k-2}}{n_{k-1}} \tag{3.13}$$

which is easily established upon expansion of the right-hand side.

For sets of elements that are partitioned into two subsets, the multinomial coefficient is simply

$$\frac{n!}{n_1!n_2!} = \binom{n}{n_1} = \binom{n}{n_2}, \tag{3.14}$$

since $n_1 + n_2 = n$. This special result can be interpreted as the number of ways of positioning the indistinguishable elements of either subset. The elements of the other subset must necessarily be placed in the remaining positions.

▶ ***Example 3-26***

If one segment of a necklace is to contain 5 blue, 6 red, and 5 yellow beads, identical except for color, how many different color patterns are possible? Using (3.13) and Table A.2, we see that the number of different color patterns is

$$\frac{16!}{5!6!5!} = \binom{16}{5}\binom{11}{6} = (4368)(462) = 2{,}018{,}016. \blacktriangleleft$$

(Problems 27, 29, 33)

PROBLEMS

1. An experiment consists of drawing two marbles at random from a box containing 6 red (R), 4 blue (B), and 6 white (W) marbles, identical except for color.
(a) List the outcomes of this experiment.
(b) List the outcomes for the event "the two selected marbles are the same color."
(c) List the outcomes for the event "at least one blue marble is selected."

2. Suppose you toss a dart once at the following target.

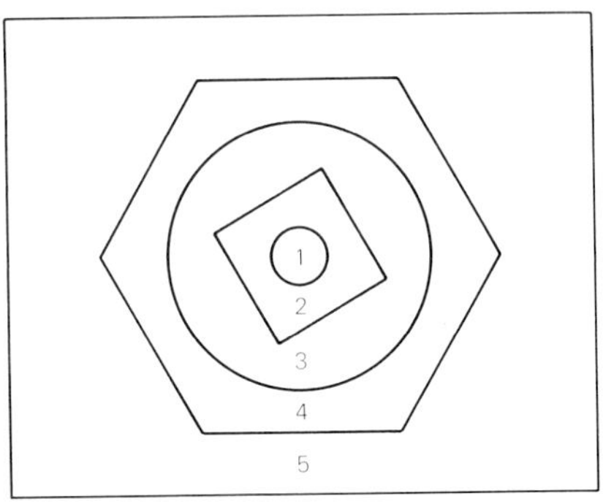

Your score is the number of the area in which the dart strikes. If the dart strikes a boundary, your score is the average of the two adjacent values (for example, 1.5 if the dart lands on the smaller circle). Assume the dart does not land on or outside the boundary of area 5.
(a) List the outcomes of this experiment.
(b) List the outcomes of the event "boundary strikes."
(c) List the outcomes of the event "nonboundary strikes."

3. Refer to Example 3-3 and list all outcomes belonging to events A, C, E, and F.

4. Suppose one dart is tossed twice at the target in Problem 2 above. For this experiment, how many outcomes are there?

5. Let $\mathcal{S} = \{0, 1, 2, 4, 6, 7, 8\}$ denote a sample space and define the following subsets of $\mathcal{S}$:

$$A = \{0, 2, 6\} \qquad B = \{1, 7\}$$

$$C = \{0, 1, 2, 4\} \qquad D = \{4, 6, 7, 8\}$$

Find

(a) $A \cup B$ (b) $A \cup C$ (c) $B \cup C$
(d) $\bar{A}$ (e) $A \cap B$ (f) $\overline{A \cap C}$
(g) $A \cup B \cup C$ (h) $A \cap B \cap C$ (i) $\overline{(B \cup D)}$
(j) $B \cup C \cup D$ (k) $B \cap C \cap D$ (m) $\overline{(B \cap D)}$
(n) $\bar{B}$ (p) $\bar{C}$

6. Repeat Problem 5 with the following substitution of sets:

$$\mathcal{S} = \{0, 1, 2, 3, 7, 8, 9\} \qquad A = \{0, 2, 3\}$$
$$C = \{0, 1, 2\} \qquad D = \{7, 8\}$$

7. Give a Venn diagram representation of the events in
(a) Problem 1(b).
(b) Problem 1(c).

8. Refer to the dart experiment in Problem 2 and give a Venn diagram representation of the sample space and the two events "boundary strikes" and "nonboundary strikes."

9. Verify that events defined in Problem 2(b) and (c) form a partition of the sample space.

10. Do the events in Problem 1(b) and (c) form a partition of the sample space? Why or why not?

11. Refer to Example 3-3 and find the number of outcomes in
(a) $A \cup B$ (b) $B \cup C$ (c) $C \cup D$ (d) $A \cup E$

12. Given that $7! = 5040$, find $8!$, $9!$, and $10!$ using equation (3.6).

13. Use Table A.2 to evaluate

(a) $\binom{10}{3}$ (b) $\binom{14}{4}$ (c) $\binom{14}{11}$ (d) $\binom{25}{21}$

(e) $\binom{17}{6}$ (f) $\binom{19}{12}$ (g) $\binom{13}{10}$ (h) $\binom{15}{5}$

14. Use Table A.2 to evaluate
(a) ${}_{16}P_4$ (b) ${}_{21}P_3$ (c) ${}_{18}P_{12}$ (d) ${}_{23}P_4$

Note from (3.8) and (3.9) that ${}_nP_r = r!\binom{n}{r}$.

15. Consider the following nine digits as a sample space:

$$\mathcal{S} = \{1, 2, 3, 4, 5, 6, 7, 8, 9\}$$

Experiment: Draw one digit at random from $\mathcal{S}$.
Events: A: even digit
B: digit at least 4
C: digit at least 7
D: digit no greater than 5

(a) Which, if any, of the above pairs of events are mutually exclusive?
(b) Which, if any, of the above pairs of events are exhaustive?
(c) What is the complement of $A \cup B$?
(d) What is the complement of $A \cap B$?

16. Repeat Problem 15 using as the sample space:

$$\mathcal{S} = \{1, 2, 4, 5, 7, 9\}$$

17. (a) How many different license plates can be made using seven symbols? The first two should be letters of the English alphabet and the last five should be digits, the first being non-zero.
(b) How many plates would there be if the two letters can both appear at the beginning, both at the end, or one at the beginning and one at the end?

18. Determine the value of n for which

(a) $({}_nP_4) = 10({}_nP_3)$ (b) $9\binom{n}{5} = 6\binom{n+1}{6}$

19. With the aid of Table A.2, evaluate

(a) $\binom{15}{3} \Big/ \binom{15}{7}$ (b) $\binom{10}{2}\binom{3}{2} \Big/ \binom{13}{4}$

(c) $\dfrac{\binom{10}{6}\binom{9}{6} + \binom{10}{5}\binom{9}{7}}{\binom{19}{12}}$

(d) $\dfrac{\binom{12}{4}\binom{8}{3}\binom{4}{1}}{\binom{24}{8}}$ (e) $\binom{25}{4} + \binom{25}{5}$

20. A string of Christmas tree lights has 12 sockets, and bulbs are available in 5 different colors. In how many ways can the string be filled if
(a) any color of bulbs can be used?
(b) at least two colors must be used?
(c) exactly two colors must be used?

21. Suppose that 10 students are eligible for 6 scholarships to be awarded. In how many different ways could the scholarships be awarded if
(a) each scholarship is worth $500?
(b) each scholarship is worth a different amount?
(c) two scholarships are worth $750 each and four are worth $375 each?

22. How many different permutations are there when using all letters of the word PROBABILISTICALLY?

23. A commercial airliner has 60 coach seats. In how many different ways could the coach passengers on a particular flight be seated if
(a) 60 coach tickets have been sold?
(b) 50 coach tickets have been sold?

24. Ten men sign up to work overtime on a given project. The employer needs only four men to finish the project during the overtime period. Assuming the ten men are equally qualified, how many choices does the employer have in determining the overtime workers?

25. Suppose 16 cattle are entered for judging, with first, second, third, and fourth places to be awarded. How many different possibilities must the judges consider?

26. Eight people are to have their picture taken in front of a building. How many different pictures might be taken if all eight people are to appear in the picture, and they are to stand in a row?

27. A person decides to plant 4 blue spruce, 3 ponderosa pine, and 3 Douglas fir trees along his property line. How many different arrangements might he consider?

28. An examination consists of 10 different problems. Students are to omit two of the first 4 problems and two of the last 6 problems. How many different solutions must a student consider?

29. For hands of 5 cards dealt from an ordinary deck of 52 cards, specify the number of outcomes comprising each of the following events:
(a) exactly 2 aces
(b) exactly 3 spades
(c) at least 3 spades
(d) all face cards (K, Q, or J).
Set up only; do not evaluate.

30. How many different bridge hands (13 cards from a deck of 52) are there which exhibit:
(a) all honors cards (A, K, Q, J, or 10)
(b) all red cards
(c) 8 spades, 5 hearts
(d) at least 10 spades?
Set up only; do not evaluate.

31. Two married couples decide to play cards. How many different seating arrangements may they take around a circular table if
(a) each person may sit at any location?
(b) no person may sit adjacent to his or her mate?

32. Two married couples and a third lady decide to play cards. How many different seating arrangements may they take around a circular table if
(a) each person may sit at any location?
(b) no person may sit adjacent to his or her mate?
If the five people are already seated when the third husband arrives, in how many locations may he sit which are not adjacent to his wife?

33. In his personal library, a student has 6 statistics books, 5 mathematics books, and 9 miscellaneous books (each book by a different author). In how many different ways can the books be arranged in a single row across his desk if
(a) no particular order is desired?
(b) the miscellaneous books are to separate the mathematics and statistics books, but the arrangement within these three groups is immaterial?
(c) the books are to be arranged alphabetically by authors within each of the three groups?

34. Volunteers for a psychology project were classified as follows:

Class	Men	Women
Freshmen	6	14
Sophomores	10	15
Juniors	21	12
Seniors	18	14

In how many ways can the project leader select
(a) 10 men and 10 women for the project?
(b) 5 volunteers from each class?
(c) 20 volunteers if 5 must be seniors?

35. A young man has in his wardrobe 6 slacks, 8 shirts, 5 ties, and 4 pairs of shoes. How many different outfits (one item of each type) can he consider if
(a) all items of each type are distinguishable?
(b) four of the shirts are identical (white), but all other items are distinguishable?
(c) two of the slacks are identical and four of the shirts are identical, but all other items are distinguishable?

36. A vegetable gardener has 20 acres of tillable land. From past experience, he knows that 12 different vegetables are suitable to the soil and climatic conditions of his farm. He wants to grow only 5 different vegetables, however.
(a) If he intends to grow 4 acres of each selected vegetable, how many programs might he consider for the upcoming season?
(b) If his tillable land is made up of 5 fields of 2, 3, 4, 5, and 6 acres and he intends to plant a single vegetable in each field, how many programs might he consider for the upcoming season?

37. A football pool is comprised of the ordered pairs (0, 0), (0, 1) through (9, 9). The winner of the pool is the holder of the pair whose entries coincide with the last digit of the final score of the teams involved. For example, in the (Army, Navy) pool, the holder of (6, 0) is the winner if the final score is any one of the following:

Army	Navy
6	0
16	0
16	10
6	10
6	20
16	20

and so on

Of course, the winner in the (Navy, Army) pool would be the holder of (0, 6). How many different sets of 5 pairs are available

(a) from the entire pool, for the first purchaser?

(b) from the pairs corresponding to tied scores, for the first purchaser?

(c) if only 20 pairs are left when a person makes a purchase?

38. A personnel manager gives each applicant a battery of six tests: A, B, C, D, E, F. The tests are unrelated and may be given in any order.

(a) How many different sequences can be used in administering these tests?

(b) For how many of these sequences does test B follow test A?

REFERENCES

[1] Beckenbach, E. F., Editor, *Applied Combinatorial Analysis*, Wiley, New York, 1964.

[2] Hall, Marshall, Jr., *Combinatorial Theory*, Blaisdell, Waltham, Mass., 1967.

[3] Riordan, J., *An Introduction to Combinatorial Analysis*, Wiley, New York, 1958.

[4] Whitworth, W. A., *Choice and Chance*, Stechert, New York, 1942.

Chapter 4

PROBABILITY

Recall that any event either occurs or does not occur. How should we describe the uncertainty of an event's occurrence? To this concept of uncertainty, most people apply the word *probability*, a word with many shades of meaning. Probability usually is calculated from a relative frequency point of view, but used or interpreted from a subjective point of view.

Few words in the English language have more synonyms than the word "probability." Phrases and terms such as "it is likely," "it is almost certain," "may," "I think," "I believe," "surely," "perhaps," "no doubt," and so on, all convey some degree of confidence which the speaker has in the occurrence of an event. Their vagueness indicates the subjective nature of probability.

This brief overview of probability is elaborated on in the remainder of this chapter. Section 4.1 contains an exposition of the ideas necessary for a full understanding of later material; a succinct summary concludes this section. The last part of the chapter, Section 4.2, gives an introduction to methods of calculating probabilities; this material is helpful but not required for later work.

4.1 PROBABILITY—A PRACTICAL EVALUATION

Because probability has become a part of the daily weather forecasts, most people have an intuitive feeling for the implications of a reported value entitled "precipitation probability." If climatic conditions are such that no storm (whether rain, snow, hail, or any other form) is foreseeable today, a precipitation probability of 0 will be reported. The value of 0 means that on all previous days having "identical" climatic conditions, no precipitation has occurred. Precipitation may occur today; the value of 0 is no guarantee against it. But, upon hearing the value of 0, most people would act as though no precipitation will occur today. On the other hand, if a storm is imminent today, a value of 1 will be reported for precipitation probability. Even though precipitation has occurred on all previous days having "identical" climatic conditions, the value of 1 does not guarantee precipitation; today could be an exception.

Because of the relative frequency aspect, a probability cannot be negative and cannot exceed 1. These extreme values of 0 and 1 have already been discussed in relation to precipitation probability. If a precipitation probability of 0.25 is reported for today, how was the value 0.25 obtained and what does it mean? For all days in the past having climatic conditions "identical" to today, precipitation occurred on 1/4 of them. We would attach chances of 1/4 to the event "precipitation today." Or, we can expect to see precipitation on 1/4 of all future days having climatic conditions "identical" to today.

Let us consider another situation. Suppose an insurance company states, "the probability that a 28-year-old automobile driver will receive a traffic citation this year is 0.38." Here the event of interest is "a 28-year-old automobile driver receives a traffic citation this year." Most likely, the value 0.38 is based on past information: the relative frequency (or proportion) of drivers over 28 years of age who received a traffic citation when they were 28 years old. The insurance company may use the value 0.38 as a guide in determining the premium to charge 28-year-old drivers for auto insurance (assuming the number of accident claims is related to the number of traffic citations).

Each time a given experiment is conducted, one outcome in the sample space must occur. This means the sample space has relative frequency of 1, so we make the assignment: $P(\mathcal{S}) = 1$. (The symbol P will be reserved for probability.) On a given day the probability "precipitation does or does not occur" is unity. The probability "a 28-year-old driver does or does not get traffic citations this year" is also unity.

4.1.1 Probability for a Union of Events

The relative frequency for the union of mutually exclusive events is the sum of the individual relative frequencies (since the number of outcomes add when forming such unions). Therefore, the probability for a union of mutually exclusive events is the sum of their individual probabilities:

$$P(A \cup B) = P(A) + P(B) \tag{4.1}$$

This result holds only when A and B are mutually exclusive. Figure 4-1 gives a Venn diagram illustration of this rule. When tossing a regular die once, the event {even spots} is the union of three mutually exclusive events: {2 spots}, {4 spots}, and {6 spots}. Thus,

$$\begin{aligned} P(\text{even spots}) &= P(2\text{ spots}) + P(4\text{ spots}) + P(6\text{ spots}) \\ &= 1/6 + 1/6 + 1/6 \\ &= 1/2 \end{aligned}$$

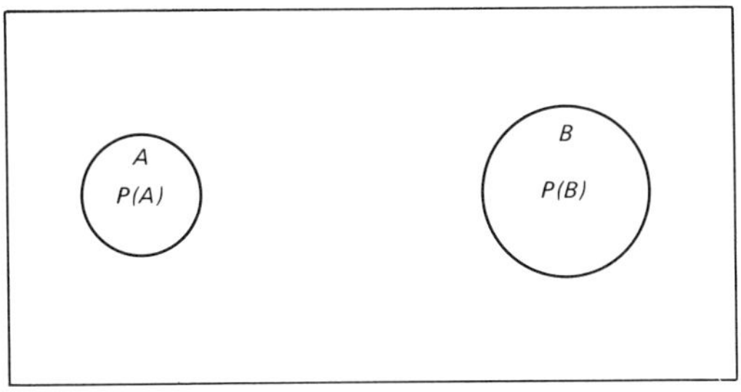

Figure 4-1 $P(A \cup B)$ for Mutually Exclusive Events A and B

The components of a partition of $\mathcal{S}$ are mutually exclusive; hence their probabilities add. And since a partition is exhaustive of $\mathcal{S}$, their probabilities must add to 1. Figure 4-2 illustrates a partition of a sample space. For today

$$P(\text{precipitation}) + P(\text{no precipitation}) = 1$$

and for any 28-year-old driver,

$$P(\text{traffic citation this year}) + P(\text{no traffic citation this year}) = 1$$

If a precipitation probability of 0.25 is reported for today,

$$\begin{aligned} P(\text{no precipitation today}) &= 1 - P(\text{precipitation today}) \\ &= 1 - 0.25 \\ &= 0.75 \end{aligned}$$

And if the probability a 28-year-old driver gets a traffic citation this year is 0.38,

$$P(\text{he doesn't get a traffic citation this year}) = 0.62$$

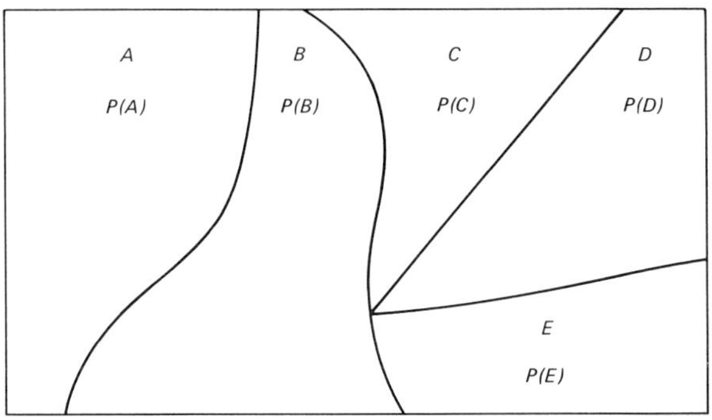

Figure 4-2 Partition of $\mathcal{S}$: $P(A) + P(B) + P(C) + P(D) + P(E) = 1$

For two events not mutually exclusive, the probability of their union is *not* the sum of their individual probabilities, because such a sum would result in a duplication of the overlap. As illustrated in Figure 4-3, this duplication is eliminated either by adding the probability of the overlap only once or by subtracting the probability of the overlap from the sum of individual probabilities. Note that $P(A) - P(A \cap B)$ represents the probability of "only A" occurring, and $P(B) - P(A \cap B)$ the probability of "only B" occurring. Adding the three components of the union listed in Figure 4-3 gives the following general rule:

$$P(A \cup B) = P(A) + P(B) - P(A \cup B), \tag{4.2}$$

when A and B are any two events; should A and B be mutually exclusive, the result in (4.2) reduces to that in (4.1).

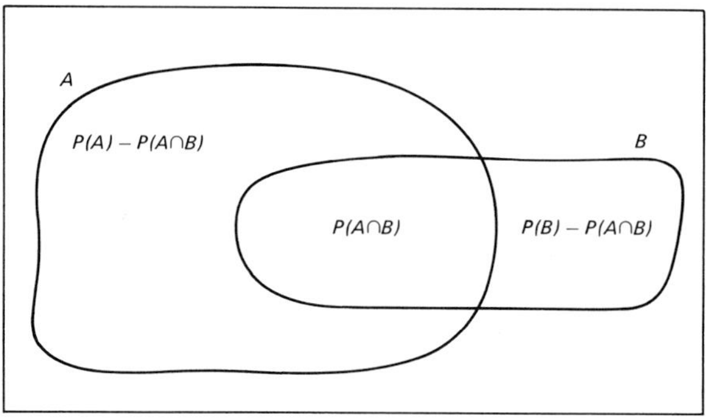

Figure 4-3 Probability of the Union of Two Events Not Mutually Exclusive

In the insurance example, consider the events "a 28-year-old driver gets a speeding citation this year" and "a 28-year-old driver gets a parking citation this year." These two events are not mutually exclusive; some 28-year-old drivers will get both types of citations this year. The probability "a 28-year-old driver gets either a speeding or a parking citation this year" equals

$$P(\text{speeding citation}) + P(\text{parking citation}) - P(\text{both citations})$$

or

$$P(\text{only speeding citation}) + P(\text{only parking citation}) \\ + P(\text{both citations})$$

In the first expression the overlap is included in each of the first two components and must be subtracted out; in the second expression there is no overlapping—the three components represent the three portions of the union as shown in Figure 4-3. If for 28-year-old drivers this year *P*(speeding citation) = 0.12, *P*(parking citation) = 0.24, and *P*(both citations) = 0.06, then

$$P(\text{either speeding or parking citation}) = 0.12 + 0.24 - 0.06 = 0.30$$

Note that *P*(both citations) is not equal to the product of the two individual probabilities; we discuss this type of probability in the next subsection. Referring to Figure 4-3, we note that

$$P(\text{only speeding citation}) = 0.12 - 0.06 = 0.06$$

$$P(\text{only parking citation}) = 0.24 - 0.06 = 0.18$$

so, from the second expression above,

$$P(\text{either speeding or parking citation}) = 0.06 + 0.18 + 0.06 = 0.30$$

4.1.2 Probability for the Intersection of Events

First, consider a simple experiment of tossing a (regular) penny and a (regular) dime. From the relative-frequency standpoint, the probability of a head showing on (the upper face of) the penny is 1/2. Likewise, *P*(head on the dime) = 1/2. What, then, is the probability of heads on both the penny and the dime? If the penny is tossed first and shows a head, that result really provides no information about the behavior of the dime: the chance of getting a head on the dime is still 1/2. One-half of the penny tosses should result in a head; on only one-half of these tosses should we also expect a head on the dime. In other words, both heads should occur one-quarter of the time, or *P*(both heads) = (1/2)(1/2) = 1/4.

For the coin-tossing experiment, the two events "head on penny" and "head on dime" have the property: knowledge of either event's occurrence or nonoccurrence gives no information about the probability of the other event's occurrence or nonoccurrence. Two events having this property are said to be *independent*. For independent events A and B, the probability of their intersection is the product of their individual probabilities:

$$P(A \cap B) = P(A)P(B) \tag{4.3}$$

When (4.3) does not hold, events A and B are said to be *dependent*. (Elsewhere, you may find these events called nonindependent.)

Let us consider two problems related to our earlier discussion. If we agree that 0.38 is the probability a 28-year-old driver will get a traffic citation this year, and if two 28-year-old drivers apply for auto insurance, what is the probability both drivers will get traffic citations this year? Or if a weather bureau reports a precipitation probability of 0.40 for today and 0.55 for tomorrow, what is the probability of precipitation for both today and tomorrow? In each case we are asking for the probability of an intersection occurring; nevertheless, the two problems differ markedly. Generally, knowledge about the occurrence of $A = \{\text{first driver gets a traffic citation this year}\}$ provides no information about the chances of $B = \{\text{second driver gets a traffic citation this year}\}$ occurring or not occurring, and we would treat A and B as independent events. But, for most localities, the occurrence or nonoccurrence of $C = \{\text{precipitation today}\}$ greatly affects the chances of $D = \{\text{precipitation tomorrow}\}$ occurring or not occurring, and we would treat C and D as dependent events.

Using (4.3), we get for the insurance example

$$P(\text{both drivers get traffic citations}) = (0.38)(0.38) = 0.1444$$

For the weather example, we cannot use (4.3), since events C and D are dependent. Our goal is to find $P(C \cap D)$, the probability that precipitation occurs both today and tomorrow. We were given $P(\text{precipitation today}) = P(C) = 0.40$. But, if precipitation occurs today, what is the probability that precipitation also occurs tomorrow? We are really asking for the probability "precipitation tomorrow given that precipitation occurs today." Probabilities such as this are called *conditional probabilities* and are written symbolically as

$$P(\text{precipitation tomorrow}|\text{precipitation today}) = P(D|C)$$

In this particular problem, event C represents the "given condition." Conditional probabilities are used in defining the probability of the intersection of two dependent events. Precisely, for dependent events C and D,

$$P(C \cap D) = P(C)P(D|C) \tag{4.4}$$

For the weather example, there is no way to calculate $P(C \cap D)$ without the value of $P(D|C)$.

To illustrate the use of (4.4), consider the selection of two apples from a bag of 20, four of which are wormy. Label the events as

$$C = \{\text{first one selected is wormy}\}$$

$$D = \{\text{second one selected is wormy}\}$$

If the apples are selected at random

$$P(C) = 4/20$$

$$P(\text{second one is wormy}|\text{first one is wormy}) = P(D|C) = 3/19$$

$$P(\text{two wormy apples}) = P(C \cap D) = (4/20)(3/19)$$

4.1.3 A Summary

Probability is a term applied to an event's uncertainty. Calculated from relative-frequency considerations, probability is a number indicative of an event's chances of occurring. As such, any probability must lie between

0 and 1, inclusive. To impossible events, we assign a probability of 0; to the sample space, $\mathcal{S}$, we assign a probability of 1.

Certain probabilities have simple structures. The probability for the union of mutually exclusive events is merely the sum of probabilities of the individual events forming the union. When finding the probability for the union of events not mutually exclusive, the probability of their overlap must be allowed to enter only once. The probability of this overlap (the intersection) is given by the product of certain probabilities, depending on whether the events are independent or dependent. The probability for the intersection of independent events is the product of probabilities of the individual events forming the intersection. The probability for the intersection of dependent events involves conditional probability.

(This first section of the chapter has emphasized concepts of probability rather than their calculations. Should the reader wish to calculate probabilities for a few simple problems without doing further reading, Problems 1 and 3 are especially recommended.)

*4.2 *CALCULATION OF PROBABILITY*

As indicated earlier, most calculated probabilities are based on relative frequency. Our discussion will be limited to a number of basic rules for calculating probability according to this basis. Some problems may require repeated application of a single rule, while others may require application of several rules.

*4.2.1 Properties of Probability

The structure of probability is governed primarily by two factors: (1) every conceivable event should have a probability, and (2) the set of all probabilities should be consistent in the relative frequency sense. In view of the first factor, probability may be described as a function relating the experimental sample space to a set of real numbers. And in view of the second factor, the following properties must hold:

(4.5)
(i) For any event A of the sample space $\mathcal{S}$, $0 \leqslant P(A) \leqslant 1$,
(ii) $P(\mathcal{S}) = 1$ and $P(\varnothing) = 0$, where $\varnothing$ is the null event,
(iii) For any set of mutually exclusive events, say $A_1, A_2, \ldots, A_k$,

$$P(A_1 \cup A_2 \cup \cdots \cup A_k) = P(A_1) + P(A_2) + \cdots + P(A_k)$$

Property (iii) is also called the *additive property of probability.*

*4.2.2 Sample Spaces Having a Finite Number of Outcomes

By their very nature, outcomes of any sample space $\mathcal{S}$ are mutually exclusive. Hence properties (ii) and (iii) of (4.5) require that the probabilities assigned to the outcomes of $\mathcal{S}$ add to unity. Furthermore, the probability for any event of $\mathcal{S}$ may be obtained by summing the probabilities assigned to those outcomes forming the event. Of course, when a large number of outcomes are involved, this summing process becomes tedious.

When selecting a digit at random from the set $\{1, 2, 3, 5, 6, 7, 8\}$, each digit should occur with relative frequency of 1/7. If we make the assignment $P(\text{any digit}) = 1/7$, then

$$P(\text{digit is divisible by 3}) = P(\text{selected digit is 3 or 6}) = 2/7$$

and

$$P(\text{odd digit}) = 4/7$$

For some experiments certain symmetries make it reasonable to assume that each outcome is equally likely to occur. In these cases the *Principle of Insufficient Reason* governs the assignment of probabilities; that is, when there is no particular reason to suppose that any one outcome is more likely than any other, each outcome is assigned the same probability. In the above example, each of the seven digits appears once in the set; no one digit should be favored if the selection is done randomly: each digit should occur with a frequency of 1/7. When tossing a well-balanced coin, the two outcomes "heads" and "tails" are assumed to be equally likely, and each is assigned a probability of 1/2.

The following examples further illustrate the Principle of Insufficient Reason and the summing process.

► ***Example 4-1***

If a well-balanced die is tossed once, each outcome is expected to occur 1/6 of the time. Thus, event A: uppermost face shows an odd number of spots has probability $P(A) = 3/6$. ◄

► ***Example 4-2***

If one card is drawn at random from a bridge deck, there are 52 possible outcomes, and each should occur with the same relative frequency. Then, events

A: red card
B: spade
C: red card or spade
D: face card (king, queen, or jack)

have probabilities

$P(A) = 26/52$
$P(B) = 13/52$
$P(C) = 39/52$
$P(D) = 12/52$. ◄

▶ ***Example 4-3***

Experiment $\mathscr{E}_5$, drawing a two-digit random number, has 100 equally likely outcomes. Events

A: 20 or greater
B: even number (00 taken as even)
C: second digit exceeds first

have probabilities

$P(A) = 0.80$
$P(B) = 0.50$
$P(C) = 0.45$. ◀

(Problems 5, 7, 9)

*4.2.3 Empirical Probabilities

When the Principle of Insufficient Reason is applicable, the assignment of probabilities to outcomes is straightforward. But how should we assign probabilities when the outcomes are not equally likely? Some experiments in this category are: (1) tossing a bent coin, (2) tossing a thumbtack, (3) recording the number of tomato seeds that germinate when six are planted in a peat pot, and (4) allowing a laboratory animal to run a maze with 10 possible paths.

And how should we assign probabilities when the outcomes are infinite in number? Experiments of this type include: (1) tossing a coin until a head appears, (2) selecting a real number at random from the interval $[0, 1]$, (3) recording the water pressure required to break a steel pipe, and (4) recording the time required for a laboratory animal to run a maze.

For experiments where the outcomes are not equally likely, or where the outcomes are infinite in number, consider the following procedure:

(1) perform the experiment a number of times, say n, and
(2) count the number of experiments where event A occurs; let this number be n_A.

Then the ratio n_A/n would represent an estimate of the true, unknown relative frequency of event A. This ratio is an empirical probability, a measure based on sampling or experimentation. Intuitively, larger and larger values of n should give more stable values for n_A/n. Thus, 5,000 tosses of a bent coin would hopefully give a "better" estimate of $P(\text{head})$ than would 1,000 tosses.

For most everyday situations, it simply isn't practical to obtain empirical probabilities because of the time and expense involved in repeating an experiment a large number of times. It is quite practical, though, to consider empirical probabilities when a substantial amount of empirical data is available from various records—as kept by weather bureaus, insurance companies, universities, and so on. In these cases only the time and expense of calculations are of concern.

Such issues as: (i) the size of n to use for "good" empirical probabilities, (ii) the stabilization of the ratio n_A/n, and (iii) the philosophical arguments for and against the use of empirical probabilities will not be pursued here. The interested reader should consult the references at the end of this chapter.

*4.2.4 Unions of Events

Sometimes we are asked to calculate the probability of a union of events. We may already have the probabilities of the individual events. Or, if the probability of an event E cannot be readily evaluated, we might be able to calculate the probabilities of other events whose union is E. We need to know how these individual probabilities are related to the probability of their union. There are two cases to consider: when the events are not mutually exclusive and when they are.

•• *Rule P-1:* For any events, A and B, of $\mathcal{S}$, their union has probability

$$P(A \cup B) = P(A) + P(B) - P(A \cap B) \tag{4.6}$$

▶ ***Example 4-4***

One card is drawn at random from a bridge deck and the following events are defined:

A: spade
B: honor card (ace, king, queen, jack, or ten)
C: black card

Then

$A \cap B$: spade honor card
$A \cap C$: spade
$B \cap C$: black honor card,

and by the Principle of Insufficient Reason,

$$\begin{aligned}
P(A \cup B) &= P(\text{spade or honor card}) \\
&= \tfrac{13}{52} + \tfrac{20}{52} - \tfrac{5}{52} = \tfrac{28}{52} \\
P(A \cup C) &= P(A) + P(C) - P(A \cap C) \\
&= \tfrac{13}{52} + \tfrac{26}{52} - \tfrac{13}{52} = \tfrac{26}{52} \\
P(B \cup C) &= \tfrac{20}{52} + \tfrac{26}{52} - \tfrac{10}{52} = \tfrac{36}{52}
\end{aligned}$$

These three probabilities could also be calculated by directly counting the number of outcomes in the unions:

$A \cup B$ is the set of 13 spades plus 15 nonspade honor cards, for 28 outcomes;
$A \cup C$ is the set of black cards which already contains the spades, or 26 outcomes;
$B \cup C$ is the set of 26 black cards plus 10 red honor cards, or 36 outcomes. ◀

▶ ***Example 4-5***

At an agricultural experiment station of a midwestern university, 16 plots of land are available for graduate students. The plots are numbered and a road runs around the entire set of plots as indicated.

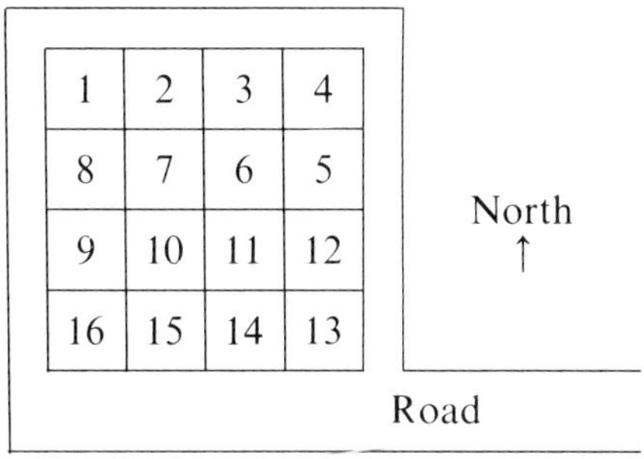

For a particular project, 10 plots are needed, and these are selected at random by 10 different teams of students. Suppose the first team to select a plot considers the following events

A: plot adjacent to road
B: plot in southern half
C: even-numbered plot

For this team

$$P(A \cup B) = 7/8$$
$$P(A \cup C) = 7/8$$
$$P(B \cup C) = 3/4.$$ ◀

When more than two events are involved, the probability of their union can be found in several ways: by direct summing of outcome probabilities (as discussed in Section 4.2.2), by repeated application of Rule P-1, or by an extension of Rule P-1.

For three events, A, B, and C, repeated application of Rule P-1 takes the following form:

(1) Find $P(A \cup B)$ by Rule P-1,
(2) Let $D = A \cup B$,
(3) Find $P(D \cup C)$ by Rule P-1,
(4) $P(D \cup C)$ is the desired probability, since $D \cup C = A \cup B \cup C$.

Other possibilities exist; we could have started with $A \cup C$ or with $B \cup C$.

The extension of Rule P-1 to the union of three events, A, B, and C, is:

$$P(A \cup B \cup C) = P(A) + P(B) + P(C) - P(A \cap B) - P(A \cap C) - P(B \cap C) + P(A \cap B \cap C) \tag{4.7}$$

The reader should verify for himself, by one or more of the above methods, that $P(A \cup B \cup C) = 9/13$ for Example 4-4.

When events A and B are mutually exclusive, the probability of their union is simply a sum of probabilities.

•• *Rule P-2:* If A and B are mutually exclusive events of $\mathcal{S}$, their union has probability

$$P(A \cup B) = P(A) + P(B) \tag{4.8}$$

Inductively, the additive structure of (4.8) is retained when Rule P-2 is extended to more than two mutually exclusive events. This extension is property (iii) of (4.5).

▶ ***Example 4-6***

When drawing two random digits, the events

B: The second digit exceeds the first.
C: Both digits are the same.
E: The sum of the two digits is 4.
F: The two digits differ by 3.

are such that $B \cap C = \varnothing$ and $E \cap F = \varnothing$. Thus

$$P(B \cup C) = \tfrac{45}{100} + \tfrac{10}{100} = 0.55$$

$$P(E \cup F) = \tfrac{5}{100} + \tfrac{14}{100} = 0.19 \blacktriangleleft$$

For some events, calculating the probability of the complement is an easier task. Since an event and its complement form a partition of $\mathcal{S}$—that is, $A \cup \bar{A} = \mathcal{S}$ and $A \cap \bar{A} = \varnothing$—the following special case of Rule P-2 is extremely useful:

•• *Rule P-3:* For any event A of $\mathcal{S}$

$$P(A) = 1 - P(\bar{A}) \tag{4.9}$$

▶ ***Example 4-7***

If a quarter is tossed until a head appears, the event

A: At least two tosses are required to get a head.

has probability

$$P(A) = 1 - P(\text{only one toss is required}) = 1 - 1/2. \blacktriangleleft$$

▶ ***Example 4-8***

When dealing a 13-card hand from a bridge deck, the probability that the hand contains at least two spades is given by

$$P(2 \text{ spades}) + P(3 \text{ spades}) + \cdots + P(13 \text{ spades}).$$

But, by Rule P-3, this sum is easily found as

$$1 - P(\text{no spades}) - P(1 \text{ spade}).$$

A solution of this problem is given in Example 4-16. ◀

(Problems 1(c), (d), (e), (f), (g), (h), 3(a))

*4.2.5 Conditional Probability

The symbol $P(A)$ stands for the probability of an event A which is defined for some experiment $\mathscr{E}$. Subsequent calculation of a value for $P(A)$ must be done with reference to the sample space $\mathcal{S}$ of $\mathscr{E}$. The underlying experimental procedure and its sample space are implicit parts of any probability, yet this fact is in no way revealed by the notation. Better notation for the probability of event A of $\mathcal{S}$ would be $P(A|\,\mathcal{S})$, which is read "the probability of event A given the sample space $\mathcal{S}$" or "the probability of event A given the event $\mathcal{S}$."

Consider an engineer who is trying to determine the time to failure of a component. One component that he has tested has operated successfully for 100 hours, and he wishes to know the chances that this component will operate successfully for another 100 hours. Thus, for the events

A: component lasts 200 hours
B: component lasts 100 hours,

the engineer would like to ascertain $P(A|B)$.

Or consider a secretive person who tosses a fair die and, without letting you see the die, tells you the number of spots on the uppermost face was at least two. What is the probability there were an odd number of spots on the uppermost face? In this case the desired probability is $P(\text{odd spots}|\text{at least two spots})$.

In any problem of this type, one might wonder whether knowledge that an event has already occurred lessens or betters the chances of another event's occurrence. The engineer might know that 100 hours of operation indicates a serious deterioration of the component; therefore, the probability of another hundred hours of operation would be greatly diminished. On the other hand, if the first hundred hours of operation represent a critical break-in period, it could well be that the chances of an additional hundred hours of operation are near certainty.

When $P(A) \neq 0$, the ratio $P(A \cap B)/P(A)$ is called the *conditional probability of event B given event A*, and is denoted by $P(B|A)$:

$$P(B|A) = P(A \cap B)/P(A) \quad \text{if } P(A) \neq 0 \tag{4.10}$$

Likewise,

$$P(A|B) = P(A \cap B)/P(B) \quad \text{if } P(B) \neq 0 \tag{4.11}$$

is called the *conditional probability of event A given event B*. Should one of the individual probabilities be 0, the corresponding conditional probability is undefined, of course.

Any conditional probability restricts attention to a "reduced sample space"—that portion of $\mathcal{S}$ corresponding to the given event. To measure the uncertainty of a second event in the reduced sample space where the given event occurs, we must determine the chances of both events occurring (the event of interest and the given event) relative to the reduced sample space. Hence the probability $P(A \cap B)$ must be adjusted; this is the purpose of dividing $P(A \cap B)$ by the probability of the given event.

In some instances it is easier to determine the conditional probability directly, instead of finding the probability of the intersection and dividing this by the probability of the given event. Thus, when drawing two cards without replacement from a bridge deck

$$P(\text{spade on second draw}|\text{spade on first draw}) = 12/51$$

since, at the second draw, 51 cards are available, 12 of which are spades—one spade was removed on the first draw.

► ***Example 4-9***

In the toss of a fair die, $P(\text{at least 2 spots}) = 5/6$, $P(\text{odd spots} \cap \text{at least 2 spots}) = 2/6$, whereby

$$P(\text{odd spots}|\text{at least 2 spots}) = \frac{2/6}{5/6} = 2/5.$$

And

$$P(\text{at least 2 spots}|\text{odd spots}) = 2/3. \blacktriangleleft$$

► ***Example 4-10***

Referring to Example 4-5,

$$P(A|C) = \frac{P(A \cap C)}{P(C)} = \frac{6/16}{8/16} = 3/4$$

$$P(B|A) = \frac{P(A \cap B)}{P(A)} = \frac{6/16}{12/16} = 1/2. \blacktriangleleft$$

The next rule gives the prescription for calculating the probability of an intersection of two events. This rule is a direct consequence of our assignment of conditional probabilities.

•• *Rule P-4:* If A and B are any two events of $\mathcal{S}$, their intersection has probability

$$P(A \cap B) = P(A)P(B|A) = P(B)P(A|B) \tag{4.12}$$

▶ ***Example 4-11***

Two cards are drawn at random without replacement from a bridge deck. Define

A: first card red
B: second card red

Then

$$P(\text{both cards red}) = P(A)P(B|A) = (1/2)(25/51). \blacktriangleleft$$

(Problems 3(d), (e), 11, 13)

*4.2.6 Independent Events

Note in Examples 4-5 and 4-10 that $P(A|C) = P(A) = 3/4$. Thus, the conditional probability of A given C is the same as the unconditional probability of A. This means that a knowledge of C occurring is immaterial as far as the probability of A is concerned. This is an important concept of probability.

Definition 4-1

Two events, A and B, are said to be *independent* (statistically) if the occurrence or nonoccurrence of either event gives no information about the probability of occurrence or nonoccurrence of the other event.

A pair of *independent events* are events for which the conditional and unconditional probabilities coincide. Thus, when A and B are independent, the conditional probabilities appearing in (4.12) may be replaced by the unconditional probabilities. This substitution is the basis of the next rule.

•• *Rule P-5:* If A and B are independent events, their intersection has probability

$$P(A \cap B) = P(A)P(B) \tag{4.13}$$

and conversely.

Rule P-5 is sometimes taken as the definition of independence.

If the relationship in (4.13) does not hold, the events A and B are said to be *dependent* (or *not independent*). Recall from the discussion at the beginning of this section that events A and C of Example 4-5 are independent. It is easy to show that each pair of events given in that example are independent.

If elements are drawn at random and with replacement from a finite set, events defined with respect to different draws are independent. When the set has an infinite number of elements, events defined on different draws are independent if the sampling is randomly done, with or without replacement.

► ***Example 4-12***

If a fair die is tossed twice, what is the probability of getting a pair of sixes? The two tosses are made independently of each other, whereby

$$\begin{aligned} P(\text{pair of sixes}) &= P(\text{six spots on first toss})P(\text{six spots on second toss}) \\ &= (1/6)(1/6) = 1/36. \blacktriangleleft \end{aligned}$$

► ***Example 4-13***

A psychologist studying the learning ability of a particular monkey defines events

A: task completed on the first day within 5 minutes
B: same task completed on the second day within 5 minutes

It is doubtful that these would be taken as independent events, since anything learned on the first day should be beneficial to the second day's performance. However, if the monkey is given different tasks on two successive days, the benefit received from the first day's performance may or may not affect the completion of the second task. ◄

► ***Example 4-14***

In the selection of two random digits, four events defined previously are

A: The first digit is a 2.
B: The second digit exceeds the first.
C: Both digits are the same.
E: The sum of the two digits is 4.

whereby,

$$\begin{aligned} P(A) &= 10/100 \\ P(B) &= 45/100 \\ P(C) &= 10/100 \\ P(E) &= 5/100 \end{aligned}$$

and

$$P(A \cap B) = 7/100 \neq P(A)P(B)$$
$$P(A \cap C) = 1/100 = P(A)P(C)$$
$$P(A \cap E) = 1/100 \neq P(A)P(E)$$
$$P(B \cap C) = 0 \neq P(B)P(C)$$
$$P(B \cap E) = 2/100 \neq P(B)P(E)$$
$$P(C \cap E) = 1/100 \neq P(C)P(E)$$

Consequently, the only events which are independent are A and C. ◀

In Example 4-14, we examined a set of four events and found only one pair, A and C, to be independent. We say that A and C are *pairwise independent*.

For a set of more than two events, say $A_1, A_2, \ldots, A_n$, the concept of their independence is *not* the simple, obvious extension of equation (4.13), which is

$$P(A_1 \cap A_2 \cap \cdots \cap A_n) = P(A_1)P(A_2)\cdots P(A_n) \tag{4.14}$$

Examples can be constructed where each pair of events satisfy (4.13) but (4.14) does not hold. Likewise, there are sets of events for which (4.14) holds but the events are not pairwise independent.

Definition 4-2

Events $A_1, A_2, \ldots, A_n$ are *independent* (also called mutually independent) if the intersection of events in *every* subset of $A_1, A_2, \ldots, A_n$ has probability given by the product of probabilities for the events in the subset.

Definition 4-2 implies that a number of relationships must hold:

$$\left.\begin{array}{l} P(A_1 \cap A_2) = P(A_1)P(A_2) \\ P(A_1 \cap A_3) = P(A_1)P(A_3) \\ \quad\vdots \\ P(A_1 \cap A_n) = P(A_1)P(A_n) \\ P(A_2 \cap A_3) = P(A_2)P(A_3) \\ \quad\vdots \\ P(A_{n-1} \cap A_n) = P(A_{n-1})P(A_n) \end{array}\right\} \text{all possible pairs}$$

$$\left.\begin{aligned} P(A_1 \cap A_2 \cap A_3) &= P(A_1)P(A_2)P(A_3) \\ &\vdots \\ P(A_2 \cap A_3 \cap A_4) &= P(A_2)P(A_3)P(A_4) \end{aligned}\right\} \text{all possible trios}$$

$$P(A_1 \cap A_2 \cap A_3 \cap A_4) = P(A_1)P(A_2)P(A_3)P(A_4) \Big\} \text{ all possible subsets of four events}$$

$$\vdots \qquad\qquad\qquad\qquad \vdots$$

$$P(A_1 \cap A_2 \cap \cdots \cap A_n) = P(A_1)P(A_2) \cdots P(A_n) \qquad \text{all } n \text{ events}$$

(Problems 7, 21)

*4.2.7 Probabilities for Partitioned Events

Section 3.7 discussed counting procedures for partitioned events. We now combine these counting techniques with the Principle of Insufficient Reason in the calculation of probabilities. Some rather complex probability problems can be solved by this method.

With the selection of n elements from a finite set of N elements, the number of outcomes is finite. If the elements are drawn at random, the $\binom{N}{n}$ different subsets are equally likely candidates, and the Principle of Insufficient Reason can be applied. It is therefore necessary to count the number of different subsets that yield a particular event.

Basically the set of N elements is partitioned according to the event in question, with possibly one subset containing all elements not specified by the event. In general terms, the event would be stated as follows:

A: The sample of n elements contains
n_1 elements from category 1
n_2 elements from category 2
$\vdots$
n_k elements from category k

Some, but not all, of the n_i may be zero; these categories are usually combined into one. Of course, $n_1 + n_2 + \cdots + n_k = n$. With respect to event A, the N elements are partitioned as in Figure 4-4.

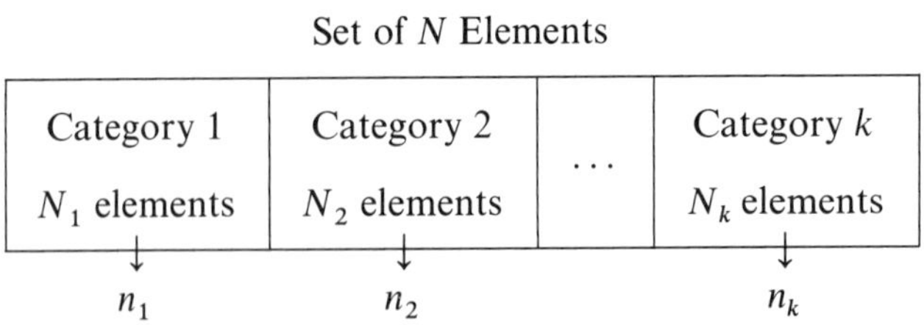

Figure 4-4 Partitioning of a Finite Set

The n_1 elements may be drawn, without regard to order, from the first category in $\binom{N_1}{n_1}$ ways. Similarly, there are $\binom{N_2}{n_2}$ ways of drawing the n_2 elements from the second category. Hence, by Rule C-3 (Section 3.8.2), there are $\binom{N_1}{n_1}\binom{N_2}{n_2}$ ways of drawing the elements from these two categories. Continuing in this fashion gives

$$(4.15)\quad P(A) = \frac{\binom{N_1}{n_1}\binom{N_2}{n_2}\cdots\binom{N_k}{n_k}}{\binom{N}{n}};\ \text{each } n_i \geq 0 \text{ with } n_1 + n_2 + \cdots + n_k = n.$$

Note that (i) two events may require two different partitionings of the set and (ii) a single event may define several partitionings. The following examples illustrate these points.

► ***Example 4-15***

Suppose that in an undergraduate course there are 14 men and 8 women who are also classified as 1 freshman, 7 sophomores, 10 juniors, and 4 seniors. The professor selects 6 students at random to put problems on the board. What is the probability that the selected students are
(a) half men?
(b) two sophomores and 4 juniors?
(c) all of the same class?
To find the first two probabilities, the 22 students would be partitioned as:

(a)

Men	Women
14	8
↓	↓
3	3

(b)

Sophomores	Juniors	Others
7	10	5
↓	↓	↓
2	4	0

$$P(\text{half men}) = \frac{\binom{14}{3}\binom{8}{3}}{\binom{22}{6}} = 0.2732$$

$$P(2 \text{ sophomores, } 4 \text{ juniors}) = \frac{\binom{7}{2}\binom{10}{4}\binom{5}{0}}{\binom{22}{6}} = 0.0591.$$

Since the selected students might be all sophomores or all juniors, the event defined in (c) does not define a unique partition. Thus,

Sophomores	Others
7	15
↓	↓
6	0

Juniors	Others
10	12
↓	↓
6	0

and from Rule P-2

$$P(\text{all of same class}) = \frac{\binom{7}{6}\binom{15}{0} + \binom{10}{6}\binom{12}{0}}{\binom{22}{6}} = 0.0029. \blacktriangleleft$$

▶ ***Example 4-16***

When dealing a 13-card hand from a well-shuffled deck of 52 cards, three events of interest are:

A: The hand contains 4 clubs and 6 hearts.
B: The hand contains 10 cards of one suit.
C: The hand contains at least 2 spades.

$$P(A) = \frac{\binom{13}{4}\binom{13}{6}\binom{26}{3}}{\binom{52}{13}}.$$

Note: If a 13-card hand contains 4 clubs and 6 hearts, it must also contain 3 cards which are neither clubs nor hearts.

Event B does not define a unique partitioning, but

$$\begin{aligned} P(B) &= P(10 \text{ clubs} \cup 10 \text{ hearts} \cup 10 \text{ diamonds} \cup 10 \text{ spades}) \\ &= 4\,P(10 \text{ clubs}) \cdots \text{by symmetry and Rule P-2} \\ &= 4\frac{\binom{13}{10}\binom{39}{3}}{\binom{52}{13}}. \end{aligned}$$

Using Rule P-3,

$$P(C) = 1 - \frac{\binom{13}{0}\binom{39}{13} + \binom{13}{1}\binom{39}{12}}{\binom{52}{13}}. \blacktriangleleft$$

(Problems 11, 15, 17)

PROBLEMS

1. Suppose a person throws a dart at this target

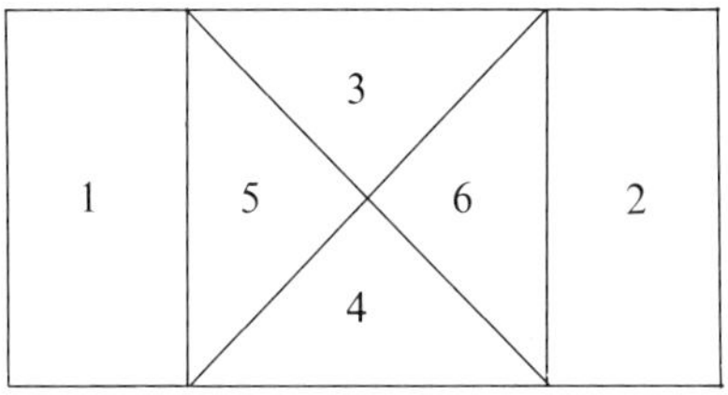

Segments 1 and 2 are equal in size; their combined area represents one-half the total area. Segments 3 through 6 also are equal in size. Assume that a dart is likely to land anywhere on the target—if the dart lands on a line or outside the target enclosure, the dart is thrown over. Find the probability for a single toss that the dart strikes

(a) segment 1. (b) segment 5. (c) segment 1 or 5.
(d) a segment having an even number.
(e) a segment having a number divisible by 3.
(f) a segment having a number at least 4.
(g) a segment having a number no greater than 4.
(h) a segment numbered 10.

2. One element is to be drawn at random from a set of 12 elements which are identical except for their numbers. They are numbered 1 through 10, with the number 2 occurring three times and all other numbers occurring once. Find the probability the selected element has a number

(a) equal to 3. (b) equal to 2.
(c) equal to 2 or 3. (d) that is even.
(e) greater than 5. (f) less than or equal to 7.

3. For an experiment, two events, A and B, and their probabilities are specified by the Venn diagram on the next page.

(a) Are A and B exhaustive events of this experiment? Why or why not?
(b) Are A and B mutually exclusive events? Why or why not?
(c) Are A and B independent events? Why or why not?
(d) Find $P(B|A)$.
(e) Find $P(\bar{A} \cap \bar{B})$.

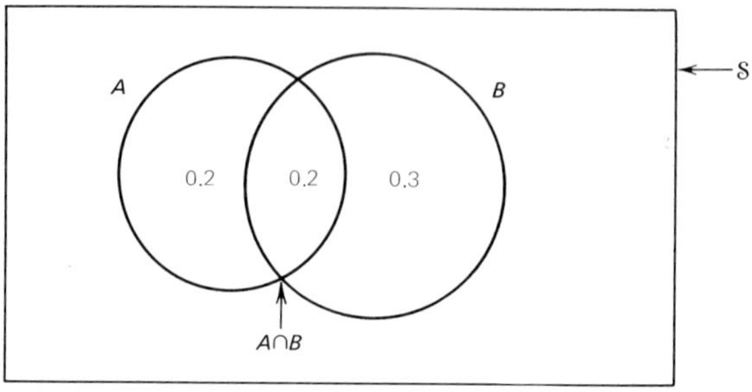

4. Repeat Problem 3 using the following Venn diagram:

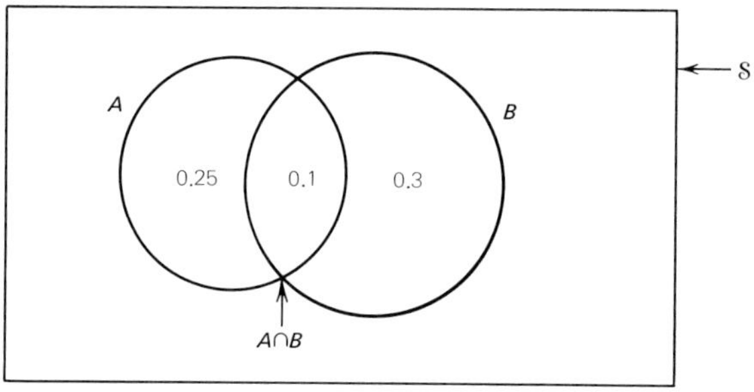

5. An engineer randomly arranges 15 cement molds in a curing furnace. Five contain mixture *A*, five contain mixture *B*, and five contain mixture *C*.
(a) How many different randomizations are there?
(b) What is the probability that the five molds containing mixture *A* are closest to the door?

6. Given the following spinner having three equal sectors

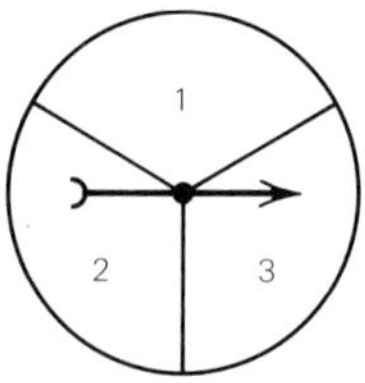

(a) the pointer is spun once. Set up the sample space for this experiment.
(b) the pointer is spun twice. Set up the sample space for this experiment.
(c) the pointer is spun six times. How many outcomes are there in the sample space of this experiment?

7. Assign probabilities to the outcomes in the sample spaces of the previous problem.

8. Find the probability for each of the following events:

Experiment	Events
(a) Two fair (well-balanced) coins are tossed.	Both show H (head). Exactly one shows H. At least one shows H.
(b) Five fair (well-balanced) coins are tossed.	All show H. Exactly three show H. At least three show H. At least one shows H.
(c) One fair (well-balanced) die is rolled.	Face up shows even spots. Face up shows at least four spots. Even and at least four spots Even or at least four spots.
(d) One card is drawn at random from an ordinary deck of 52 cards.	Red card. Ace. Red ace. Red or ace. Red or queen or spade. Red and queen and spade.
(e) A 5-card hand is dealt from a deck of 52 cards.	One pair. Two pairs. Full house. Four of a kind.

9. Some bowling alleys use one red and nine white pins on each alley. If a person makes a strike (knocks down all ten pins with the first ball) when the red pin is the head pin (occupies the number one spot), a prize is awarded. If the pins are randomly set by the pinsetter,
(a) how many different arrangements are there?
(b) what proportion of the time should a person have the opportunity of trying for the prize?

10. If two marbles are drawn simultaneously from a bag containing six white, four blue, and three yellow marbles, what is the probability that the two selected are
(a) of the same color?
(b) of different colors?
(c) one blue and one yellow?

11. If four marbles, instead of two, are drawn from the bag in the previous problem, what is the probability of getting
(a) all white?
(b) all of the same color?
(c) exactly two white?
(d) two white and two of the same color (not white)?

12. One letter is chosen at random from each of the two words P R O B A B I L I T Y and S T A T I S T I C S.

(a) What is the probability that the same letter is drawn from each word?

(b) Given that the two letters drawn were the same, what is the probability they were both T's?

13. At the start an urn contains four chips numbered 1, 2, 3, 4. One chip is drawn at random, its number recorded, and *two* chips (the one drawn plus another) with this number are placed in the urn. Then one chip is drawn at random from the urn, its number recorded, and *three* chips (the one drawn plus two others) with this number are placed in the urn. Finally, one chip is drawn at random from the urn and its number recorded. Find

(a) $P(1, 2, 4 \text{ in this order})$

(b) $P(1, 2, 4 \text{ in any order})$

(c) $P(4|1, 2 \text{ on the first two draws})$

14. An enemy code is sent to two code experts for "cracking." Suppose the probability that A cracks the code is 1/3 and the probability that B cracks the code is 1/5.

(a) Set up a suitable sample space describing the possible outcomes and assign probabilities.

(b) What is the probability that the code is cracked?

(c) If the code is cracked, what is the probability that A cracked it?

15. A class of forty is composed of 10 undergraduate students and 30 graduate students.

(a) If two students are selected at random to put problems on the board, what is the probability they are both undergraduates?

(b) If 10 students are selected at random to put problems on the board, what is the probability 8 graduate students are selected?

16. Consider the following nine digits as a sample space:

$$\mathcal{S} = \{1, 2, 3, 4, 5, 6, 7, 8, 9\}$$

Experiment: Draw one digit at random from $\mathcal{S}$.

Events: A: even digit

B: digit at least 4

C: digit at least 7

D: digit no greater than 5

Find

(a) $P(A)$ (b) $P(B)$ (c) $P(A|B)$

(d) $P(B|A)$ (e) $P(C|A)$ (f) $P(C|D)$

17. A library has 12 copies of a certain book. One copy has two missing pages, and two copies each have three missing pages. If all copies are on the shelf, and a person selects two copies at random, what is the probability the two selected copies have

(a) no missing pages?

(b) five missing pages?

(c) some missing pages?

18. A person decides to plant 4 blue spruce, 3 ponderosa pine, and 3 Douglas fir trees along his property line. If the person picks a random arrangement for planting the trees, what is the probability the 4 blue spruce trees would appear in the 4 locations closest to the house? (You may assume that the house is at one end of the property line.)

19. For hands of 5 cards dealt from a deck of 52 cards, find the probability of each of the following events:
(a) exactly 2 aces
(b) exactly 3 spades
(c) at least 3 spades
(d) all face cards (kings, queens, or jacks).
Set up only, do not evaluate.

20. Find the probability of obtaining each of the following bridge hands (13 cards from a deck of 52):
(a) all honors cards (aces, kings, queens, jacks, or 10's)
(b) all red cards
(c) 8 spades and 5 hearts
(d) at least 10 spades
Set up only, do not evaluate.

21. A certain basketball player makes 2/3 of his free throws. If he attempts 5 free throws during the first quarter of play, what is the probability he makes
(a) exactly 4 of them?
(b) none of them?
(c) all of them?
(d) at least 3 of them?

22. Unknown to you upon your arrival at the bakery section of a grocery store, there are 40 loaves of fresh bread and 20 loaves of day-old bread; the day-old bread has been dispersed throughout the fresh. You select 4 loaves at random. (Squeezing will not help in detecting freshness, as the loaves are triple-wrapped.) What is the probability of getting
(a) 4 fresh loaves?
(b) 4 loaves of day-old bread?
(c) 2 fresh and 2 day-old loaves?
If the first two loaves that you open when you get home are fresh, what is the probability that the remaining two loaves are also fresh?

23. A grabbag contains 30 identically wrapped boxes, 10 of which are empty, 10 contain items worth less than \$1.00, and 10 contain items worth over \$5.00. If the 30 boxes are available when a customer arrives to buy 3 items, what is the probability of getting
(a) one from each category?
(b) all empty boxes?
(c) all items worth more than \$5.00?
(d) all items having some value?

24. Four different varieties of wheat (similar in appearance but genetically different) were used in an experiment. Ten plants of each variety were randomly located in the experimental section of the greenhouse. Mice invaded the greenhouse and chewed labels from 3 variety A plants, 2 variety B plants, and 2 variety D plants. What are the chances of correctly identifying these 7 plants if

(a) they are identical in appearance and nothing further is known about them?

(b) they are identical in appearance but one variety D plant can be positively identified from its location?

REFERENCES

[1] Feller, W., *An Introduction to Probability Theory and Its Applications*, Vol. I, 3rd ed., Wiley, New York, 1968.

[2] Guenther, W. C., *Concepts of Probability*, McGraw-Hill, New York, 1968.

[3] Kolmogorov, A. N., *Foundations of the Theory of Probability*, Second English Edition, Chelsea, New York, 1956 (translation of the original 1933 German monograph and a 1936 Russian translation).

[4] Mullins, E. R., Jr., and D. Rosen, *Probability and Calculus*, Bogden & Quigley, Tarrytown, N.Y., 1971.

[5] Parzen, E., *Modern Probability Theory and Its Applications*, Wiley, New York, 1960.

Chapter 5

PRESENTATION OF SAMPLE DATA

After an experiment is conducted, the results and conclusions are usually summarized and presented in a bulletin, technical report, journal article, or some other "write-up." Because of space and cost considerations, and because readers find detailed tables boring, most experimental summaries do not include a complete listing of the individual sample observations, known as the *raw data*. A condensed version of the sample data is often included, however.

The remainder of this chapter is devoted to some techniques of presenting sample data in condensed form. In addition to their role in experimental summaries, these techniques will prove useful in our later examinations of distributions and probability.

5.1 THE FREQUENCY DISTRIBUTION

One of the most common forms of data presentation is the *frequency table* or *frequency distribution*, a tabulation showing the number of observations occurring in each of a number of classes. As such, the frequency distribution summarizes the sample data by grouping.

Before the advent of high-speed computers, frequency distributions were extremely valuable for simplifying calculations. Even now in cases where the sample size is large and computing facilities are not readily available, the frequency distribution can result in a considerable saving of time and effort in many calculations; this will be discussed in Section 7.10.

The first step in constructing a frequency distribution is to determine the sample range, the largest observation minus the smallest observation. In conjunction with the range, we can specify two additional (and arbitrary) quantities: the number of classes and the class designations. The classes must have total range no smaller than the sample range. The number of classes is normally between 6 and 25, depending upon the sample size and the nature of the data. If we use too many classes, some may contain no observations; using too few may distort the actual distribution. An indication of the number of classes to use is furnished by a rule devised by Sturges [1]:

$$k^* = 1 + 3.3 \log(n), \tag{5.1}$$

where k^* is the indicated number of classes, n is the sample size, and the logarithm is to the base 10. Actually, in [1] Sturges gave a formula for the class interval in terms of the sample range and k^*, indicating, as we already have, that these three quantities are related.

Once the classes have been defined, we can determine the *class limits*, the smallest and largest observed value belonging to each class. *Class boundaries* are dividing points between adjacent class limits while the *class intervals* are found by subtracting the lower boundary from the corresponding upper boundary. If equal-sized classes are not used whenever possible, certain modifications are necessary.

Classes should be constructed so that no observation can belong to more than one class. Since the sample observations are always measured to a fixed number of decimal places, the sample data can easily be divided into nonoverlapping classes by adding an extra decimal to the class boundaries.

After the classes are defined, we can determine the *class frequency*, the number of sample elements belonging to a given class. The actual number of elements in a given class is also called the *absolute frequency* or simply the *frequency*. If we divide the absolute frequency by the sample size, we get the *relative* frequencies which can be converted to *percentage* frequencies upon multiplication by 100. In a given problem, we might use one, two, or all three types of frequencies.

In the frequency distribution, all sample elements belonging to a given class are replaced by a typical or representative element called the *class mark*. Unless there is some reason for doing otherwise, the midpoint of the class limits (or boundaries) is taken to be the class mark.

In general, then, the individual observations cannot be identified in a frequency distribution. The original sample cannot be regenerated from the frequency distribution. Two exceptions to this are (i) when the classes each have a single nonnumerical designation or (ii) when the upper and lower limits coincide for each class.

Let k denote the number of classes to be used in the frequency table. Then, for $i = 1, 2, \ldots, k$, the symbols f_i and X_i will denote, respectively, the class frequency and the class mark of the i-th class.

To summarize, the frequency distribution is the list of classes together with corresponding measures of frequency; the values of X_i may or may not be given. A title and the source of data (if available) are implicit parts of any data presentation.

▶ ***Example 5-1***

A class of 112 students enrolled in an introductory statistics course were polled on the first day to determine their college distribution. The results are given in the following frequency distribution. The classes have non-numerical designations.

Distribution of Students in Introductory Statistics

College	Number of Students (f_i)	Relative Frequency
Arts and Science	34	0.304
Education	13	0.116
Engineering	26	0.232
Agriculture	17	0.152
Business	20	0.178
Law	2	0.018

◀

▶ ***Example 5-2***

A person takes inventory of the books he purchased for his undergraduate training and subsequently kept. He has 40 books whose costs were:

$8.25	8.75	1.45	3.60	8.00
8.95	14.95	4.75	6.95	2.75
7.25	6.70	1.50	6.50	10.50
10.50	7.50	3.40	8.75	5.95
9.95	8.25	5.40	8.75	7.95
7.50	11.00	7.00	3.25	14.75
4.75	1.75	1.75	1.40	1.65
7.95	3.50	9.50	7.50	12.50

The sample range is $14.95 − 1.40 = 13.55. By Sturges' rule, $k^* = 1 + 3.3\log(40) \doteq 6.3$; about 6 classes should be used. With the sample range of 13.55, seven classes each of size 2.00 would be a convenient choice. One possible grouping of the data would be the following frequency distribution.

Value of Books in a Personal Library

Cost	Number of Books (f_i)	X_i
\$1.00– 2.99	7	1.995
3.00– 4.99	6	3.995
5.00– 6.99	5	5.995
7.00– 8.99	14	7.995
9.00–10.99	4	9.995
11.00–12.99	2	11.995
13.00–14.99	2	13.995

Boundaries of the above classes are: 0.995–2.995; 2.995–4.995; and so on. ◄

(Problems 1(a), 3(a), 9)

5.2 *CUMULATIVE FREQUENCY DISTRIBUTIONS*

Suppose the person referred to in Example 5-2 wishes to know what proportion of his books cost \$5.00 or more. He can accumulate the frequencies of the last five classes in the frequency distribution from which he can get the desired proportion. Similarly, an accumulation of the frequencies of the first four classes gives information about the books costing less than \$9.

Tabulations showing the accumulation of frequencies for the classes either above or below each class boundary are called *cumulative frequency distributions.* The accumulations below each class boundary, also called the *less than cumulative distribution,* is more common and more useful than the *or more cumulative distribution.* Many theoretical concepts are defined and investigated by way of less than cumulative distributions. As in the case of the frequency tables, the cumulative frequencies may be given in absolute, relative, or percentage form.

► ***Example 5-3***

The less-than-cumulative distribution for the book data of the last example is:

Value of Books in a Personal Library

Cost	Number of Books
Less than or equal to \$0.995	0
2.995	7
4.995	13
6.995	18
8.995	32
10.995	36
12.995	38
14.995	40

◄

(Problems 1(b), 3(b))

5.3 GRAPHICAL REPRESENTATIONS

Graphs are often used in place of, or in addition to, the tabular presentations discussed in the two previous sections. Graphs corresponding to the frequency and cumulative frequency distributions are identified in the next two definitions.

Definition 5-1

A *histogram* is the graphic form of a frequency distribution consisting of rectangles erected over the class intervals such that the *areas* of the rectangles are proportional to the class frequencies.

Note that each rectangle extends from the lower class boundary to the upper boundary, leaving no gaps between the rectangles. Whenever equal class intervals are used, the heights of the rectangles are proportional to the class frequencies. Otherwise, the heights must be adjusted so that the *areas* of the rectangles are proportional to the class frequencies.

A vertical scale depicting either absolute, relative, or percentage frequencies may be used. Relative frequency histograms will prove useful later in studying distributions and related topics.

Definition 5-2

A line graph connecting the cumulative frequencies plotted over the class boundaries is called an *ogive* (pronounced o-jive).

Since there are two types of cumulative distributions, there are two types of ogives. Each may be presented in absolute, relative, or percentage form.

► ***Example 5-4***

The histogram of the frequency table given in Example 5-2 is:

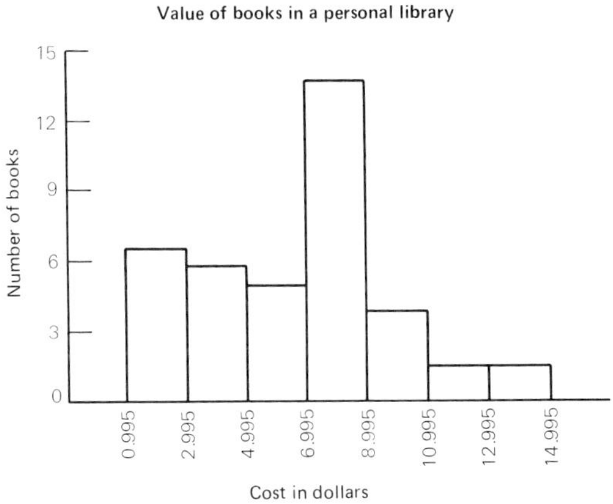

Dividing each value on the vertical scale by $n = 40$ would give the proportion of books in each interval; that is, the relative frequency histogram. ◀

▶ ***Example 5-5***

The ogive for the cumulative distribution given in Example 5-3 is:

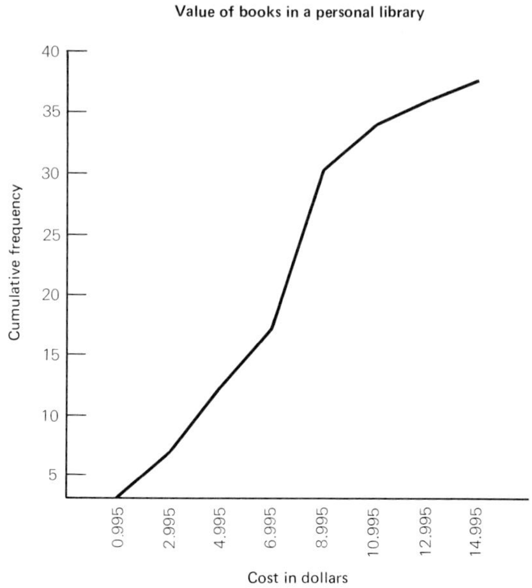

◀

PROBLEMS

1. The following represent live weights, to the nearest quarter-pound, of whitefish caught in the Madison River:

1.25	0.75	1.00	1.00	1.50	1.25
1.00	1.50	0.75	0.75	1.75	1.25
2.50	2.00	1.00	1.25	1.00	1.25
0.75	1.50	1.50	1.75	1.00	1.50
1.50	2.25	2.75	1.25	1.50	2.00
1.50	1.00	1.75	1.50	1.00	2.25

(a) Construct a frequency distribution using class limits.
(b) Construct a cumulative frequency distribution.

2. (a) Draw the histogram corresponding to the distribution found in Problem 1(a).
(b) Draw the ogive corresponding to the distribution found in Problem 1(b).

3. Traveling a certain route to work, a person found the travel time, in seconds, over a 6-week period to be:

350	533	715	445	295	674
426	565	591	616	380	563
815	488	436	566	621	581
744	846	810	632	687	387
612	268	542	474	439	650

(a) Give the frequency distribution using equal-sized classes.
(b) Give the cumulative frequency distribution.

4. (a) Convert the tabulation in Problem 3(a) to a relative frequency distribution.
(b) Convert the tabulation in Problem 3(b) to a relative frequency form.

5. A field infested with grasshoppers was sprayed with an insecticide. After 20 minutes, the number of dead grasshoppers found on 40 randomly chosen areas (each 9 square feet) was:

3	16	19	21	41	9	18	22
18	31	15	11	26	14	5	8
7	8	23	7	23	29	34	16
22	14	12	15	16	21	17	9
29	26	7	36	19	13	26	38

(a) Construct a frequency distribution using 5 equal-sized classes.
(b) Give the less-than-cumulative distribution using the results of (a).

6. (a) Present the histogram related to Problem 5(a).
(b) Present the ogive related to Problem 5(b).

7. Final exam grades for 60 students completing an elementary statistics course were:

53	86	67	82	85	80	92	85	93	79
91	78	42	90	83	88	70	84	80	76
64	78	88	47	54	70	62	74	73	77
91	61	60	79	76	95	94	82	90	92
94	62	90	95	60	91	84	85	92	82
82	90	86	48	39	81	79	76	55	78

(a) Set up a frequency distribution for these data. Use equal-sized classes.
(b) Give the cumulative distribution for final exam grades. Use the results of (a).

8. (a) Draw the histogram for Problem 7(a).
(b) Draw the ogive for Problem 7(b).

9. The following frequency distribution is available to you.

Class	Frequency
0– 49	14
50– 99	12
100–149	9
150–199	5
200–249	5

(a) How many observations were there?
(b) Give the class limits.

(c) Give the class boundaries.
(d) Give the class marks.
(e) Give the relative frequencies.

10. Repeat Problem 9 using the following frequency distribution:

Class	Frequency
0.5–10.5	8
10.5–20.5	15
20.5–30.5	6
30.5–40.5	7
40.5–50.5	20
50.5–60.5	4

REFERENCE

[1] Sturges, H. A., "The Choice of a Class Interval," *Journal of American Statistical Association* (March 1926).

Chapter 6

DISTRIBUTIONS

The concepts discussed so far—functions, variables, randomness, experiments, sample spaces, outcomes, and events—can now be coordinated with other basic ideas to form the foundation of statistical analysis.

6.1 RANDOM VARIABLES

Each experiment provides information about a multitude of variables. For five cars selected at random from those passing a traffic checkpoint, we might evaluate one or more of such variables as: (1) the number of cars having faulty equipment, (2) the number of foreign-made cars, (3) the number of cars having only one occupant, and so on. Or when a person is selected at random from a population, we might determine: (1) his weight, (2) his height, (3) his IQ score, (4) his annual salary, (5) the number of his brothers and sisters, and so on.

Each of the variables mentioned above assigns a real number to the experimental outcome. And, since randomness is an implicit part of experimentation (Chapter 3), the values of each variable occur according to underlying probabilities associated with each variable. These facts are summarized in the following definition.

Definition 6-1

A random variable is any (real-valued) function defined on the sample space of an experiment.

Random variables are also called *chance variables* or *stochastic variables.*

As we see from Definition 6-1, the domain of any random variable is the sample space of an experiment; the range is a set of real numbers. The range of each variable defined for the car example is a finite set; the number in each case must be 0, 1, 2, 3, 4, or 5. For height and weight variables of the person, the range is an uncountable set; we might obtain any real number in some interval. The range provides a means of classifying random variables.

Definition 6-2

A random variable† is said to be

(i) *discrete* if its range is either a finite or a countable set,

and

(ii) *continuous* if its range is an uncountable set.

A discrete random variable can take on at most a countable number of different values, while a continuous random variable can take on an uncountable number of different values, any value in an interval or set of intervals.

Verbal descriptions of random variables will be denoted by capital letters, with the corresponding lower-case letter denoting the *values* of the random variable. Thus X is a random variable, and x is some value which X takes on. Or X represents the entity that we can measure repeatedly, while x represents a specific value that we obtain from one observation (or measurement) of the entity. This distinction between random variables and their values is also necessary in order to have symbolism for the probability that a random variable takes on certain of its values; $P(X = x)$ is the probability that X takes on the value x.

In the car example some variables and their values are:

X = the number of cars having faulty equipment
x = 0, 1, 2, 3, 4, or 5

Y = the number of foreign-made cars
y = 0, 1, 2, 3, 4, or 5.

And in the example of selecting one person, the following variables and values could be assigned:

Z = the number of his brothers and sisters
z = 0, 1, 2, 3, . . .

W = his height
w = any value between, say, 20 and 90 inches.

† A third case where the random variable is a mixture of the discrete and continuous types will not be discussed here because it rarely occurs in applications.

There is no restriction on either the number or the type of variables that a researcher might consider in a given experiment. All variables defined in the car example above were discrete variables. But notice for the selection of a person, both continuous and discrete variables were defined (the first three continuous, the last two discrete). In the car example, we could obtain information on the continuous variables

U = length of wheelbase for each car

and

V = weight of each car

as well as information on the three variables stated earlier.

(Problems 1, 3)

6.2 ESSENTIALS OF DISTRIBUTIONS

In a given experiment the researcher restricts his attention to those random variables which give him the most useful information. Each random variable he chooses to use has a unique distribution. By "the distribution of a random variable," we mean the set of all possible values of the random variable and the associated probability structure. We would like to know how probability is distributed over the outcomes of our experiment.

The term "distribution" will be used only with reference to the entire population of values that the random variable takes on. On the other hand, an *empirical distribution* is the term used to mean a tabulation based on sample observations. Empirical distributions were discussed in Chapter 5.

The distribution of a random variable having only a finite number of values can be specified by listing the values and their associated probabilities. When the number of values of a random variable is infinite or finite but very large, the listing procedure would not be used. Alternatives include the use of equations, formulas, and graphs. Two examples illustrate these methods.

▶ *Example 6-1*

If a random variable, X, can take on the eight values, 5, 6, 7, 8, 9, 10, 11, and 12, and if all values are equally likely, we can specify the distribution of X in any one of the following ways:

(1) X has eight equally likely values:
$x = 5, 6, 7, 8, 9, 10, 11$, and 12

(2) Probability is distributed equally over the eight X values:
$x = 5, 6, 7, 8, 9, 10, 11$, and 12

(3) $P(X = x) = 1/8$, for $x = 5, 6, 7, 8, 9, 10, 11, 12$

The phrase *discrete uniform distribution* is used to describe a distribution having equally likely values. We say, in this case, "X has a discrete uniform distribution." ◀

▶ ***Example 6-2***

Analogous to the discrete uniform distributions (Example 6-1) are the *continuous uniform distributions.* Suppose buses arrive at (and leave) a bus stop every 10 minutes, and a person arrives at the bus stop at a random time. Then, the random variable

$$Y = \text{time he must wait for a bus}$$

has a uniform distribution given by

$$f(y) = 1/10; \qquad 0 \leqslant y \leqslant 10.$$

In graphic form

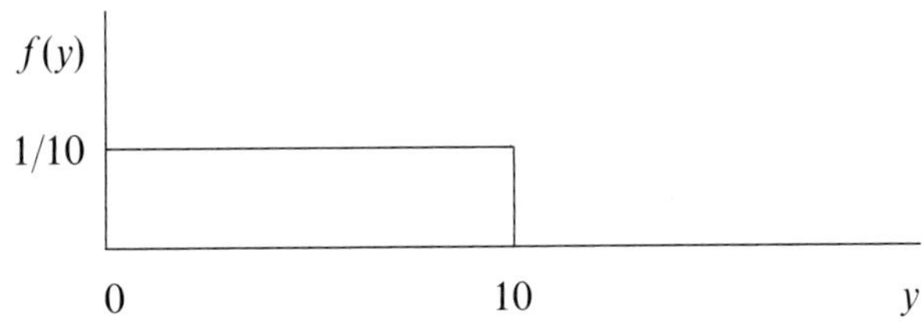

He may arrive at the same instant as a bus does, or he may arrive as a bus is leaving, in which case he must wait 10 minutes for the next bus. Here "uniform" refers to the same probability for all intervals of a given size. There is the same chance of waiting between 1 and 2 minutes for a bus as there is of waiting between 4.5 and 5.5 minutes, and so on. ◀

In dealing with a random variable U having 100 equally likely values (as yet unspecified), a complete listing of the values, like that in Example 6-1 may be unnecessary. Should the values be successive integers beginning with 12, say, the distribution of U can be specified by saying, "U has $N_1 = 100$ equally likely values which are successive integers beginning with $m_1 = 12$." Of course, for the variable X of Example 6-1, we could say: "X has $N_2 = 8$ equally likely values which are successive integers beginning with $m_2 = 5$." There are many discrete distributions of the type "N equally likely values which are successive integers beginning with the integer m." Likewise, the expression

$$f(y) = 1/10; \qquad a \leqslant y \leqslant a + 10$$

specifies many continuous uniform distributions; the one for $a = 0$ appears in Example 6-2.

By changing the values of N and m in the first general expression above, and the value of a in the second general expression, we can generate two classes of different distributions. We call quantities like N, m, and a *parameters.*

A *constant* is a numerical quantity, such as 1, π, $\sqrt{2}$, which remains the same in *every* discussion. A *parameter* is a numerical quantity which might vary from one discussion to the next but is constant throughout any one discussion. A *variable* is a numerical quantity which might vary even throughout a single discussion.

Many times a number of distributions can be considered under a general framework because they have the same structure but differ only in the values of their parameter(s). When this is the case, we say that the collection of distributions "depends upon" or "can be indexed" by a certain set of parameters. A collection of distributions that is indexed by a certain set of parameters is called a *family of distributions.* The phrase "N equally likely values which are successive integers beginning with the integer m" defines a family of distributions that is indexed by two parameters, N and m. Each different assignment of values to N and m gives a different distribution, but all have the same probability structure—namely, $P(X = x) = 1/N$. We call this collection of distributions "a two-parameter family of discrete distributions." Earlier, we discussed two members of this family—those having parameters $N_1 = 100$, $m_1 = 12$ and $N_2 = 8$, $m_2 = 5$.

A second collection of distributions that we discussed was a one-parameter family specified by

$$f(y) = 1/10; \qquad a \leqslant y \leqslant a + 10$$

Here the parameter is a, a quantity which fixes the location of the distribution.

At this time we should emphasize the fact that distributions have parameters other than indexing parameters. Certain parameters, whether they are indexing parameters or not, are important in describing specific features of a population. Two important classes of parameters in this regard are: (i) measures of location† and (ii) measures of dispersion.

Location parameters are quantities which describe "where the distribution is situated," usually by specifying the middle of the distribution, or an endpoint. ("Middle" may refer to the center in some sense, to an average value, and so on.) The most common measure of location is the *population mean,* denoted by μ (Greek letter mu), which is the average value of the population elements. Two other measures of location are the population median and the population mode. More will be said about these parameters and their estimation in ensuing chapters.

Dispersion parameters are quantities which describe "how the population elements distribute about the middle value." The most common measure

† Elsewhere, you may find these parameters called "measures of central tendency."

of dispersion is the *population variance*, denoted by σ^2 (the square of the Greek letter sigma). If we consider the deviation of each population element from the mean μ, the variance σ^2 is the average of these squared deviations. The positive square root of the variance σ^2 is called the *population standard deviation*, and is denoted by σ. Even though σ may be obtained from σ^2, and vice versa, we retain both parameters because each occurs in various phases of statistical analysis. Less common measures of dispersion include the population range and the population interquartile range.

A distribution can be specified by giving, in addition to the distributional family, either values of the indexing parameters or values of other parameters. To determine the precise values of such parameters, we would have to evaluate the entire population. Consequently, we are usually forced into accepting *estimates* of parameters based on a random sample of the population. These calculated quantities are called *statistics*. Specifically:

Definition 6-3

Any function whose value can be determined from sample observations is called a *statistic*.

Notice from this definition that a statistic cannot contain unknown quantities, such as unspecified parameters. Statistics used to estimate parameters are frequently called *estimators*.

Parametric inference is a primary objective of most experimentation, and statistics are the vehicles used in accomplishing this objective. Now, here is the crux of the problem:

> Every statistic is a random variable whose values are governed by the probability structure of the population being sampled.
> Therefore, every statistic has a distribution which is determined by the probability structure of the underlying population.

The distributions of statistics are called *sampling distributions* and are fundamental in statistical inference. They are needed to investigate and compare properties of different statistics, for determining the degree of confidence in our inferences, and so on.

Finding the distribution for any random variable is generally difficult. The distribution for a random variable satisfying a fixed set of assumptions can often be determined with recourse to probability theory. Except for a few simple cases, this distribution theory is beyond the scope of this book. Refer to [1], [2], [3], [4], and [5].

We conclude this section with a simple example to show what is involved in finding a distribution.

▶ ***Example 6-3***

Take the following set of numbers to be equally likely values of a random variable $X:\{2, 6, 8, 10, 12\}$. The population mean is $\mu = 38/5$. Ignorant of this fact, a person wishes to estimate μ from a random sample of size 2 (drawn without replacement from the population). Suppose the person decides to use the sample average as the estimator of μ, denoting this average by Y. Then the distribution of Y is the set of all values that Y can take on and the probability of each value occurring.

Because the population is small and known to us, we can easily determine the distribution of Y from a listing of the different samples:

Sample	Sample Values	Sample Average $= Y$
1	2, 6	$y_1 = 4\ [=(2+6)/2]$
2	2, 8	$y_2 = 5$
3	2, 10	$y_3 = 6$
4	2, 12	$y_4 = 7$
5	6, 8	$y_5 = 7$
6	6, 10	$y_6 = 8$
7	6, 12	$y_7 = 9$
8	8, 10	$y_8 = 9$
9	8, 12	$y_9 = 10$
10	10, 12	$y_{10} = 11$

Random sampling assures us that each sample occurs with probability of 1/10. Hence, each individual value of Y listed in the third column has probability of 1/10 of occurring. There are only 8 distinct values of Y, since two values each occur twice: $y_4 = y_5$ and $y_7 = y_8$. The distribution of Y is specified by

$$
\begin{aligned}
P(Y = y) &= 1/10 \quad \text{if } y = 4, 5, 6, 8, 10, \text{ or } 11 \\
&= 2/10 \quad \text{if } y = 7, \text{ or } 9.
\end{aligned}
$$ ◀

(Problems 5(a), 7(a), (b), (c))

6.3 *GENERAL TERMINOLOGY*

As indicated earlier, most of the mathematical foundations of distributions are beyond the scope of the present material. We will simply state and rationalize certain basic results of distribution theory. By concentrating on the rationale of each result and ignoring the mathematical details of their proof, we hope to bring more clearly into focus the type of situations where each "borrowed" result is applicable.

The members of a family of distributions need identification. An arbitrary member of a family of distributions will be denoted by $f(x; \theta_1, \theta_2, \ldots)$, where

$\theta_1, \theta_2, \ldots$ (Greek letter theta), are the parameters required for indexing. For a single parameter family the notation reduces to $f(x;\theta)$. Should it be necessary to indicate the random variable, we will use the notation $f_X(x;\theta)$, or $f_X(x;\theta_1,\theta_2,\ldots)$. Note that a semicolon separates the variables and parameters.

▶ ***Example 6-4***

A one parameter family of distributions is the set defined by

$$f(x;\theta) = (\theta + 1)x^{\theta}; 0 \leqslant x \leqslant 1 \text{ and } \theta > -1$$
$$= 0; \text{ otherwise}$$

Each value of θ greater than minus one defines a different distribution. (The family is actually a subset of a much larger family called "the Beta distributions.") ◀

The cumulative distribution of X is defined for all real values, x, by

(6.1) $$F_X(x;\theta) = F(x;\theta) = P(X \leqslant x;\theta)$$

and gives the probability that the random variable X has a value less than or equal to x when the distribution is specified by the parameter θ. Again, we will use the simpler expression $F(x;\theta)$ unless the random variable needs to be indicated. When more than one parameter is involved, θ appearing in (6.1) is replaced by the set of parameters. We will consider only those distributions for which $F(x;\theta)$ is either a step function or a continuous function. The former correspond to discrete distributions, the latter to continuous distributions.

In the discrete case, $f(x;\theta)$ gives the probability that X takes on the value x; that is,

(6.2) $$f(x;\theta) = P(X = x;\theta)$$

whereby (6.1) can be written, in the discrete case, as

(6.3) $$F_X(x;\theta) = \text{sum of all } f_X(t;\theta) \text{ for which } t \leqslant x$$

In the continuous case $f(x;\theta)$ is called a *probability density*. The probability is 0 at any point, but $f(x;\theta)$, when not 0, gives the density (height of the curve) of X at the point x. Equation (6.1) is valid in the continuous case, but the right-hand side of (6.3) would be replaced by an integral. Interpreting the integral as an area under a curve, it follows in the continuous case that probabilities are related to areas and are non-zero only for intervals which are not single points.

► ***Example 6-5***

Suppose X has a discrete distribution specified by the following:

$$f(0.5) = P(X = 0.5) = 2/10$$
$$f(1) = P(X = 1) = 3/10$$
$$f(2) = P(X = 2) = 4/10$$
$$f(4) = P(X = 4) = 1/10$$

Then the cumulative distribution of X is the step function

$$\begin{aligned} F(x) &= 0, && \text{for } x < 0.5 \\ &= 2/10, && \text{for } 0.5 \leqslant x < 1 \\ &= 5/10, && \text{for } 1 \leqslant x < 2 \\ &= 9/10, && \text{for } 2 \leqslant x < 4 \\ &= 1, && \text{for } x \geqslant 4. \end{aligned}$$ ◄

(Problems 5(b), 7(d), (e), (f))

6.4 *EXPECTED VALUES*

The *expected value* of a random variable is defined as a special average value. For a random variable X, we denote its expected value by $E[X]$, where the capital letter E is an operator used to denote the averaging process. Values of any random variable may occur with different frequencies; by weighting each value of a random variable by its relative frequency, each value contributes its fair share to the expected value. Thus, if two values occur with the same frequency, each should contribute equally to the expected value. But if $P(X = 2) = 1/3$ and $P(X = 8) = 2/3$, we "expect" to see the value 8 twice as often as the value 2. The expected value of X is

$$E[X] = 2(1/3) + 8(2/3) = 6$$

► ***Example 6-6***

If we define X to be the number of heads showing when four balanced coins are tossed, X has values $x = 0, 1, 2, 3,$ and 4, with respective probabilities 1/16, 4/16, 6/16, 4/16, 1/16. The expected number of heads on any toss of these coins is

$$\begin{aligned} E[X] &= 0(1/16) + 1(4/16) + 2(6/16) + 3(4/16) + 4(1/16) \\ &= 2 \end{aligned}$$

We "expect" half the coins to show heads on any toss. So on 4 tosses, we "expect" 2 heads to show. This agrees with our calculations. ◄

Generally, the expected value of a function of a random variable is defined as the average value which the function takes on. Again, each value is weighted by its relative frequency. If $h(X)$ denotes an arbitrary function of X, then the expected value of the function $h(X)$ is denoted by $E[h(X)]$. Two very important functions of X are $h(X) = X$, the identity function, and $h(X) = (X - \mu)^2$. By definition, these expected values are

$$E[X] = \mu = \text{population mean of } X \tag{6.4}$$

$$E[(X - \mu)^2] = \sigma^2 = \text{population variance of } X \tag{6.5}$$

In the discrete case

$$\begin{aligned} \mu &= \text{sum of all terms of the form } xf(x;\theta) \\ &\quad\ \text{for which } f(x;\theta) \neq 0 \\ \sigma^2 &= \text{sum of all terms of the form } (x-\mu)^2 f(x;\theta) \\ &\quad\ \text{for which } f(x;\theta) \neq 0 \end{aligned} \tag{6.6}$$

give, respectively, the mean and variance of X having distribution $f(x;\theta)$. (In the continuous case, the sums would be replaced by integrals.)

Note that the variance σ^2 has a value of 0 only when each and every deviation from the mean is 0. This implies that all values of X occur at the mean value, μ.

► ***Example 6-7***

For the sampling scheme described in Example 6-3, the expected value of Y is

$$\begin{aligned} E[Y] &= 4(1/10) + 5(1/10) + 6(1/10) + 7(2/10) \\ &\quad + 8(1/10) + 9(2/10) + 10(1/10) + 11(1/10) \\ &= 76/10 = 38/5 \end{aligned}$$

We see that the average value of Y (the sample average) is equal to μ, the parameter being estimated. No sample gives an estimate equal to μ, but, on the average, the statistic Y is on target in estimating μ.

The variance of Y is

$$\begin{aligned} E[(Y - 38/5)^2] = \sigma^2 &= (4 - 38/5)^2(1/10) + (5 - 38/5)^2(1/10) \\ &\quad + (6 - 38/5)^2(1/10) + (7 - 38/5)^2(2/10) \\ &\quad + (8 - 38/5)^2(1/10) + (9 - 38/5)^2(2/10) \\ &\quad + (10 - 38/5)^2(1/10) + (11 - 38/5)^2(1/10) \\ &= 4.44 \end{aligned}$$ ◄

(Problem 11)

6.5 *A FEW WORDS ABOUT MULTIVARIATE DISTRIBUTIONS*

If more than one random variable is defined in a single experiment, there is usually more interest in the *joint distribution* of all variables than in the distributions of the individual variables. Thus, when studying the costs incurred in constructing houses, the joint distribution of

X_1 = overall cost

X_2 = number of square feet of floor space

X_3 = number of bedrooms

would probably be of more concern than the distribution of any single X_i.

The notation for a bivariate distribution that can be indexed by one parameter is $f(x_1, x_2 : \theta)$ if the variables are denoted by X_1 and X_2; it is $f(x, y; \theta)$ if the variables are denoted by X and Y. In the discrete case, $f(x, y; \theta)$ represents a probability of a joint occurrence, namely,

(6.7) $$f(x, y; \theta) = P(X = x, Y = y; \theta)$$

When v parameters are required for indexing a bivariate distribution, the notation will be

$$f(x, y; \theta_1, \theta_2, \ldots, \theta_v) \text{ or } f(x_1, x_2; \theta_1, \theta_2, \ldots, \theta_v)$$

Extensions to more than two variables follow the same general pattern. Thus

$$f(x_1, x_2, \ldots, x_k; \theta_1, \theta_2, \ldots, \theta_v)$$

is the notation for the joint distribution of $k \geqslant 2$ variables that requires $v \geqslant 2$ parameters for indexing.

Expected values were defined for univariate distributions in Section 6.4. For joint distributions, expected values can be defined for any function of the variables, say $h(X_1, X_2, \ldots, X_k)$ in a k-variate experiment. Some important expected values in joint distributions are the means and variances of the individual variables; these are defined and may be obtained as in Section 6.4. Another very special set of expected values is the set of *covariances* of pairs of variables. These parameters measure the joint variation of the pair of variables involved and are defined as the expected value of

$$h(X_1, X_2, \ldots, X_k) = (X_i - \mu_i)(X_j - \mu_j) \quad \text{for } i, j = 1, 2, \ldots, k$$

with $i \neq j$. A common notation for the *covariance of* X_i *and* X_j is

(6.8) $$\sigma_{ij} = E[(X_i - \mu_i)(X_j - \mu_j)]$$

And for the *covariance of X and Y* the notation is

(6.9) $$\sigma_{XY} = E[(X - \mu_X)(Y - \mu_Y)]$$

(Problem 13)

PROBLEMS

1. Tell whether each of the following describes a discrete or continuous random variable:
(a) the annual growth of a tree
(b) the number of cancellations of a European cruise
(c) the number of tomatoes on a given plant
(d) the number of times you must toss a penny until a head appears on the upper face
(e) the weight of a penny
(f) the number of scratches on a penny

2. Tell whether each of the following describes a discrete or continuous random variable:
(a) the number of cars sold next month by a particular car dealer
(b) the number of fish in Wall Lake
(c) the weight of the next fish caught at Wall Lake
(d) the length of time for paint to dry
(e) the width of a sheet of paper

3. Give possible values for each variable in Problem 1.

4. Give possible values for each variable in Problem 2.

5. A box contains 8 electrical fuses, 6 of which are good and 2 defective. Three fuses are drawn at random and without replacement from the box. Define the random variable: X = number of good fuses among the 3 drawn.
(a) Give the distribution of X.
(b) Give the cumulative distribution of X.

6. Suppose that 3 dimes and 3 pennies (all fair coins) are tossed. Let X_1 denote the number of heads among the dimes and X_2 denote the number of heads among the pennies. Give the distribution of $Y = X_1 - X_2$.

7. For the finite population: $\{1, 2, 4, 6, 12\}$, find, in samples of size 3 drawn at random without replacement, the distribution of
(a) the sample average, say Y.
(b) the smallest sample value, say W.
Give the cumulative distribution of
(c) Y. (d) W.

8. Each of the following spinners is spun once. Let X_1 denote the number obtained from the first spinner, and let X_2 denote the number obtained from the second spinner. Find the distribution of $Y = X_1 + X_2$.

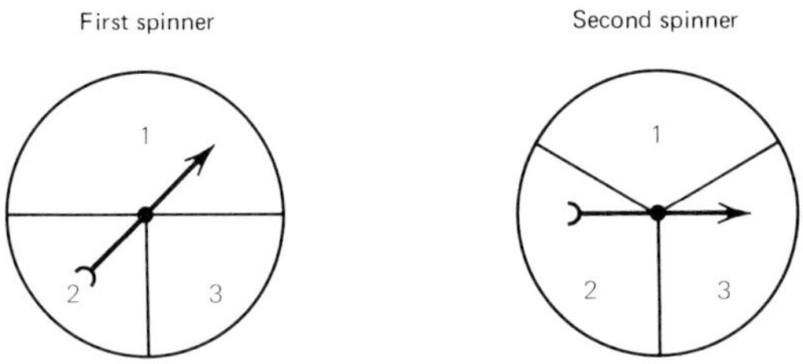

9. Refer to Problem 8 and find the distribution of
(a) $Y_1 = X_1 - X_2$. (b) $Y_2 = 2X_1 - X_2$.

10. Refer to Problem 6 and find the expected value (mean value) of Y.

11. Refer to Problem 5 and find
(a) the mean number (expected value) of good fuses drawn.
(b) the variance of X.

12. Refer to Problem 7 above and find the expected value of the two statistics mentioned.

13. Refer to Problem 8 and find the distribution of X_1 and X_2.

14. Refer to Problem 6 and find the distribution of X_1 and X_2.

15. The distribution of a random variable X is specified by

x	$f(x)$
0	0.1
1	0.2
2	0.25
4	0.20
6	0.25

(a) Find the mean μ of this distribution.
(b) Find the variance σ^2 of this distribution.

16. The distribution of a random variable Y is specified by

y	$g(y)$
−2	0.3
0	0.2
1	0.1
3	0.25
5	0.15

(a) Find the mean μ of this distribution.
(b) Find the variance σ^2 of this distribution.

17. (a) For the random variable X of Problem 15, give the cumulative distribution.
Find the probability that X has a value
(b) less than 5.
(c) greater than 2.
(d) equal to the mean value, μ.

18. Refer to the distribution of Y in Problem 16 and find
  (a) the cumulative distribution.
  (b) the probability that Y is nonnegative.
  (c) the probability that Y is positive.
  (d) the probability that Y does not have the value 1.

REFERENCES

[1] Hogg, R. V., and A. T. Craig, *Introduction to Mathematical Statistics*, 3rd ed., Macmillan, New York, 1970.

[2] Larson, H. J., *Introduction to Probability Theory and Statistical Inference*, Wiley, New York, 1969.

[3] Mood, A. M., and F. A. Graybill, *Introduction to the Theory of Statistics*, 2nd ed., McGraw-Hill, New York, 1963.

[4] Wilks, S. S., *Mathematical Statistics*, Wiley, New York, 1962.

[5] Zehna, P. W., *Probability Distribution and Statistics*, Allyn and Bacon, Boston, Mass., 1969.

Chapter 7

DESCRIPTIVE STATISTICS

The purpose of random sampling is to obtain information about populations. Depending upon research objectives, specific information must be extracted from the sample data. Operations performed on the sample data reduce the original observations to a single numerical value, a statistic. Ideally, a single statistic will provide all desired information. This ideal rarely occurs, however; usually several statistics are needed for each objective. Judiciously chosen, a few statistics can provide a wealth of information about the populations of interest.

This chapter is concerned with various operations performed on the sample data and subsequent calculation of some common descriptive measures or descriptive statistics. Further uses and analyses to which these statistics are subjected will be studied in following chapters.

7.1 SAMPLE VALUES

For each element selected from a population, we record information on those characteristics that pertain to our objectives and then define random variables for each characteristic. For example, when observing ten cars, we might record the results of an equipment check and the number of occupants in each car. We might be interested in the random variables

$$X = \text{number of defective items per car}$$

$$Y = \text{number of occupants per car}$$

Or, if we observed twenty people, we might record information about the variables

$$Z = \text{height of each person}$$

$$W = \text{weight of each person}$$

From a random sample of size n, we denote the values of a random variable X by

(7.1) $$x_1, x_2, \ldots, x_n$$

where the subscript indicates the sample observation used in obtaining the value. Similarly, for a random variable Y, we denote the values by

(7.2) $$y_1, y_2, \ldots, y_n$$

Thus, for the first car observed, we might find $x_1 = 2$ and $y_1 = 5$; for the second car, we might find $x_2 = 0$ and $y_2 = 1$.

By introducing an index

$$i = 1, 2, \ldots, n$$

we can represent typical values of X and Y by x_i and y_i, respectively. Instead of the displays (7.1) and (7.2), the values of X and Y can also be symbolized by

$$\{x_i : i = 1, 2, \ldots, n\}$$

and

$$\{y_i : i = 1, 2, \ldots, n\}$$

respectively.

7.2 ORDER STATISTICS

Let $x_1, x_2, \ldots, x_n$ denote the n sample values of a random variable X. If the sample values are ordered from low to high and the element occupying the i-th position is denoted by $x_{(i)}$, we have

(7.3) $$x_{(1)} \leqslant x_{(2)} \leqslant \cdots \leqslant x_{(n)}$$

The set of ordered values

(7.4) $$x_{(1)}, x_{(2)}, \ldots, x_{(n)}$$

is called the *array* of X values and $x_{(i)}$ is called the i-th *order statistic*. Middle values in the array are used in estimating location parameters; other values in the array are used to estimate dispersion parameters.

(Problems 1(i), 11(a))

7.3 *THE SUMMATION OPERATOR*

Many statistical calculations can be written as sums of similar terms. Verbal descriptions of sums are often cumbersome, as evidenced by the expressions in (6.3) and (6.6). Or, if we want the total number of occupants in ten cars observed, we are asking for

$$x_1 + x_2 + x_3 + x_4 + x_5 + x_6 + x_7 + x_8 + x_9 + x_{10}$$

More generally, the total of a random sample of X values is

$$x_1 + x_2 + \cdots + x_i + \cdots + x_n \tag{7.5}$$

Another frequently used quantity is the sum of squares of all observations on X:

$$x_1^2 + x_2^2 + \cdots + x_n^2 \tag{7.6}$$

If a set of terms can be indexed, their sum can be represented in "shorthand notation" by using the summation operator: $\sum_{i=m}^{n}(\ )$. The upper-case Greek letter sigma, $\sum$, tells one to *add* all terms similar to the one appearing in parentheses; the terms to be summed are those specified by the *range of summation*, $i = m, m + 1, \ldots, n$, and the letter i is called the *index of summation.*

The sums appearing in (7.5) and (7.6) can be expressed as

$$\begin{aligned} \sum_{i=1}^{n} x_i &= x_1 + x_2 + \cdots + x_n \\ \sum_{i=1}^{n} x_i^2 &= x_1^2 + x_2^2 + \cdots + x_n^2 \end{aligned} \tag{7.7}$$

▶ ***Example 7-1***

Suppose $n = 4$ observations of a random variable X gave: $x_1 = 5$, $x_2 = 4$, $x_3 = 2$, and $x_4 = 6$. Then

$$\sum_{i=1}^{4} x_i = x_1 + x_2 + x_3 + x_4 = 5 + 4 + 2 + 6 = 17$$

and

$$\sum_{i=1}^{4} x_i^2 = x_1^2 + x_2^2 + x_3^2 + x_4^2 = 25 + 16 + 4 + 36 = 81$$ ◀

▶ ***Example 7-2***

A sample of $n = 10$ cars revealed the following numbers of occupants: $x_1 = 3$, $x_2 = 4$, $x_3 = 1$, $x_4 = 1$, $x_5 = 6$, $x_6 = 2$, $x_7 = 2$, $x_8 = 5$, $x_9 = 8$, and $x_{10} = 4$. Then,

$$\sum_{i=1}^{10} x_i = 36 \quad \text{and} \quad \sum_{i=1}^{10} x_i^2 = 176$$

Suppose the first four cars were compact cars and the last three were station wagons. Then, the number of occupants in the compact cars was

$$\sum_{i=1}^{4} x_i = 3 + 4 + 1 + 1 = 9$$

and the number of occupants in the station wagons was

$$\sum_{i=8}^{10} x_i = x_8 + x_9 + x_{10} = 5 + 8 + 4 = 17 \blacktriangleleft$$

▶ ***Example 7-3***

A function is defined by $f(i) = 2(i - 1) + 5$, for $i = 1, 2, 3, 4$. Then

$$\sum_{i=2}^{3} f(i) = f(2) + f(3) = 7 + 9 = 16$$

and

$$\sum_{i=1}^{4} f(i) = 5 + 7 + 9 + 11 = 32 \blacktriangleleft$$

(Problems 1(a), (c), (e), (h))

7.4 RULES OF SUMMATION

Let c be a constant; m, n, a, and b be integers with $m \leqslant n$ and $a \leqslant b$; and $X, Y, Z, \ldots$ denote general characteristics.

•• *Rule S-1:* The index of summation is arbitrary; that is,

$$\sum_{i=m}^{n} x_i = \sum_{j=m}^{n} x_j = \sum_{k=m}^{n} x_k = \cdots \tag{7.8}$$

••• *Reason:* Any index of summation is merely a "dummy symbol" which disappears when the operation is performed.

•• *Rule S-2:* The summation of a constant, c, is the constant multiplied by the number of terms in the summation:

$$\sum_{i=1}^{n} c = nc \tag{7.9}$$

••• *Reason:* In the first equation of (7.7), set each x_i equal to the constant c.

When the summation in (7.9) extends over the range $i = m$ to $i = n$, the resulting sum is $(n - m + 1)c$.

▶ ***Example 7-4***

$$\sum_{i=1}^{9} \$1.00 = \$9.00$$

$$\sum_{i=4}^{9} \$1.00 = (9 - 4 + 1)\$1.00 = \$6.00 ◀$$

●● *Rule S-3:* If each term in a summation has a constant multiplier, the constant can be factored outside the summation:

(7.10)
$$\sum_{i=m}^{n} cy_i = c\left(\sum_{i=m}^{n} y_i\right)$$

●●● *Reason:*
$$\begin{aligned}\sum_{i=m}^{n} cy_i &= cy_m + cy_{m+1} + \cdots + cy_n \\ &= c(y_m + y_{m+1} + \cdots + y_n) \\ &= c \sum_{i=m}^{n} y_i\end{aligned}$$

▶ ***Example 7-5***

Let $c = 3$ and $y_1 = 2$, $y_2 = 4$, $y_3 = -1$, and $y_4 = 1$. Then

$$\sum_{i=1}^{4} y_i = 2 + 4 + (-1) + 1 = 6$$

and

$$3y_1 = 6, \qquad 3y_2 = 12, \qquad 3y_3 = -3, \qquad 3y_4 = 3$$

Hence

$$\sum_{i=1}^{4} (3y_i) = 6 + 12 + (-3) + 3 = 18 = 3\left(\sum_{i=1}^{4} y_i\right) ◀$$

●● *Rule S-4:* The summation of a "linear" sum of terms equals the same "linear" sum of summations of the individual terms; for example,

(7.11)
$$\sum_{i=m}^{n} (x_i - x^i y_i + cz_i^2) = \sum_{i=m}^{n} x_i - \sum_{i=m}^{n} x^i y_i + c \sum_{i=m}^{n} z_i^2$$

●●● *Reason:* Expand the left-hand side, group like terms, and use previous rules.

► ***Example 7-6***

If $x_1 = 2$, $x_2 = 3$, $x_3 = 1$ and $y_1 = 1$, $y_2 = 4$, and $y_1 = 2$, then

$$\sum_{i=1}^{3} x_i = 6 \qquad \sum_{i=1}^{3} y_i = 7$$

$$\sum_{i=1}^{3} x_i y_i = 2(1) + 3(4) + 1(2) = 16$$

$$\sum_{i=1}^{3} (x_i - y_i + x_i y_i) = (2 - 1 + 2) + (3 - 4 + 12) + (1 - 2 + 2) = 15$$

$$\sum_{i=1}^{3} x_i - \sum_{i=1}^{3} y_i + \sum_{i=1}^{3} x_i y_i = 6 - 7 + 16 = 15 \blacktriangleleft$$

Rule S-4 can be applied to certain "nonlinear" summations such as $\sum_{i=m}^{n} (x_i + c)^2$ by first making the terms in the summation "linear"; that is,

(7.12) $$\sum_{i=m}^{n} (x_i + c)^2 = \sum_{i=m}^{n} (x_i^2 + 2cx_i + c^2) = \sum_{i=m}^{n} x_i^2 + 2c \sum_{i=m}^{n} x_i + nc^2$$

•• *Rule S-5:* Double summations with independent ranges can be interchanged:

(7.13) $$\sum_{i=m}^{n} \sum_{j=a}^{b} x_{ij} = \sum_{j=a}^{b} \sum_{i=m}^{n} x_{ij}$$

••• *Reason:* Expand either side of (7.13); then rearrange and sum the terms via the "other index" first.

Many summations occurring in statistical work involve every observation in the sample. Therefore, we frequently omit the range of summation when the sum is taken over the entire sample:

$$\sum_{i=1}^{n} (\ ) = \sum (\ )$$

We will always give the range of summation when the sum does *not* extend over the entire sample or when the range is not completely clear from the context. Likewise, we will include the index when it is not completely clear from the context.

*7.5 *PRODUCT OPERATOR*

Some statistical analyses involve products of several terms: factorials, permutations, and combinations, for example. The *product operator*, $\prod$, represents this operation:

(7.14) $$\prod_{i=m}^{n} x_i = x_m x_{m+1} \cdots x_n$$

Some properties of the product operator are given in the following equations. The reasons or derivations for these rules are left as exercises for the reader.

$$\prod_{i=m}^{n} c = c^{(n-m+1)} \tag{7.15}$$

$$\prod_{i=m}^{n} cy_i = c^{(n-m+1)}\left(\prod_{i=m}^{n} y_i\right) \tag{7.16}$$

$$\prod_{i=m}^{n} (1/x_i) = \left(\prod_{i=m}^{n} x_i\right)^{-1} \tag{7.17}$$

$$\prod_{i=m}^{n} x_i y_i = \left(\prod_{i=m}^{n} x_i\right)\left(\prod_{i=m}^{n} y_i\right) \tag{7.18}$$

► ***Example** 7-7*

For the data given in Example 7-5,

$$\prod_{i=1}^{4} y_i = (2)(4)(-1)(1) = -8$$

$$\prod_{i=1}^{4} 3y_i = (6)(12)(-3)(3) = -648 = 3^4\left(\prod_{i=1}^{4} y_i\right)$$

$$\prod_{i=1}^{4} (1/y_i) = (1/2)(1/4)(-1)(1) = -1/8 = 1\bigg/\left(\prod_{i=1}^{4} y_i\right)$$ ◄

7.6 ESTIMATION OF LOCATION PARAMETERS

Location parameters were introduced in Chapter 6. In the following subsections we will expand on these parameters and consider their estimation. Recall that the population mean, μ, was described as an average value (the expected value) of a random variable. Another location parameter is the *population median*, denoted by ω (Greek letter omega). For a random variable X, ω is that point for which $P(X \leqslant \omega) = 1/2$; that is, ω is that value which divides the distribution into two equal parts. Interstate travelers are familiar with the median strip, on either side of which lies half the highway.

7.6.1 The Arithmetic Mean

The most frequently used estimator of the population mean, μ, is the *arithmetic mean* (the simple average) of the sample values. The bar over a symbol customarily denotes an average; that is,

$$\overline{X} = \text{arithmetic mean of } X = \frac{\text{sample total of } X \text{ values}}{\text{number of } X \text{ values}} \tag{7.19}$$

Since different samples may well give different realizations of X, the sample mean can vary from sample to sample. Thus, we conclude that the arithmetic mean $\overline{X}$ is a random variable in random samples of a population. Equation (7.19) represents a verbal description of this random variable (denoted by a capital letter). The value of $\overline{X}$ realized in a given sample would be an estimate of μ. For a random sample of size n, the observed value of $\overline{X}$ is represented by the corresponding lower-case letter:

(7.20)
$$\bar{x} = \frac{1}{n}\sum_{i=1}^{n} x_i = \frac{1}{n}\sum x_i$$

Hereafter, $\overline{X}$ and its values $\bar{x}$ will be called the *sample mean* or simply *the mean*.

▶ ***Example 7-8***

Twelve randomly selected pear trees had yields, to the nearest pound, of 331, 236, 284, 422, 306, 374, 268, 292, 326, 381, 346, and 343. Thus, $x_1 = 331$, $x_2 = 236, \ldots, x_{12} = 343$. We have given a complete listing of this data since we will refer to it in other examples. The average yield for these pear trees is

$$\bar{x} = \frac{1}{n}\sum x_i = 3909/12 = 325.75 \text{ pounds} \blacktriangleleft$$

▶ ***Example 7-9***

Using the costs of 40 books forming a personal library given in Example 5-2, the average value of a book in this library is

$$\bar{x} = \$273.00/40 = \$6.825 \blacktriangleleft$$

As we will see later, the estimator $\overline{X}$ possesses many desirable properties, more than any other estimator of μ.

(Problems 1(d), 3(a), 5(a), 7(a), 9(a))

7.6.2 The Median

The usual estimator of the population median, ω, is the *sample median* defined by

(7.21)
$$\begin{aligned}\widetilde{X} &= \text{sample median of } X\\ &= \text{middle value in the array of } X\end{aligned}$$

Sample values of $\widetilde{X}$ are

(7.22)
$$\begin{aligned}\tilde{x} &= x_{\left(\frac{n+1}{2}\right)} && \text{if } n \text{ is odd}\\ &= (1/2)\left[x_{\left(\frac{n}{2}\right)} + x_{\left(\frac{n+2}{2}\right)}\right] && \text{if } n \text{ is even}\end{aligned}$$

► ***Example 7-10***

For the yields given in Example 7-8, the order statistics are:

$$x_{(1)} = 236,\ x_{(2)} = 268,\ x_{(3)} = 284,\ x_{(4)} = 292,$$

$$x_{(5)} = 306,\ x_{(6)} = 326,\ x_{(7)} = 331,\ \ldots,\ x_{(12)} = 422$$

The sample median is

$$\tilde{x} = \tfrac{1}{2}[x_{(6)} + x_{(7)}] = (326 + 331)/2 = 328.5 \ ◄$$

When the population elements are symmetrically distributed, the point of symmetry is the mean μ, which coincides with the median ω. Thus, for the symmetric distributions the sample median estimates μ and ω simultaneously. We point out that the sample median is usually a poorer estimate of μ than the sample mean $\bar{X}$.

(Problems 1(j), 11(b), 17(e), (f))

7.6.3 The Weighted Mean

Let $x_1, x_2, \ldots, x_n$ denote a set of n observations, and let $w_1, w_2, \ldots, w_n$ denote respective weights of the observations. The weights may indicate costs, quantities, or other measures of worth of the observations.

Consider the following situation. A person bowling on a league has an average after 9 weeks of 172 (based on three lines, games per week), but on the tenth evening the person bowls a fantastic average of 262. His new average is not equal to the simple average of 172 and 262, which is 217, because his tenth-week average of 262 is worth only one-tenth as much as his nine-week average of 172. The two averages $\bar{x}_1 = 172$ and $\bar{x}_2 = 262$ have relative weights of $w_1 = 0.9$ and $w_2 = 0.1$, respectively. His correct average after 10 weeks is $0.9(172) + 0.1(262) = 181$. We could also have obtained the average by finding his total pins in the 30 lines. The average of 172 is based on 27 lines, for a total of 4,644 pins, while the average of 262 is based on 3 lines, for 768 pins. Thus, for the 30 lines he had 5,430 pins, or 181 pins per line.

Formally, then, we define the *weighted mean* of $x_1, x_2, \ldots, x_n$ having respective weights $w_1, w_2, \ldots, w_n$ as

$$\bar{x}_w = \frac{\sum w_i x_i}{\sum w_i} \tag{7.23}$$

The second way of calculating the bowling average also used a weighted mean, with $w_1 = 27$ and $w_2 = 3$:

$$\bar{x}_w = \frac{27(172) + 3(262)}{27 + 3} = 181$$

The *arithmetic mean* is a special case of the weighted mean where each observation is given equal weight, conveniently taken to be 1 or $1/n$.

7.6.4 Other Measures of Location

Many other measures of location exist, such as the geometric mean, the harmonic mean, modal values, the midrange, and quartiles. These measures are used primarily in specialized situations; none is used as extensively as either the sample mean or sample median.

The relative merits of the sample mean and sample median are easy to establish; the same is not true of the other measures of location. Consequently, we have difficulty in saying which estimator is "best" in a given situation. However, for estimation of the mean μ, we can show that the sample mean $\bar{X}$ generally is better than all other measures of location.

We will not present a treatment of these additional measures of location but will refer occasionally to two of them: the *sample mode* and the *sample midrange*. The sample mode is the most frequent value occurring in the sample: it may not exist, or it may not be unique. The sample midrange is defined by

(7.24)
$$\begin{aligned} MR &= \text{sample midrange} \\ &= \text{average of extreme values in the sample} \end{aligned}$$

Values of the random variable MR are

(7.25)
$$mr = \tfrac{1}{2}[x_{(1)} + x_{(n)}]$$

Referring to Examples 7-8 and 7-10, the midrange of the yields from pear trees is

$$mr = \tfrac{1}{2}(236 + 422) = 329$$

(Problem 1(k))

7.7 *DOT NOTATION*

Previously, we simplified summations over the entire sample by omitting the range. Summations which represent basic totals can be further simplified by using *dot* notation. For example, the sample total of X is

(7.26)
$$x_. = \Sigma x_i$$

At this point one might ask, "Why not eliminate the dot for further simplification?" We could do so now, but not in later work. At present the dots merely emphasize that a total is represented. Later we will encounter variables that are indexed by two or more subscripts and will need totals for each combination of indices. If our sample data is

$$\{x_{ij}: i = 1, 2, \ldots, a;\ j = 1, 2, \ldots, b\}$$

we might need

$$(7.27)\qquad \begin{aligned} x_{i\cdot} &= \sum_j x_{ij} \qquad \text{for } i = 1, 2, \ldots, a \\ x_{\cdot j} &= \sum_i x_{ij} \qquad \text{for } j = 1, 2, \ldots, b \\ x_{\cdot\cdot} &= \sum_i \sum_j x_{ij} = \sum_{ij} x_{ij} \end{aligned}$$

Without dots there would be no way to tell $x_{3\cdot}$ and $x_{\cdot 3}$ apart, $x_{4\cdot}$ and $x_{\cdot 4}$ apart, and so on.

It is indeed unfortunate that some of the common quantities, such as $\sum x_i^2$ and $\sum (x_i - \bar{x})^2$, do not have representations in dot notation. However by adopting the convention that "the dots take precedence," many functions of totals can be written in dot notation. Thus

$$(7.28)\qquad \begin{aligned} x_{\cdot}^2 &= \left(\sum x_i\right)^2 \qquad \text{not} \qquad \sum x_i^2 \\ \sqrt{x_{\cdot}} &= \left(\sum x_i\right)^{1/2} \qquad \text{not} \qquad \sum \sqrt{x_i} \\ x_{i\cdot}^2 &= \left(\sum_j x_{ij}\right)^2 \qquad \text{not} \qquad \sum_j x_{ij}^2 \end{aligned}$$

A definition summarizes the material on dot notation.

Definition 7-1

Each dot appearing as a subscript on a variable means to sum the variable over the entire range of the index previously occupying the position of the dot.

(Problems 1(b), (f), (g), 3(a))

7.8 *ESTIMATION OF DISPERSION PARAMETERS*

Variability occurs because population elements are not identical. In highly controlled processes, such as drug manufacturing and nuclear reactions, the variability is usually very slight. When this is the case, only a few observations are needed to get a clear picture of the population structure. On the other hand, variability may be so extensive that inferences will have little or no meaning unless hundreds of observations are used. As we will see later, the population variability has a great impact on the value of any inference.

It might seem that a small variability among elements of any set is a highly desirable feature. Usually this is true, but there are cases where small variability is detrimental. Consider various examining processes, such as Civil Service exams, tests given by college professors, and promotional exams given by a personnel manager. If all scores on such exams are dispersed over a very small range, the exam has very little differentiation power. In other situations, such as manufacturing processes, the presence of even a small amount of variability is a warning that something is about to go wrong, or already has.

Measures of variability should be influenced only by the differences among elements, not by their location. To eliminate location effects, most measures of variability are defined as differences among observations or as differences from some reference point, usually the mean value. In addition, the values of any variability measure should change inversely with the concentration of elements near the reference point; that is, a small value should reflect a high concentration and larger values should indicate lower concentration at the reference point.

7.8.1 The Sample Range

One of the simplest measures of dispersion is the *sample range* defined by

$$\begin{aligned} R &= \text{sample range} \\ &= \text{distance between the extreme values in a sample} \end{aligned} \tag{7.29}$$

Values of the random variable R in samples of size n are given by

$$r = x_{(n)} - x_{(1)} \tag{7.30}$$

For the yields of pear trees in Example 7-8,

$$x_{(1)} = 236 \quad \text{and} \quad x_{(12)} = 422$$

giving a sample range of

$$r = 422 - 236 = 186$$

The (sample) range is a very crude measure, indicating only the difference between the smallest and largest observations. In small samples the range is easy to calculate and is a good estimator of the population variability. In fact, when $n < 10$, the range is used extensively in quality-control operations as a measure of dispersion. But when $n > 10$, the range is an increasingly poorer measure of the population variability.

Rather than reporting the value of R, it is much more informative to give the minimum and maximum order statistics, $x_{(1)}$ and $x_{(n)}$. These two values not only provide a value of r, they also indicate the location of the sample values; $x_{(1)}$ and $x_{(n)}$ actually define the interval over which the sample values are dispersed.

7.8.2 Deviations from the Mean

For a sample of observations, it is natural to inquire about the average value of the deviations $(x_i - \bar{x})$ as a measure of variability. However, we see that

(7.31) $$\frac{1}{n}\sum (x_i - \bar{x}) = \frac{1}{n}(x. - n\bar{x}) = 0$$

It can be shown that an analogous relationship holds for the population: the expected value of $X_i - \mu$ is always 0; that is, $E[(X - \mu)] = 0$.

The cancellation effect present in (7.31) can be eliminated by taking the absolute value of each deviation. The average of these quantities is called the *average deviation*, denoted by AD, and has sample values given by:

(7.32) $$ad = \frac{1}{n}\sum |x_i - \bar{x}|$$

The popularity of AD has never been overwhelming because it is difficult to calculate; the sum in (7.32) does not simplify. Properties of AD are not well known, since the absolute value operator is not conducive to all mathematical analyses.

► ***Example 7-11***

Referring to Example 7-2, the number of occupants in each of 10 cars, the mean number of occupants per car is $\bar{x} = 36/10 = 3.6$. The deviations from the mean are $(x_i - \bar{x})$: -0.6, 0.4, -2.6, -2.6, 2.4, -1.6, -1.6, 1.4, 4.4, 0.4. Notice that $\sum (x_i - \bar{x}) = 0$. The average deviation is

$$ad = \tfrac{1}{10}[0.6 + 0.4 + 2.6 + \cdots + 0.4]$$
$$= 18/10 = 1.8$$ ◄

(Problem 1(c))

7.8.3 Variance and Standard Deviation

Another way to eliminate the cancellation effect is to square the deviations from the mean. The sum of these squared deviations is defined for a random variable X by

(7.33) SS of X = sum of squared deviations of X values from the mean

called the *sum of squares*, or SS for brevity. Sample values of SS are denoted by

(7.34) $$SS_X = \sum (x_i - \bar{x})^2$$

(The symbolism SS_X is used because it is standard, although it does represent a deviation from our convention of using small letters for values of a random variable.)

The "average" valuc of the squared deviations is an important measure of dispersion called the *sample variance*:

(7.35) $$S^2 = \text{sample variance} = \frac{\text{sum of squares}}{n - 1}$$

whose sample values are

(7.36) $$s^2 = \frac{1}{n-1}\sum (x_i - \bar{x})^2 = \frac{SS_X}{n-1}$$

When necessary for proper identification, subscripts will be attached to S^2 and s^2.

The division of the SS in (7.35) and (7.36) by $n - 1$ rather than by n will be explained later. For the present, we can consider this modification as being one which makes s^2 a better estimate of the population variance, σ^2 (better than the estimate having a divisor of n).

The nature of $\bar{x}$ may result in fractional deviations, $x_i - \bar{x}$, and cause difficulties in computing s^2 from (7.36). For example, 4.75 is the average of 3, 7, 5, and 4; two decimals are required for accurate calculation of s^2 from (7.36). Sample sizes like 7 or 11 or 23 may result in means like 1/7 or 2/7 or 1/11; the numbers must be rounded or fractions must be used. Thus efficiency and accuracy demand an alternative version of the SS. A computational form of the SS, derived in Section 7.11, permits the calculation of the sample variance as follows:

(7.37) $$s^2 = \frac{1}{n-1}\left[\sum x_i^2 - x.^2/n\right]$$

The term $\sum x_i^2$ is called the *raw or uncorrected SS*, while SS_X is called the *corrected SS*. The quantity $x.^2/n$ is called the *correction term*, or sometimes the *correction for the mean.*

► *Example 7-12*

For simplicity, consider the following eight observations to be a sample:

$$x_i\colon 6, 9, 7, 5, 6, 7, 4, 4$$

Then we can obtain the following:

sample total: $x. = 48$
sample mean: $\bar{x} = 48/8 = 6$
raw SS: $\sum x_i^2 = 6^2 + 9^2 + \cdots + 4^2 = 308$
correction term: $x.^2/n = (48)^2/8 = 288$
corrected SS: $SS_X = 308 - 288 = 20$
deviations $x_i - \bar{x}$: 0, 3, 1, −1, 0, 1, −2, −2
$SS_X = \sum (x_i - \bar{x})^2 = 0 + 9 + 1 + 1 + 0 + 1 + 4 + 4 = 20$
sample variance: $s^2 = 20/7 = 2.86$ ◄

▶ *Example 7-13*

With the aid of a desk calculator, we see for the 12 yields in Example 7-8

$$\sum x_i^2 = (331)^2 + (236)^2 + \cdots + (343)^2 = 1{,}303{,}399$$

$$x_{\cdot}^2/n = (3909)^2/12 = 1{,}273{,}356.75$$

$$s^2 = [1{,}303{,}399 - 1{,}273{,}356.75]/11 = 30{,}042.25/11$$

$$= 2731.11$$

$$s = 52.3$$ ◀

▶ *Example 7-14*

For the costs of books given in Example 5-2, variance and standard deviation calculations are

$$\sum x_i^2 = (8.25)^2 + (8.95)^2 + \cdots + (12.50)^2 = 2349.8850$$

$$x_{\cdot}^2/40 = (273)^2/40 = 1863.2250$$

$$s^2 = \tfrac{1}{39}[2349.8850 - 1863.2250] = 486.66/39 = 12.48$$

$$s = \$\sqrt{12.48} = \$3.53$$ ◀

The variances, s^2 and σ^2, are defined in terms of squared deviations. We would like to measure variation in reference to distances but because of the "squaring effect," variances are not compatible with distances. Each squared deviation actually represents a squared distance (of x_i from its mean). The units attached to variances are the square of units attached to individual values.

The standard deviations, $s = \sqrt{s^2}$ and $\sigma = \sqrt{\sigma^2}$, are more informative measures of dispersion than the variances because the squaring effect has been reduced. We are not saying that the squaring effect has been eliminated; if it were, the standard deviations would measure average distance from the mean—they do not. (The average deviation, *AD*, actually does measure average distance, but recall the undesirable features of this estimator.) The units of the standard deviations are the same as those of the individual values.

The following example illustrates the unrealistic magnitude of the variance and the practical aspects of the standard deviation.

▶ *Example 7-15*

If a sample of four people have annual salaries of \$8,000, \$9,000, \$8,400, and \$10,600, their average annual salary is $\bar{x} = \$9{,}000$, whereby

$$x_i - \bar{x}: -\$1{,}000,\ 0,\ -\$600,\ \$1{,}600,$$

and

$$s^2 = (\$)^2(3{,}920{,}000)/3 \doteq 1{,}306{,}667(\$)^2$$

Noting that the salaries cover a range of \$2,600, the figure 1,306,667 has little meaning in relation to the four salaries or their range. The standard deviation $s = \$\sqrt{1,306,667} \doteq \$1,143$ gives a clearer indication of dispersion. ◀

(Problems 3(b), (c), (d), 5(b), 7(b), 9(b))

7.8.4 Tchebycheff's Inequality

Some further justification for the standard deviation as a realistic measure of dispersion is provided by an inequality developed by the Russian mathematician, Tchebycheff [4]. It gives the proportion of elements within c standard deviations of the mean. We state the inequality without proof; it holds for samples as well as populations, for discrete as well as continuous random variables.

> *Tchebycheff's Inequality*: Let c be any positive constant. The proportion of elements *within* c standard deviations of their mean value is at least $1 - 1/c^2$.

If $0 < c \leqslant 1$, the inequality gives no information, since any proportion must be nonnegative. Table 7-1 gives, for selected values of c, the proportions stated by Tchebycheff's Inequality. At least 3/4 of the values must be within two standard deviations of their mean value, at least 8/9 within three standard deviations, and so on.

Table 7-1 Proportions from Tchebycheff's Inequality

c	Interval	Proportion
1.25	$\bar{x} \pm 1.25s$ (or $\mu \pm 1.25\sigma$)	0.36
1.5	$\bar{x} \pm 1.5s$	0.56
2	$\bar{x} \pm 2s$	0.75
2.5	$\bar{x} \pm 2.5s$	0.84
3	$\bar{x} \pm 3s$	0.89
4	$\bar{x} \pm 4s$	0.94

▶ ***Example 7-16***

The data in Example 7-12 have a standard deviation of $s = \sqrt{2.86} \doteq 1.7$ and a mean of $\bar{x} = 6$. The interval defined by $\bar{x} \pm 1.5s = 6 \pm 2.55$ must contain at least 5/9 of the observations while the interval defined by $\bar{x} \pm 2s = 6 \pm 3.4$ must contain at least 3/4 of the observations. It is seen that the first interval contains 7/8 of the observations while the second contains all observations. ◀

(Problem 22)

7.8.5 The Concept of Degrees of Freedom

In a two-dimensional plane whose axes are denoted by X and Y, an arbitrary point may be denoted by the ordered pair of real numbers (x, y); the point has an x coordinate and a y coordinate. If the point is allowed to vary (or move about) without any restrictions, both x and y are arbitrary, and we say that the point has two degrees of freedom. However, the point is restricted to the X axis (by imposing the condition: $y = 0$), it has lost one degree of freedom; since the x coordinate can vary, the point now has only one degree of freedom. Similarly, imposing the condition that $x = y$ removes one degree of freedom in determining the point. Placing another restriction on the x and y coordinates, that the point must also be on the Y axis, removes another degree of freedom, leaving none. The point is now fixed; it is the origin, (0, 0).

When no restrictions were placed on the point (x, y), each coordinate could be assigned a value independently of the other. Each restriction placed on the coordinates of the point eliminates this independence, or freedom, to a certain degree.

In statistics *degrees of freedom* denotes the number of independent components needed to calculate a given quantity ("independent" is used here in a functional rather than probability sense). Usually abbreviated as *df* or *DF*, the term is used in connection with sums of squares. Consider the *SS* of a variable X in a sample of size n:

$$SS_X = \sum_{i=1}^{n} (x_i - \bar{x})^2$$

The n deviations $x_i - \bar{x}$ define SS_X, but not all n deviations are necessary or required in the calculation of SS_X since (7.31) states that $\sum_{i=1}^{n} (x_i - \bar{x}) = 0$. This latter equation imposes a restriction on the deviations: The n-th deviation is automatically fixed by the other $n - 1$ deviations. The deviations are not governed by any other conditions or restrictions, and thus SS_X has $n - 1$ degrees of freedom.

Now looking at our definition of sample variance, we see that it is the average of the independent squared deviations from the mean. Or, the estimate of σ^2 is *SS* divided by *df*.

In many statistical distributions degrees of freedom enters as an indexing parameter. The notation for degrees of freedom as a parameter is ν (Greek letter nu).

7.8.6 Variability of the Sample Mean

Since the sample mean $\overline{X}$ is a random variable in random samples from any population, $\overline{X}$ has a distribution, a mean, a variance, and other parameters. The distribution of $\overline{X}$ will be studied later.

For random samples of size n drawn from any infinite population (equivalently, sampling with replacement from a finite population), the mean and variance of $\overline{X}$ are, respectively,

$$\mu_{\overline{X}} = \mu_X \qquad \text{and} \qquad \sigma^2_{\overline{X}} = \sigma^2_X/n, \tag{7.38}$$

where the subscript denotes the random variable. The standard deviation of $\overline{X}$ given by $\sigma_{\overline{X}} = \sigma/\sqrt{n}$ is also called the *standard error of the mean.* When an estimator is involved, standard error is synonymous with standard deviation.

For estimates of $\sigma^2_{\overline{X}}$ and $\sigma_{\overline{X}}$, we use the sample analogues

$$s^2_{\overline{X}} = s^2_X/n \qquad \text{and} \qquad s_{\overline{X}} = s_X/\sqrt{n} \tag{7.39}$$

which are called, respectively, the *sample variance of the mean* and the *sample standard error of the mean.*

When sampling without replacement from a finite population of size N, the variance of $\overline{X}$ is

$$\sigma^2_{\overline{X}} = \left(\frac{N-n}{N-1}\right)\frac{\sigma^2_X}{n} \tag{7.40}$$

▶ ***Example 7-17***

From the data in Example 7-12, the estimated variance of the mean $\overline{X}$ is seen to be

$$s^2_{\overline{X}} = 2.86/8 \doteq 0.36$$

and the standard error of the mean is

$$s_{\overline{X}} = \sqrt{0.36} = 0.6$$

From Example 7-13, the estimated variance of the mean yield per pear tree is

$$s^2_{\overline{X}} = 2731.11/12 = 227.59$$

and the standard error is

$$s_{\overline{X}} = 52.3/\sqrt{12} = 15.09 \blacktriangleleft$$

▶ ***Example 7-18***

In Example 6-3 we found the sampling distribution of $\overline{X}$ in random samples of size $n = 2$ drawn from a finite population of size $N = 5$. Since the values of X

are equally likely, $f(x) = 1/5$ for all x; hence, by (6.6),

$$\sigma_{\bar{X}}^2 = \frac{1}{5}\sum(x - 38/5)^2 = 11.84$$

Then, by (7.40),

$$\sigma_{\bar{X}}^2 = \left(\frac{5-2}{5-1}\right)\frac{11.84}{2} = 4.44$$

which agrees with the value obtained in a different manner in Example 6-7. ◀

(Problems 3(e), 7(b), 13(c))

7.8.7 Range Estimation of Variability

When drawing small samples from a symmetric distribution, a fairly good and easily obtained estimate of the population standard deviation is a fraction of the sample range. The results were derived for random sampling from normal distributions (to be discussed in Section 8.7), but serve as approximations for other symmetric distributions.

Specifically, Table A.8 gives constants d_n needed to estimate σ through the range R in samples of size $n = 2(1)10(2)20(10)50$.† For a sample of size n, the range estimate of σ is the product of r and d_n:

(7.41) $$\hat{\sigma} = r \cdot d_n$$

The other two columns of Table A.8 are related. The second indicates "how good $\hat{\sigma}$ is relative to s" as an estimator of σ; entries in the last column are reciprocals of those in the relative efficiency column and indicate the ratio of sample sizes to make the two estimators "equally good." Thus, in samples of size $n = 12$ the range estimator is about as good as s would be in samples of size 10; the range estimator of σ based on 50 observations is about as good as s based on 25 observations.

▶ ***Example 7-19***

The yields from 12 pear trees in Example 7-8 are ordered in Example 7-10, from which we obtain $r = x_{(12)} - x_{(1)} = 422 - 236 = 186$. From Table A.8, $d_{12} = 0.307$ which gives $\hat{\sigma} = 186(0.307) = 57.1$. In Example 7-13, we saw that $s = 52.3$. ◀

We emphasize that $\hat{\sigma}$ and s are different estimators (of the parameter σ); they are different functions of the sample observations. We shouldn't expect these statistics to give the same estimate of σ, and in fact they didn't in Example 7-19.

(Problem 24(d))

† This notation is defined on page 33.

*7.8.8 Bounds on the Variance

It would be advantageous to have a check on the calculations made in finding the variance. Instead of an exact check, which would mean independent calculations, bounds are available. Quick and easy bounds are given in the following (for reasons, see [1], [2], and [5]):

- *Theorem 7-1: Least upper and greatest lower bounds on the sample variance are*

(7.42)
$$\frac{r^2}{2(n-1)} \leqslant s^2 \leqslant \frac{Mr^2}{4},$$

where r is the sample range and $M = n/(n-1)$ *or* $(n+1)/n$ *according to whether n is even or odd.*

▶ ***Example 7-20***

For the data in Example 7-12, $r = 5$ and $n = 8$, giving $1.79 \leqslant s^2 \leqslant 7.14$.

For the yields in Example 7-8, $r = 186$ and $n = 12$, giving $1572.5 \leqslant s^2 \leqslant 9435.3$. ◀

(Problem 35)

7.9 *THE TECHNIQUE OF CODING*

When the sample values contain more than a few digits, the calculation of most statistics becomes time-consuming and tedious. No matter how the computations are made, merely handling cumbersome values increases the chance of human errors. To simplify matters, the raw data may be *coded*, that is, transformed to a new scale. The transformed values are treated as a "new sample." After the desired statistics have been calculated for the transformed data, a transformation back to the original scale, or an "uncoding," is required.

Only linear transformations will be studied here, with the discussion limited to means and variances. In general, suppose the sample X values are transformed to a Y scale by the linear transformation

(7.43)
$$y_i = Ax_i + B,$$

where A and B are arbitrary constants with $A \neq 0$. Certain choices of A and B will make the calculation of $\bar{y}$ and s_Y^2 easier than others. Through experience, guidelines can be established for making these choices.

Relevant statistics are actually the mean and variance of the X values, $\bar{x}$ and s_X^2. Thus it is necessary to know the relationship between the means and variances of two sets of data related by (7.43). It is shown in Section 7.11 that these four statistics are linked by the equations

$$\bar{y} = A\bar{x} + B \tag{7.44}$$

$$s_Y^2 = A^2 s_X^2 \qquad \text{or} \qquad s_Y = |A| s_X \tag{7.45}$$

We see from (7.44) that the means are related in the same way as the variables. Equation (7.45) implies that the additive constant B has no effect on the variance, while the multiplicative constant A enters as a squared factor (not surprising, since the variance is calculated from square quantities).

In calculating the variance s_Y^2, the computational form of SS_Y would generally be used. This is obtained from (7.37) by replacing x_i with y_i.

► ***Example 7-21***

Suppose that 10 people applying for a fellowship have the following cumulative grade points:

3.74	3.33	3.83	3.35	4.00
3.24	3.61	3.63	3.40	3.33

Calling these values the X data, consider the values obtained under the transformation $y_i = 100x_i - 350$:

24	−17	33	−15	50
−26	11	13	−10	−17

Now

$$y. = 46 \qquad \text{and} \qquad \bar{y} = 4.6 \quad \text{or} \quad 4.6 = 100\bar{x} - 350 \quad \text{or} \quad \bar{x} = 3.546,$$

$$\sum y_i^2 = 6034 \qquad \text{and} \qquad s_Y^2 = \tfrac{1}{9}[6034 - (46)^2/10]$$

$$= 5822.4/9 = 646.9,$$

and

$$\sqrt{646.9} = 100s_X \quad \text{or} \quad s_X = 0.254. \blacktriangleleft$$

(Problems 9(c), (d), 10)

7.10 CALCULATIONS FOR GROUPED DATA

This section discusses the calculation of the sample mean and variance for grouped data, with and without coding. Other measures of location will not be discussed since their calculation from grouped data is not universally accepted.

Calculations made from frequency distributions are based on the assumption that all data of each class occurs at the class mark. This is a reasonable assumption if there is a uniform distribution (Examples 6-1 and 6-2) of observations throughout each class. Violation of this assumption introduces a so-called *grouping error* in the calculations. When calculating the mean, the error is usually quite small, since some cancellation effect results from observations falling above and below the midpoint. However, when calculating the variance, these differences enter as squared quantities and do not cancel. The grouping error in calculating the variance is related to the size of the class interval, increasing as the length of the interval increases; a correction for this error will be discussed later.

Under the above assumption, instead of all n sample values entering the calculations, only the k class marks and k class frequencies are needed. The sample mean, calculated from a frequency distribution, is simply a weighted mean of the class marks with the frequencies as the weights. Thus, by (7.23), we have

(7.46) $$\bar{x} = \frac{1}{n}\sum_{i=1}^{k} f_i x_i ,$$

since $n = \sum_{i=1}^{k} f_i$. The sample variance would be a weighted mean of the squared deviations $(x_i - \bar{x})^2$ if the divisor of the SS was n instead of $n - 1$. Because of this modification, the sample variance for a set of grouped data is

(7.47) $$s^2 = \frac{1}{n-1}\sum_{i=1}^{k} f_i(x_i - \bar{x})^2,$$

which can be simplified for calculation purposes (see next section) to

(7.48) $$s^2 = \frac{1}{n-1}\left[\sum_{i=1}^{k} f_i x_i^2 - \left(\sum_{i=1}^{k} f_i x_i\right)^2 \Big/ n\right]$$

Note that in (7.48) the first term written within the brackets is the weighted sum of squared class marks, *not* the sum of $f_i^2 x_i^2 = (f_i x_i)^2$. The parenthetical quantity in (7.48) is, of course, the total used in defining the mean; see (7.46).

When the class intervals are large (and equal in size), a partial compensation for the grouping error in s^2 is given by Sheppard's correction (see [3]). If the common class interval is denoted by h, the correction simply amounts to subtracting $h^2/12$ from the sample variance. The corrected variance is

(7.49) $$s_c^2 = s^2 - h^2/12$$

Coding is easily accomplished when making calculations from grouped data, because the class marks enter the analysis. Simply code the class marks which, when equally spaced, can always be reduced to integers symmetric about 0.

▶ ***Example 7-22***

From the frequency distribution given in Example 5-2, calculations of the mean, variance, and standard deviation are:

$$\sum_{i=1}^{7} f_i x_i = 7(1.995) + 6(3.995) + \cdots + 2(13.995) = 271.80$$

$$\sum_{i=1}^{7} f_i x_i^2 = 7(1.995)^2 + 6(3.995)^2 + \cdots + 2(13.995)^2 = 2277.281$$

correction term: $(271.80)^2/40 = 1846.881$

$$\bar{x} = \$271.80/40 = \$6.795$$

$$s^2 = \tfrac{1}{39}[2277.281 - 1846.881] = (430.4)/39 = 11.04$$

$$s = \sqrt{11.04} = \$3.32$$

These statistics were calculated from the raw data in Examples 7-9 and 7-14 and were $\bar{x} = \$6.825$, $s^2 = 12.48$, and $s = \$3.53$. While the two means are very close, only \$.03 apart, the two variances differ by a considerable amount. This demonstrates that grouping errors have a bigger impact on the variance than on the mean. Grouping has in this case caused an underestimate of the variability, whereas it generally does the opposite. Sheppard's correction, amounting to only $h^2/12 = 1/3$, would widen the gap even further. ◀

▶ ***Example 7-23***

If the data used in the previous example are subjected to the transformation $y_i = (x_i - 7.995)/2$, we have

f_i	y_i	$f_i y_i$	$f_i y_i^2$
7	−3	−21	63
6	−2	−12	24
5	−1	−5	5
14	0	0	0
4	1	4	4
2	2	4	8
2	3	6	18
40		−24	122

$$\bar{y} = -24/40 = -0.6$$

$$\bar{y} = (\bar{x} - 7.995)/2 \quad \text{or} \quad \bar{x} = 2\bar{y} + 7.995$$

$$\bar{x} = 7.995 - 1.20 = 6.795$$

$$s_Y^2 = \tfrac{1}{39}[122 - (-24)^2/40] = \tfrac{1}{39}[122 - 14.4] = 107.6/39$$

$$s_Y^2 = \tfrac{1}{4}s_X^2 \quad \text{or} \quad s_X^2 = 4s_Y^2 = 430.4/39 = 11.04$$ ◀

(Problems 8, 32)

*7.11 *DERIVATION OF SOME FORMULAS*

Using the summation rules of Section 7.4, it is a straightforward process to derive the expanded form of the *SS* appearing in (7.37):

$$\begin{aligned} SS_X = \sum (x_i - \bar{x})^2 &= \sum (x_i - \bar{x})x_i - \sum (x_i - \bar{x})\bar{x} \\ &= \sum (x_i - \bar{x})x_i \qquad \text{by (7.31)} \\ &= \sum x_i^2 - \bar{x} \sum x_i \\ &= \sum x_i^2 - x_.^2/n \end{aligned}$$

When the sample data are coded by the linear transformation (7.43), the mean of the transformed data is given by

$$\bar{y} = y_./n = \frac{1}{n} \sum y_i$$

$$= \frac{1}{n} \sum (Ax_i + B)$$

(7.50)
$$= \frac{1}{n}(A \sum x_i + nB)$$

$$= \frac{1}{n}(Ax_. + nB)$$

$$= A\bar{x} + B$$

With the result obtained in (7.50), the variance on the transformed scale is seen to be

$$s_Y^2 = \frac{1}{n-1} \sum (y_i - \bar{y})^2$$

$$= \frac{1}{n-1} \sum [(Ax_i + B) - (A\bar{x} + B)]^2$$

(7.51)
$$= \frac{1}{n-1} \sum [Ax_i - A\bar{x}]^2$$

$$= \frac{A^2}{n-1} \sum (x_i - \bar{x})^2 = A^2 s_X^2$$

For grouped data, the expanded form of the *SS* appearing in (7.48) is easily found as follows:

$$\begin{aligned} SS_X = \sum_{i=1}^{k} f_i(x_i - \bar{x})^2 &= \sum_{i=1}^{k} f_i(x_i - \bar{x})x_i - \sum_{i=1}^{k} f_i(x_i - \bar{x})\bar{x} \\ &= \sum_{i=1}^{k} f_i(x_i - \bar{x})x_i, \end{aligned}$$

since

$$\sum_{i=1}^{k} f_i(x_i - \bar{x}) = \sum_{i=1}^{k} f_i x_i - \bar{x} \sum_{i=1}^{k} f_i = 0$$

Therefore,

$$SS_X = \sum_{i=1}^{k} f_i x_i^2 - \bar{x} \sum_{i=1}^{k} f_i x_i$$
$$= \sum_{i=1}^{k} f_i x_i^2 - \left(\sum_{i=1}^{k} f_i x_i \right)^2 / n$$

Performing operations similar to those in (7.50) and (7.51) but beginning with

$$\bar{y} = \frac{1}{n} \sum_{i=1}^{k} f_i y_i \qquad \text{and} \qquad s_Y^2 = \frac{1}{n-1} \sum_{i=1}^{k} f_i (y_i - \bar{y})^2,$$

it is easily shown that the results stated in (7.44) and (7.45) also hold for grouped data.

PROBLEMS

1. Let the following numbers represent a sample of size $n = 10$:

$$x_1 = 3 \quad x_4 = 4 \quad x_7 = 5 \quad x_{10} = 3$$
$$x_2 = 5 \quad x_5 = 2 \quad x_8 = 3$$
$$x_3 = 4 \quad x_6 = 7 \quad x_9 = 4$$

Find:

(a) $\sum_{i=3}^{8} x_i$ (b) $x.$ (c) $\sum_{i=1}^{10} (x_i - 4)$ (d) $\bar{x}$

(e) $\sum_{i=1}^{10} x_i^2$ (f) $x.^2$ (g) $\sum_{i=1}^{10} x_i^2 - x.^2/10$ (h) $\sum_{i=1}^{10} (x_i - 4)^2$

(i) the array of this data
(j) $\tilde{x}$, the median of this data
(k) *mr*, the midrange of this data

2. Consider the following as a random sample of size $n = 9$:

$$x_1 = 6 \quad x_4 = 9 \quad x_7 = 6$$
$$x_2 = 5 \quad x_5 = 6 \quad x_8 = 8$$
$$x_3 = 7 \quad x_6 = 7 \quad x_9 = 9$$

Find:

(a) $x.$ (b) $\sum_{i=2}^{7} x_i$ (c) $\bar{x}$

(d) $\sum_{i=1}^{9} (x_i - 6)$ (e) $x.^2$ (f) $\sum_{i=1}^{9} x_i^2$

(g) $\sum_{i=1}^{9} (x_i - 7)$ (h) $\sum_{i=1}^{9} x_i^2 - x.^2/9$ (i) $\sum_{i=1}^{9} (x_i - 7)^2$

(j) $\tilde{x}$, the median of this data
(k) s_X^2, the variance of this data
(l) $s_{\bar{X}}$, the standard error of the mean

3. From a random sample of size 15, the following was recorded:

$$x. = 210 \qquad \sum x_i^2 = 3{,}780$$

For this sample, calculate
(a) the mean (b) the sum of squares, SS_X (c) the variance
(d) the standard deviation (e) the standard error of the mean

4. Statistics recorded for a given experiment were

$$n = 25 \qquad x. = 250 \qquad \sum x_i^2 = 2{,}986$$

Determine

(a) the mean (b) SS_X (c) the variance (d) s_X (e) $s_{\bar{X}}$

5. The number of correct responses to a sixteen-stimuli paired-associate test given by a psychologist to a group of 20 subjects were

12	11	10	9	12
10	10	11	10	9
8	13	14	8	8
12	9	9	11	9

For this group of subjects, find
(a) the average number of correct responses to the test.
(b) the variance exhibited by the number of correct responses.

6. For the weights of whitefish (Problem 1 of Chapter 5), find from the raw data
(a) the mean.
(b) the variance.
(c) the standard deviation.
(d) the standard error of the mean.

7. Using the raw data in Problem 5 of Chapter 5, find
(a) the average number of grasshoppers killed per area by the insecticide.
(b) the standard error of the estimate found in (a).

8. Using the frequency distribution in Problem 7 of Chapter 5, determine
(a) the mean of the final exam grades.
(b) the variance of the final exam grades.

9. Each evening, a particular television variety show begins with a monologue. For a three-week period, the length of the monologue was, in seconds,

560	580	320
580	495	540
305	375	505
440	330	380
295	290	460

Let X denote the length of the monologue, in seconds.
(a) Calculate the mean of X.
(b) Calculate the variance of X.
(c) Calculate the mean of Y, the length of the monologue given in minutes.
(d) What is the sample variance of Y?

10. A sample of 20 observations gave $\bar{y} = 15$ and $s_Y^2 = 80$. Give the mean and variance of X if the data were coded according to

(a) $Y_i = 3X_i - 5$ (b) $Y_i = 3(X_i - 5)$ (c) $2Y_i = 5X_i + 4$

11. On a dozen fishing trips, an avid fisherman caught (but not necessarily kept) the following number of grayling:

18, 41, 32, 68, 13, 114, 55, 26,15, 24, 38, 76

(a) Give the array for the number of grayling caught on these fishing trips.
(b) What was the median number of grayling caught on these fishing trips?

12. The number of carloads of potatoes shipped by rail over a two-week period from a given collection point were 37, 23, 27, 35, 31, 40, 28, 36, 32, 25, 32, and 30. For this two-week period,
(a) what is the mean number of carloads shipped?
(b) what is the median number of carloads shipped?
(c) what is the variance of the number of carloads shipped?

13. In a psychology study, subjects were asked to mechanically trace an object. The times, in seconds, required by 30 subjects were:

83	93	95	100	91	93
92	97	85	87	94	87
104	88	90	106	86	80
79	101	91	88	88	102
84	86	94	96	95	97

(a) Calculate the mean time to trace the object.
(b) Calculate the variance of tracing times.
(c) What is the standard error of the mean found in (a)?

14. Airmail deliveries to a postal station, given in pounds X, for a ten-day period were 64, 107, 110, 86, 73, 78, 105, 126, 116, and 96. For this ten-day period, find
(a) the median poundage delivered
(b) the mean poundage delivered
(c) the standard deviation of X
(d) the standard deviation of $\bar{X}$

15. In a new housing development a contractor built 16 houses during the summer. Their sale prices, X, were

\$16,800	\$16,700	\$16,000	\$16,700
15,900	16,500	16,250	15,750
15,750	16,500	15,750	16,650
16,650	15,900	16,400	16,500

Code the data and find
(a) the average sale price of these houses
(b) the variance of these sale prices
(c) the standard error of the average sale price

16. Pounds of red ripe chili pods harvested from 10 experimental plots, denoted by X, were

1,635	1,652	1,674	1,639	1,658
1,688	1,661	1,648	1,655	1,671

After coding the data, find
(a) the mean, $\bar{x}$
(b) the variance, s_X^2

17. Over an eighteen-month period, a family spent the following amount on two of their utilities:

Month	Gas: X	Electricity: Y	Month	Gas: X	Electricity: Y
1	\$23.84	\$12.55	10	\$7.20	\$10.91
2	16.69	12.11	11	10.97	12.34
3	18.77	14.50	12	16.17	12.66
4	14.09	10.24	13	17.99	13.02
5	14.87	11.23	14	24.49	13.80
6	11.36	10.68	15	16.04	11.59
7	8.50	12.81	16	17.34	14.53
8	3.17	7.82	17	11.68	12.85
9	4.60	9.22	18	8.12	13.69

(a) What was the monthly average spent on gas?
(b) What variance is exhibited by monthly gas values?
(c) What was the monthly average spent on electricity?
(d) What variance is exhibited by monthly electricity values?
(e) What is the median amount spent on gas?
(f) What is the median amount spent on electricity?

18. One stock on the New York Stock Exchange (NYSE) and one stock on the American Stock Exchange (AMEX) were each on the most-active list every day for a three-week period. Sales, in hundreds, were

AMEX: X			NYSE: Y		
470	1,195	1,392	4,884	1,063	1,787
973	1,151	1,099	5,594	1,169	2,432
1,366	655	400	2,998	2,266	3,155
1,448	561	517	3,134	3,515	1,651
1,001	2,019	1,424	2,955	2,705	2,052

During this three-week period, find (in hundreds)
(a) the average of sales of X
(b) the standard deviation of X
(c) the average of sales of Y
(d) the standard deviation of Y
In unit sales, find
(e) the mean and standard deviation of the AMEX stock.
(f) the mean and standard deviation of the NYSE stock.

19. An industrial company has three manufacturing plants. Plant A has 125 employees with an average weekly salary of \$132.40; B has 140 employees, with an average weekly salary of \$126.82; and C has 95 employees, with an average weekly salary of \$134.65. What is the average weekly salary of all employees of these three plants?

20. The assembly of a group of components is timed on six occasions, giving, in seconds, 108, 107, 109, 107, 108, and 107. Find
(a) the median of assembly times.
(b) the mean of assembly times.
(c) the variance of assembly times.

21. The number of passengers departing daily from an airport during July were:

6, 8, 12, 5, 16, 9, 9, 6, 7, 11, 10, 5, 6, 8, 6, 7,
9, 12, 13, 5, 7, 9, 12, 8, 6, 8, 5, 10, 9, 10, 8

(a) Find the average number of passengers departing during July.
(b) What variance do the daily departures exhibit?

22. (a) What proportion of the sample observations must lie in the interval specified by the limits $\bar{x} \pm 2.5s$, according to Tchebycheff's Theorem?
Determine the actual proportion in the interval specified by the limits $\bar{x} \pm 2.5s$
(b) for the data of Problem 13.
(c) for the data of Problem 15.
(d) for the X data of Problem 17.
(e) for the Y data of Problem 17.
(f) for the data of Problem 21.

23. Before buying a shipment of walnuts, a purchasing agent weighs 9 sacks of walnuts selected at random. Their weights, in pounds, are 98, 102, 103, 100, 103, 102, 99, 98, and 104. For the entire shipment, estimate
(a) the weight per sack.
(b) the variance of weights.

24. At a large company the number X of weekly absences because of illness for the first quarter of this year were 8, 2, 3, 0, 5, 2, 18, 10, 6, 3, 0, and 4. For the first quarter of this year, calculate
(a) $\bar{x}$ (b) s_X^2 (c) $s_{\bar{X}}$ (d) the range estimator of σ_X

25. Costs per acre for water used by 15 ranchers in irrigating alfalfa land were:

\$3.02	\$2.91	\$2.74	\$3.14	\$3.21
2.88	2.48	2.55	2.81	2.96
2.67	3.06	2.76	2.62	2.57

(a) What is the average water cost per acre for all ranchers?
(b) What is the variance of the 15 acreage water costs?

26. When a rancher sheared 18 sheep, he obtained the following clean fleece weights, in pounds:

4.3	6.8	5.7	5.1	4.5	3.9
5.2	4.6	3.8	5.5	5.3	4.7
4.4	5.0	5.8	5.6	4.9	5.1

(a) What is the average clean fleece weight?
(b) What is the variance of the clean fleece weights?
(c) What is the range estimator of variance of clean fleece weights?

27. The number of camper units using a Forest Service campsite on weekends in June, July, and August were:

33	44	38
38	21	34
26	30	37
35	32	27

(a) Calculate the average number of camper units using this campsite on a summer weekend.
(b) Give the standard error of this estimate.

28. A chemical engineer is investigating temperatures at which various gases liquefy under pressure. He cools one particular gas under 3 atmospheres of pressure and records the following temperatures, in Absolute degrees (A°), at which it liquefies:

110.54	110.56
110.58	110.56
110.50	110.52
110.55	110.56
110.53	110.55

(a) On the Absolute scale, what is the average temperature at which this gas liquefies?
(b) What is the variance of these temperatures, on the Absolute scale? Temperatures on the Absolute and Centigrade scales are related by: A° − 273° = C°.
(c) Give the mean temperature at which this gas liquefied on the Centigrade scale.
(d) Give the variance of the temperatures on the Centigrade scale.

Temperatures on the Centigrade and Fahrenheit scales are related by $9C° = 5(F° - 32°)$.

(e) Give the mean at which this gas liquefies, on the Fahrenheit scale.

(f) Give the variance of the temperatures on the Fahrenheit scale.

29. Official high temperatures recorded by the weather bureau in a southwestern city over a ten-day period were 88, 102, 96, 96, 103, 95, 88, 95, 97, and 93. For this ten-day period, what is the (a) mean, (b) median, (c) variance, and (d) standard deviation of the high temperatures?

30. Tchebycheff's inequality states that 3/4 of the sample observations must lie in the interval specified by the limits $\bar{x} \pm 2s$. How does this compare with the actual proportion exhibited by the grasshopper data of Problem 5 of Chapter 5?

31. A survey was conducted at a midwestern university to determine the average cost of utilities to married student families. Data were collected for a two-year period giving:

Location	Number of Families	Average Monthly Utilities
Campus housing	140	\$48.40
Apartments—private homes	90	51.20
Apartment buildings	110	46.80

If a single monthly average of all married student families is to be given in a housing brochure, what value would be most appropriate?

32. A person traveling a certain route to work found the travel times over an eight-week period distributed as follows:

Travel Time	Number of Days
8 but less than 10 minutes	2
10 but less than 12 minutes	8
12 but less than 14 minutes	14
14 but less than 16 minutes	12
16 but less than 18 minutes	4

(a) Find the mean and standard deviation of the travel times in minutes.

(b) Find the mean and standard deviation of travel times measured in seconds.

***33.** (a) Write the formula for the number of permutations, ${}_nP_r$, in terms of the product operator.

(b) Write the formula for the number of combinations, ${}_nC_r$, in terms of the product operator.

***34.** For the data given in Problem 1, find

(a) $\prod_{i=1}^{5} x_i$ (b) $\prod_{i=1}^{5} (x_i/x_{i+5})$

(c) $\prod_{i=1}^{6} (x_i - x_{i+1})$ (d) $\prod_{i=1}^{5} x_i^2$

***35.** Determine least upper and greatest lower bounds on the sample variance using
(a) the data given in Problem 9.
(b) the data given in Problem 13.
(c) the data given in Problem 15.
(d) the *X* data given in Problem 18.
(e) the *Y* data given in Problem 18.
(f) the data given in Problem 26.

REFERENCES

[1] Guterman, H. E., "An Upper Bound for the Sample Standard Deviation," *Technometrics*, **4**, 1 (1962).
[2] Lentner, M. M., "Bounds on the Sample Variance in Terms of the Range," being prepared for publication.
[3] Sheppard, W. F., "On the Application of the Theory of Error to Cases of Normal Distribution and Correlations," *Philosophical Transactions*, A, **192** (1898).
[4] Tchebycheff, P. L., "Des Valeurs Moyennes," *Journal Mathematique Pures et Appliques*, **12** (1867).
[5] Thomson, G. W., "Bounds for the Ratio of Range to Standard Deviation," *Biometrika*, **42** (1955).

Chapter 8

COMMON STATISTICAL DISTRIBUTIONS

Past experience and the researcher's knowledge greatly aid in determining a general form for the distribution of population elements under investigation. When certain assumptions are satisfied, the researcher may use a distribution belonging to a particular family. It is generally agreed that few real-world situations coincide with theoretical results to the extent that general distributions explain the situation exactly. However, most people are willing to accept small deviations from the true situation rather than attempt to find a more exact distribution. The increased accuracy may not compensate for the additional time and effort required.

Some types of statistical distributions have been used to explain real-world problems so frequently that the techniques and theories appropriate to them have been extensively developed. And the fact that results can be readily obtained from general techniques already developed provides added incentive for using a known distribution. Thus much statistical analysis is accomplished under the assumption of certain distributional forms. The use of a particular distribution must withstand any challenge, however, since the validity of inferences depends upon the validity of assumptions made. To use known methods when appropriate saves time, effort, and money, but to use them erroneously gives results which may be worse than pure guessing.

Statistical inference requires probabilities of certain random variables. Recall from Section 6.3 that probabilities for continuous random variables are given by areas under continuous curves and those for discrete random variables by individual or accumulated point probabilities. Since areas under curves may be represented in the calculus by integrals (special summations of infinitely many terms), the probabilities for continuous random variables require evaluation of an integral. However, only a few of the integrals occurring in statistical work can be readily evaluated by elementary methods; most require evaluation by "numerical methods." And the summations required for most discrete probabilities are difficult to perform.

It is because of the difficulties encountered in evaluating summations and integrals that probabilities have been tabulated for basic members of many families. Modern high-speed computers are used in the calculation of probabilities, and in the creation of graphs which basically represent special probability tabulations. Tabulations of probabilities are now available for most of the distributional families covered in this chapter. Relevant portions of existing tables appear in Appendix A; Section 8.9 indicates additional tables available and the coverage they give. Appendix B contains graphs useful in statistical inference.

8.1 THE BINOMIAL FAMILY

This family of discrete distributions has applications in many areas of scientific work. The binomials explain exactly some random variables and serve as approximations for many others.

Consider a series of trials made under the following conditions:

(8.1)

(i) The number of trials, say n, is fixed.
(ii) The outcome of each trial belongs to one of two classes.
(iii) The n trials are mutually independent.
(iv) The probability of an outcome in either class remains constant for all trials.

Sequences of trials satisfying these conditions are called *Bernoulli trials*, after the Swiss mathematician James Bernoulli (1654–1704), who was one of the first to study them extensively. His most important findings were published posthumously in *Ars conjectandi* (*The Art of Conjecture*) in 1713.

The two binomial classes are frequently called *successes* and *failures*, terms carried over from the era when the binomial was applied mainly to gambling problems. The two binomial classes form a partition of the experimental outcomes; hence, each class may contain several subcategories. For example, we might classify people as "redheads" or as "not redheads." In essence, condition (iv) states that the chances of a success (or failure) must be the same for each trial, and condition (iii) states that a specified number of successes (or failures) must have the same chance of occurring in any combination of the trials.

The following are examples of variables which have an exact binomial distribution: (1) the number of red cards observed when 10 cards are selected with replacement from a well-shuffled bridge deck, (2) the number of 1's showing when 4 fair dice are tossed, and (3) the number of 5's appearing in sets of 10 random numbers. The binomial distribution also serves as a good approximation to many variables, such as the following: (1) number of boys in families of a certain size, (2) number of pink-flowered plants in a dozen second-generation offspring, and (3) the number of cars having defects in a random sample of 20 that are safety checked.

Define the following quantities for a series of Bernoulli trials:

$$\begin{aligned} n &= \text{number of trials} \\ X_1 &= \text{number of observations in class 1, successes} \\ X_2 &= \text{number of observations in class 2, failures} \\ p_1 &= \text{probability of a success on any trial} \\ p_2 &= \text{probability of a failure on any trial} \end{aligned}$$

Because the two classes form a partition, the number of observations in the two classes must add to $n: X_1 + X_2 = n$. And the two probabilities must add to unity: $p_1 + p_2 = 1$. With these two conditions, we can eliminate one variable and one probability from consideration. It is common to specify the distribution of the *number of successes*, denoted simply by X; that is, with the symbolism

$$X = X_1; n - X = X_2; p = p_1; q = 1 - p_1 = p_2$$

the distribution of X is

$$f(x; n, p) = \frac{n!}{x!(n-x)!} p^x q^{n-x} = \binom{n}{x} p^x q^{n-x}; \qquad x = 0, 1, \ldots, n \tag{8.2}$$

The distribution specified in (8.2) is a member of the binomial family. The binomials are a two-parameter family of discrete distributions, the parameters being the number of trials, n, and the proportion of successes, p. For each positive integer n and each real value p in the interval $[0, 1]$, a different distribution is obtained. A binomial distribution is symmetric if and only if $p = q = 1/2$, for any n.

We write

$$X \sim \mathrm{B}(n, p) \tag{8.3}$$

to mean "X is distributed as a binomial with parameters n and p." The number $f(x; n, p)$ is really $P(X = x; n, p)$—the probability that X has the value x when the parameters are n and p. Cumulative probabilities are given by

$$\begin{aligned} P(X \leqslant x; n, p) &= F(x; n, p) \\ &= f(0; n, p) + f(1; n, p) + \cdots + f(x; n, p) \end{aligned} \tag{8.4}$$

Table A.3 gives values of $f(x; n, p)$ and $F(x; n, p)$ for selected values of n and p; additional tables are listed in Section 8.9.

► ***Example 8-1***

The number of heads showing when 10 balanced coins are tossed is binomial with $n = 10$ and $p = 1/2$. The chances of seeing 8 heads on one such toss is $f(8; 10, \frac{1}{2}) = 0.0439$ and the chances of seeing at least 4 heads on one such toss is

$$P(X \geqslant 4) = 1 - P(X \leqslant 3) = 1 - F(3; 10, \tfrac{1}{2})$$
$$= 1 - 0.1719 = 0.8281$$ ◄

► ***Example 8-2***

Suppose that 40% of all cars in a certain city have at least one defect. In a random sample of 20 of these cars, what is the probability that exactly 6 cars have defects? no more than 8 cars have defects? between 6 and 15 cars have defects? Equivalently, we might ask: what proportion of all random samples of 20 cars in the given city have these numbers of defective cars?

If a car with defects is considered a success—in statistical terms only—the distribution of

$$X = \text{number of cars among the 20 having defects}$$

is given approximately by (8.2), with $n = 20$ and $p = 0.4$,

$$f(x; 20, 0.4) = \binom{20}{x}(0.4)^x(0.6)^{20-x}; x = 0, 1, \ldots, 20$$

whereby

$$P(X = 6) = f(6; 20, 0.4) = 0.1244$$
$$P(X \leqslant 8) = F(8; 20, 0.4) = 0.5956$$
$$P(6 < X < 15) = F(14; 20, 0.4) - F(6; 20, 0.4)$$
$$= 0.9984 - 0.2500 = 0.7484$$

Thus random samples of 20 cars would have 6 with defects about 1/8 of the time and would have between 6 and 15 with defects about 3/4 of the time. ◄

Since X counts the number of successes in n trials and since $P(\text{success}) = p$ for any trial, we expect to see successes on a proportion p of the trials. For n trials we expect to see np successes. Therefore, repetitions of a binomial experiment having fixed n, and p near zero, should consistently produce only a few successes and many failures. In other words, the variation of X values should be small when p is small. Similarly, when p is near unity, we expect to see a large number of successes (and only a few failures) consistently—again, small variations in the X values. Specifically, when $X \sim \mathrm{B}(n, p)$ the mean and variance of X are:

(8.5) $$\mu_X = np \qquad \text{and} \qquad \sigma_X^2 = npq$$

When $n = 10$ balanced coins are tossed, we expect to see $\mu_X = 5$ heads. On repeated tosses, the number of heads should exhibit a variance of $\sigma_X^2 = 10(1/2)(1/2) = 2.5$, or a standard deviation of $\sigma_X = \sqrt{2.5} = 1.6$.

Most binomial tables have entries only for $p \leqslant 1/2$. Should a binomial be encountered with $p > 1/2$, simply consider the distribution of failures which is binomial with parameters n and $q \leqslant 1/2$. This amounts to interchanging the roles of the two classes.

Let Y denote the number of failures in a series of n Bernoulli trials. Then, since $X \sim \mathrm{B}(n, p)$ and $Y = n - X \sim \mathrm{B}(n, q)$, we see from (8.2) that

(8.6) $$\begin{aligned} P(X = n - y) = f_X(n - y; n, p) &= \binom{n}{n-y} p^{n-y} q^y \\ &= \binom{n}{y} q^y p^{n-y} \\ &= f_Y(y; n, q) = P(Y = y) \end{aligned}$$

or the probability of $x = n - y$ successes is the same as the probability of y failures in n Bernoulli trials. Similarly, the relationship between the cumulative distributions of X and Y is easily established by noting that

(8.7) $$\begin{aligned} F_Y(y; n, q) &= P(Y \leqslant y) \\ &= 1 - P(Y > y) \\ &= 1 - P(n - X > y) \\ &= 1 - P(X < n - y) \\ &= 1 - P(X \leqslant n - y - 1) \\ &= 1 - F_X(n - y - 1; n, p) \end{aligned}$$

▶ ***Example 8-3***

About 90% of all truckers who eat at a given truck stop order the luncheon special. What is the probability that the next 15 truckers all ask for the luncheon special? no more than 10 of the next 15 ask for the luncheon special?

Assuming the truckers order their lunches independently and each has probability of 0.9 of ordering the luncheon special, the required probabilities are

$$P(Y = 15) = f_Y(15; 15, 0.9) = f_X(0; 15, 0.1) = 0.2059$$

and

$$P(Y \leqslant 10) = F_Y(10; 15, 0.9) = 1 - F_X(4; 15, 0.1) = 1 - 0.9873$$
$$= 0.0127 \blacktriangleleft$$

(Problems 3, 21, 23)

8.2 THE HYPERGEOMETRIC FAMILY

One of the assumptions necessary for the binomial to provide a valid explanation of a random variable was independent trials (as a result, independent observations). This assumption is satisfied when sampling with replacement from a finite population and when sampling with or without replacement from an infinite population. Independence does not hold in the remaining case, sampling without replacement from a finite population.

The following are examples of sampling schemes where the elements are not independent: (1) drawing cards without replacement from a bridge deck, (2) selecting a committee from a fixed group of individuals, (3) selecting 1,000 United States citizens, (4) drawing components from a fixed stockpile, (5) selecting apple trees from an orchard, (6) selecting cattle from a feedlot, and so on. In each of these and similar situations the selection of any element depends upon what has been selected previously.

Hypergeometric distributions are characterized by *sampling without replacement* from *finite, two-class populations.* Let the number of elements in the population be denoted by N, those in class 1 by N_1, and those in class 2 by $N_2 = N - N_1$. In a random sample of $n \leqslant N$ elements drawn without replacement, consider the random variable

$$X = \text{number of observations belonging to class 1}$$

The number of observations belonging to the second class is $n - X$, which is automatically known when X is specified. It is not too difficult to show that the distribution of X is

$$f(x; n, N_1, N) = \frac{\binom{N_1}{x}\binom{N - N_1}{n - x}}{\binom{N}{n}}; \quad \text{for integers such that } 0 \leqslant x \leqslant N_1, \text{ and } n + N_1 - N \leqslant x \leqslant n \tag{8.8}$$

Equation (8.8) defines the hypergeometric distributions, discrete distributions that can be indexed by three parameters: positive integers n, N_1, and N.

The mean and variance of the hypergeometric distribution specified by (8.8) are

$$(8.9)\qquad \mu_X = \frac{nN_1}{N} \quad \text{and} \quad \sigma_X^2 = \left[\frac{nN_1}{N}\right]\left[\frac{N - N_1}{N}\right]\left[\frac{N - n}{N - 1}\right]$$

Because of the many possible choices of N_1 and N_2 for each N, the tabulation of hypergeometric probabilities requires considerable space. Rather than giving a table of hypergeometric probabilities, we evaluate simple hypergeometric problems with the aid of Table A.2. Refer to Section 8.9 for information about existing tabulations.

▶ ***Example 8-4***

A class of 16 students is made up of 5 women and 11 men. Each Friday the instructor selects 4 students at random to put solved problems on the board. After several weeks, the women believe they are called upon too often to put problems on the board. The women question the random selection procedure used by the instructor. The distribution of X = number of women selected to present problems on a given Friday is specified by the functional form

$$f(x; 4, 5, 16) = \frac{\binom{5}{x}\binom{11}{4-x}}{\binom{16}{4}}; x = 0, 1, 2, 3, 4$$

Using Table A.2, or by direct enumeration, this distribution can be specified more exactly by

x	0	1	2	3	4
$f(x; 4, 5, 16)$	.1813	.4533	.3022	.0604	.0027

We see that the chance of one woman being selected on a given Friday is nearly one-half. And, the chance of one or more women being selected on a given Friday is over 0.8. If three or four women are consistently chosen each Friday, however, they should indeed question the selection procedure! ◀

Sometimes, the sample size n is very small relative to the population size N. For such cases the conditional probability of observing a success (or a failure) changes very little at each draw, thus the constant probability and independence conditions needed for using the binomial are "almost satisfied." As might be expected, the hypergeometric can be approximated by the binomial. It is recommended that $n/N < 1/10$ for a good approximation. The binomial used to approximate (8.8) is $B(n, N_1/N)$.

▶ ***Example 8-5***

A car dealer has 60 used cars of which 45 are American-made and 15 foreign-made. For a weekend special, he selects 5 cars at random and reduces their price by \$200 if American-made and by \$100 if foreign-made. Consider the random variable

$$X = \text{number of American cars in the special sale}$$

From (8.8) with $n = 5$, $N_1 = 45$, and $N = 60$:

$$f(x; 5, 45, 60) = \frac{\binom{45}{x}\binom{15}{5-x}}{\binom{60}{5}}; x = 0, 1, \ldots, 5$$

Using the Lieberman–Owen Tables [5]:

x	0	1	2	3	4	5
$f(x)$	.0005	.0112	.0825	.2728	.4092	.2237

Approximating this distribution by the binomial with $n = 5$ and $p = 45/60 = 0.75$ gives

x	0	1	2	3	4	5
$f(x)$	.0010	.0146	.0879	.2637	.3955	.2373

◀

(Problems 7, 19)

8.3 THE POISSON FAMILY

This family of discrete distributions has found widespread applications since first introduced in [10] by the French mathematician, Poisson (1781–1840), who in the original investigation, studied the distribution of deaths due to mule kicks per year in the Prussian Army. These distributions depend upon a single parameter, λ, the average or expected number of "successes" per observational unit. Any convenient unit of time, area, space, and the like may be used as an observational unit. Consider the random variable

(8.10) X = number of successes occurring within an observational unit of material, and the following conditions:

(8.11)
(i) The successes occur independently in nonoverlapping units.
(ii) The probability of one success in a small fractional unit is directly proportional to the size of the unit.
(iii) The probability of two or more successes occurring within a small fractional unit is negligible.

A variable defined as in (8.10) and satisfying the assumptions in (8.11) is said to be a *Poisson random variable*. Some random variables which are Poisson, or approximately so, include the number of:
(1) misprints on the front page of a newspaper,
(2) bacteria in a unit of culture medium,
(3) pollen particles in a unit of air,
(4) raisins, or chocolate chips, in a cookie,
(5) tornadoes in a certain locality in a week,
(6) Supreme Court vacancies,
(7) incoming, or outgoing, telephone calls at a given switchboard, and
(8) whitefish caught per hour on the Madison River.

The distribution of a Poisson variable X can be derived by differential equation methods (see [16], for example) and is

(8.12)
$$f(x;\lambda) = \frac{\lambda^x e^{-\lambda}}{x!}; x = 0, 1, 2, \ldots$$
$$= 0; \quad \text{otherwise},$$

where λ is the expected number of successes per unit, and $e = 2.71828\ldots$, the base of natural logarithms. The mean and variance of the distribution specified by (8.12) are

(8.13)
$$\mu_X = \lambda \qquad \text{and} \qquad \sigma_X^2 = \lambda$$

► ***Example 8-6***

If a cook puts raisins in a batch of cookie dough so that each cookie should have 6 raisins on the average, what fraction of the cookies have only 3 raisins? at least 5 raisins? no more than 10 raisins? Variable X, the number of raisins per cookie, has an approximate Poisson distribution with $\lambda = 6$. From Table A.4,

$$P(X = 3) = f_X(3;6) = 0.0892$$
$$P(X \geqslant 5) = 1 - F_X(4;6) = 1 - 0.2851 = 0.7149$$
$$P(X \leqslant 10) = F_X(10;6) = 0.9574$$

In other words, about 9% of the cookies have only 3 raisins, roughly 71% have 5 or more raisins, and almost 96% have 10 or fewer raisins. ◀

The Poisson distribution (8.12) can also be obtained as a limiting form of the binomial distribution $B(n, p)$ as n gets larger and p smaller, in such a way that np remains constant. Consequently, the Poisson distribution can be used to approximate the binomial under these conditions. Since the quantity np is the mean value in the $B(n, p)$ distribution, it follows that the Poisson used in the approximation is the one having mean $\lambda = np$. This approximation does not depend upon the conditions in (8.11), of course. It is recommended that $n \geqslant 20$ and $p \leqslant 1/20$ before considering the approximation; the results are very good when $n \geqslant 100$ and $np \leqslant 10$.

▶ ***Example 8-7***

To show the quality of the Poisson approximation to the binomial, consider the binomial having $n = 1{,}000$ and $p = 0.001$. From the Robertson Tables [11], we have

x	0	1	2	3	4	5	6
$f(x; n, p)$	.3677	.3681	.1840	.0613	.0153	.0030	.0005

Approximating this distribution by the Poisson with $\lambda = np = 1.0$ gives, from Table A.4,

x	0	1	2	3	4	5	6
$f(x; \lambda)$	.3679	.3679	.1839	.0613	.0153	.0030	.0005

This Poisson distribution would also approximate the binomial having parameters $n = 100$ and $p = 0.01$; these probabilities, obtained from either the Harvard or Robertson Tables, are

x	0	1	2	3	4	5	6
$f(x; n, p)$	.3660	.3698	.1848	.0610	.0149	.0029	.0005

◀

Let $X_1, X_2, \ldots, X_n$ denote the number of successes observed on n independent, identical Poisson units. Then the total number of successes on the n units, $X_. = \sum X_i$, has a Poisson distribution with parameter $\lambda^* = n\lambda$, where λ is the parameter related to an individual unit. For example, suppose pollen particles are Poisson distributed with an average per liter of air of $\lambda = 18$ particles. Then, the number of pollen particles, $X_.$, for five liters of this air has a Poisson distribution with $\lambda^* = 5(18) = 90$ particles.

Likewise, if we observe a fractional Poisson unit, the number of successes on the partial unit has a Poisson distribution with parameter having the "fractional value." Thus, the number of pollen particles in a half liter of air has a Poisson distribution with $\lambda^* = (18)/2 = 9$ particles.

(Problems 5, 25)

8.4 THE MULTINOMIAL FAMILY

We now consider an extension of the binomial distributions to cover the cases involving more than two classes. Retaining the other Bernoulli conditions—fixed n trials, independence of outcomes, and constant probability of drawing from each class—we suppose that every sample and population element belongs to one of $k \geqslant 2$ classes. Then for $i = 1, 2, \ldots, k$, let

p_i = probability of observing an element from the i-th class

X_i = number of sample elements occurring in the i-th class

The binomials form a subset of the multinomials—those for which $k = 2$. Note that in the binomial case we could have written

$$P(X_1 = x_1, X_2 = x_2) = f(x_1, x_2; n, p_1, p_2) = \frac{n!}{x_1!x_2!} p_1^{x_1} p_2^{x_2}, \tag{8.14}$$

which is really an alternative expression for the right-hand side of (8.2). When $k > 2$, the joint distribution of $X_1, X_2, \ldots, X_k$ is the k-class multinomial distribution, an extension of (8.14):

$$f(x_1, x_2, \ldots, x_k) = \frac{n!}{x_1!x_2!\ldots x_k!} p_1^{x_1} p_2^{x_2} \cdots p_k^{x_k}; \quad \text{each } x_i = 0, 1, \ldots, n \text{ with } \sum_{i=1}^{k} x_i = n \tag{8.15}$$

(We have omitted the parameters from the left-hand side for simplicity.) Two restrictions, $\sum_{i=1}^{k} x_i = n$ and $\sum_{i=1}^{k} p_i = 1$, are seen to hold. Hence we could replace x_k and p_k by

$$x_k = n - x_1 - x_2 - \cdots - x_{k-1} \quad \text{and} \quad p_k = 1 - p_1 - p_2 - \cdots - p_{k-1},$$

but this replacement would make the right-hand side of (8.15) very lengthy.

The k-class multinomials are discrete distributions depending upon k parameters: the sample size n and any $k - 1$ of the p_i, the class proportions.

▶ ***Example 8-8***

According to genetic theory, the second-generation plants of a certain parental cross of summer squash should be the 4 varieties: white disc, white sphere, yellow disc, and yellow sphere, in the 9:3:3:1 ratio; that is, 9/16 white disc, and so on. If a person sets out 100 randomly selected squash plants of such a cross, the distribution of X_i in the i-th class, $i = 1, 2, 3, 4$, is

$$f(x_1, x_2, x_3, x_4) = \frac{100!}{x_1!x_2!x_3!x_4!}\left(\frac{9}{16}\right)^{x_1}\left(\frac{3}{16}\right)^{x_2}\left(\frac{3}{16}\right)^{x_3}\left(\frac{1}{16}\right)^{x_4};$$

$$\text{each } x_i = 0, 1, \ldots, 100$$

$$\text{with } x_1 + x_2 + x_3 + x_4 = 100 \blacktriangleleft$$

*8.5 THE EXTENDED HYPERGEOMETRICS

When the independence condition was removed from the binomial, the hypergeometric distributions resulted. Removing the independence condition from the multinomial, which is an extension of the binomial, results in an extension of the hypergeometric to $k \geqslant 3$ classes. For $i = 1, 2, \ldots, k$, let

X_i = number of observations belonging to the i-th class

N_i = number of population elements in the i-th class

$N = N_1 + N_2 + \cdots + N_k$

Then the joint distribution of $X_1, X_2, \ldots, X_k$ is an extension of (8.8) and is called the extended hypergeometric distribution:

$$\text{(8.16)} \quad f(x_1, x_2, \ldots, x_k) = \frac{\binom{N_1}{x_1}\binom{N_1}{x_2}\cdots\binom{N_k}{x_k}}{\binom{N}{n}}; \qquad \text{each } x_i = 0, 1, \ldots, N_i \quad \text{with } \sum_{i=1}^{k} x_i = n$$

$$= 0; \qquad \text{otherwise}$$

This is a discrete family indexed by $k + 1$ parameters: the sample size n and the set of N_i.

▶ ***Example 8-9***

The distribution of cards by suits in a bridge hand follows the four-class hypergeometric with $n = N_1 = N_2 = N_3 = N_4 = 13$. Let X_1 correspond to clubs, X_2 to diamonds, X_3 to hearts, and X_4 to spades. Then

$$f(x_1, x_2, x_3, x_4) = \frac{\binom{13}{x_1}\binom{13}{x_2}\binom{13}{x_3}\binom{13}{x_4}}{\binom{52}{13}}; \quad \begin{array}{l}\text{each } x_i = 0, 1, \ldots, 13 \\ \text{with } x_1 + x_2 + x_3 + x_4 = 13\end{array} \blacktriangleleft$$

(Problem 13)

8.6 *THE NORMAL FAMILY*

The most frequently encountered continuous distributions in probability and statistics are the members of the *normal* family. If not appearing directly, they form the basis for other related distributions to be discussed in this chapter. Each normal distribution can be represented by a symmetric, bell-shaped curve. Two parameters are required to identify a particular member: the population mean μ (location parameter) and the population variance σ^2 (dispersion parameter). A general normal curve appears in Figure 8-1. The distance between μ and the points of inflection of the curve (the points where the curve changes its upward-downward direction) is σ, the standard deviation. The point of symmetry is necessarily the mean value, μ.

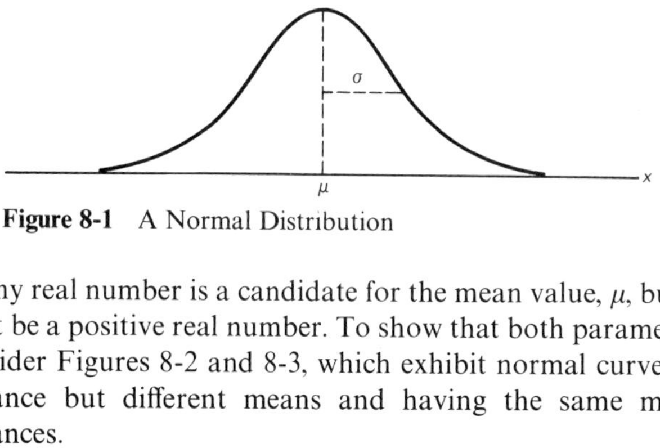

Figure 8-1 A Normal Distribution

Any real number is a candidate for the mean value, μ, but the variance, σ^2, must be a positive real number. To show that both parameters are necessary, consider Figures 8-2 and 8-3, which exhibit normal curves having the same variance but different means and having the same mean but different variances.

If X is distributed as a normal random variable with mean, μ, and variance, σ^2, we use the customary notation

(8.17) $$X \sim N(\mu, \sigma^2)$$

The mathematical equation of the $N(\mu, \sigma^2)$ curve, first obtained in 1733 by DeMoivre (1667–1754), is

(8.18) $$f(x; \mu, \sigma^2) = \frac{1}{\sigma\sqrt{2\pi}}\, exp\left[\frac{-1}{2\sigma^2}(x - \mu)^2\right]; \qquad -\infty < x < \infty$$

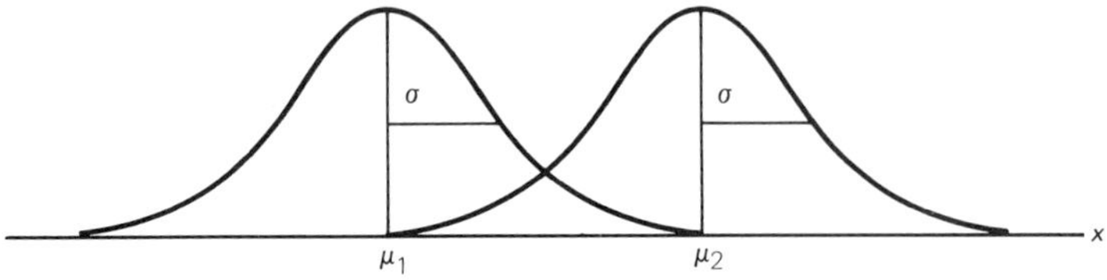

Figure 8-2 Normals with Different Means, Same Variance

where $\pi = 3.14159\ldots$, $exp\,[a] = e^a$, and $e = 2.71828\ldots$. The normal distribution (8.18) was later derived independently by Gauss (1777–1855) in his research on the theory of errors in repeated measurements.

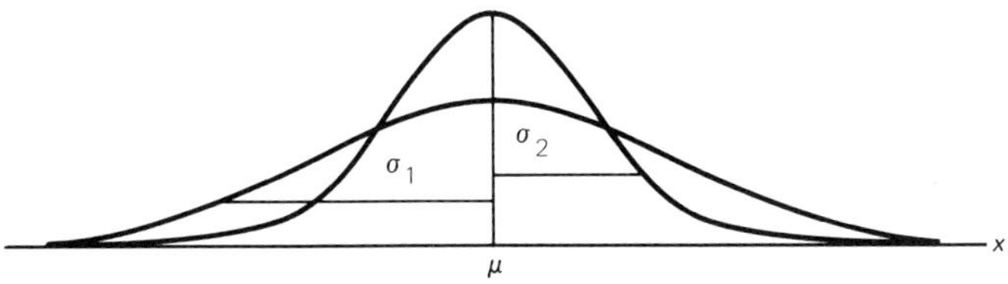

Figure 8-3 Normals with Different Variances, Same Mean

Few if any real-world variables have an exact normal distribution. However, the normal law serves as an adequate model to explain the behavior of many continuous variables, such as (1) vertical distances of hits from the bull's eye of a target, (2) heights of certain groups of people, (3) weight gains of experimental animals, (4) lengths of production times, (5) voltage fluctuations, (6) times to failure of light bulbs, (7) yields of a chemical process, (8) weight of cereal per package, (9) amount of beverage per bottle, and (10) width of sheets of paper.

It should not be assumed that all of the variables listed, or closely related variables, are approximately normal in every case. For example, the times required to assemble cardboard cartons may be approximately normal, whereas the times required to assemble metal cages may be explained better by some non-normal distribution. And, while the lifetimes of certain electrical components, such as light bulbs, may follow a normal distribution approximately, experience has revealed the times to failure of certain vacuum tubes and electronic components have distributions belonging to another family (the Gamma family).

Important features of *any* normal distribution are that

(8.19)
68.3% of its elements fall in the interval $\mu \pm \sigma$
95.4% of its elements fall in the interval $\mu \pm 2\sigma$
99.7% of its elements fall in the interval $\mu \pm 3\sigma$

These features are indicated in Figure 8-4. Recall that areas are interpreted as probabilities (proportions) in the continuous case.

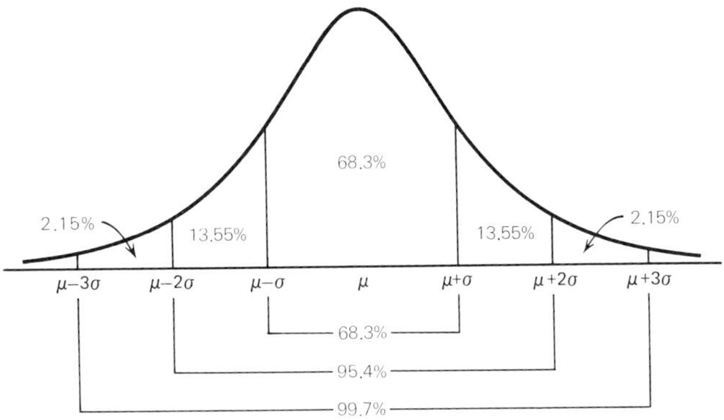

Figure 8-4 Distribution of Elements in a $N(\mu, \sigma^2)$ Population

By virtue of the properties stated in (8.19), certain probabilities can be obtained for any normal distribution. These probabilities could also be obtained from Figure 8-4. For example, $P(X > \mu + 2\sigma) = 0.023$ for any normal variable X. Although it would be helpful to have tables from which any probability of any normal distribution could be obtained, there are an uncountable number of normal distributions which makes such a complete tabulation impossible. Fortunately, a complete tabulation is unnecessary, since every member of the normal family is related in a special way to the N(0, 1) distribution; that is, the normal with zero mean and unit variance. Because of its importance, the N(0, 1) distribution is called the *standard normal*, the *normal in standard form*, or the *Z distribution* and is extensively tabulated. Table A.5 gives values of $F_Z(z)$ for $z = -4(0.01)4$.

▶ ***Example 8-10***

Letting Z denote the standard normal variable, we obtain the following from Table A.5:

$$P(Z \leqslant 0.5) = F_Z(0.5) = 0.69146$$

$$P(0 \leqslant Z \leqslant 1.5) = F_Z(1.5) - F_Z(0) = 0.93319 - 0.50000 = 0.43319$$

$$P(Z \geqslant 2) = 1 - F_Z(2) = 1 - 0.97725 = 0.02275$$

$$P(Z \leqslant -1.8) = F_Z(-1.8) = 0.03593$$ ◀

It can be shown that the random variable $X \sim N(\mu, \sigma^2)$ can be converted to the random variable $Z \sim N(0, 1)$ by the linear transformation (special coding)

(8.20) $$Z = \frac{X - \mu}{\sigma}$$

As a result, finding the probability for any interval of the $N(\mu, \sigma^2)$ distribution reduces to finding the probability for the corresponding interval on the $N(0, 1)$ distribution; that is, for $x_1 \leqslant x_2$

$$P(x_1 \leqslant X \leqslant x_2) = P\left(\frac{x_1 - \mu}{\sigma} \leqslant \frac{X - \mu}{\sigma} \leqslant \frac{x_2 - \mu}{\sigma}\right)$$

(8.21)

$$= P\left(\frac{x_1 - \mu}{\sigma} \leqslant Z \leqslant \frac{x_2 - \mu}{\sigma}\right)$$

Application of (8.20) to a normal random variable X is called the *standardization* of X.

► ***Example 8-11***

The reaction times, X, of a set of laboratory mice when subjected to a certain stimulus are assumed to follow a normal distribution with $\mu = 0.8$ second and $\sigma = 0.5$ second. What proportion of the reaction times are over 1 second? between 0.3 and 1 second? Standardizing X and using Table A.5, we get

$$P(X > 1) = P\left(Z > \frac{1 - 0.8}{0.5}\right) = P(Z > 0.4)$$

$$= 1 - F(0.4) = 1 - 0.65542 = 0.34458$$

and

$$P(0.3 \leqslant X \leqslant 1) = P\left(\frac{0.3 - 0.8}{0.5} \leqslant Z \leqslant 0.4\right) = P(-1 \leqslant Z \leqslant 0.4)$$

$$= F(0.4) - F(-1)$$

$$= 0.65542 - 0.15865 = 0.49677$$

About 34% of the reaction times are expected to exceed 1 second, while nearly 50% of the reaction times should be between 0.3 and 1 second. ◄

(Problems 9, 11, 17)

8.7 DISTRIBUTIONS IN SAMPLING FROM A NORMAL

In many experimental situations a normal distribution adequately describes the behavior of a random variable. Primary inference problems then focus on the mean, μ, and variance, σ^2. General solutions exist for the most important parametric inferences in the normal case. These solutions are based upon special functions of sample observations (that is, certain transformations of commonly used statistics), and result in a unified treatment of various inference problems.

8.7.1 Sampling Distribution of the Mean and Other Linear Forms

First we consider the distribution of $\overline{X}$ in random samples of size n drawn from a $N(\mu, \sigma^2)$ population. On several earlier occasions we have expressed our interest in using $\overline{X}$ as an estimator of μ.

- *Theorem 8-1: If the random variable X has the* $N(\mu, \sigma^2)$ *distribution, then for random samples of size n, the sample mean $\overline{X}$ has the* $N(\mu, \sigma^2/n)$ *distribution.*

Briefly, Theorem 8-1 states "if X is normal, so is $\overline{X}$." The mean and variance of $\overline{X}$, $\mu_{\overline{X}} = \mu$ and $\sigma_{\overline{X}}^2 = \sigma_X^2/n$, respectively, correspond to the results in (7.38).

Applying the technique of standardization to $\overline{X}$, when sampling from a normal, we see from Theorem 8-1 that

$$Z = \frac{\overline{X} - \mu_{\overline{X}}}{\sigma_{\overline{X}}} = \frac{\overline{X} - \mu}{\sigma/\sqrt{n}} \sim N(0, 1) \tag{8.22}$$

This special transformation on $\overline{X}$ allows us to calculate probabilities concerning $\overline{X}$, when μ and σ are known, by appealing to the Z distribution. If μ and σ are unknown, probabilities of $\overline{X}$ are indeterminate.

▶ ***Example 8-12***

Boxes of cake mix are filled by a machine which has been adjusted so that the amount X per box is $\mu = 20.1$ ounces. Suppose we are willing to assume (based on past experience) that X is approximately normal with mean $\mu = 20.1$ and standard deviation of $\sigma = 0.12$ ounces. Each half hour, nine boxes are selected at random from the production and their contents weighed. What are the chances of obtaining 9 boxes whose contents average less than the advertised weight of 20 ounces? average between 20 and 20.25 ounces?

Since $\overline{X} \sim N(20.1, 0.0016)$, the required probabilities are

$$P(\overline{X} < 20) = P\left(Z < \frac{20 - 20.1}{0.04}\right) = P(Z < -2.5) = 0.00621$$

and

$$\begin{aligned} P(20 < \overline{X} < 20.25) &= P\left(\frac{20 - 20.1}{0.04} < Z < \frac{20.25 - 20.1}{0.04}\right) \\ &= P(-2.5 < Z < 3.75) \\ &= F_Z(3.75) - F_Z(-2.5) \\ &= 0.99991 - 0.00621 = 0.99370 \end{aligned}$$

These probabilities are represented by the shaded areas of the following graph.

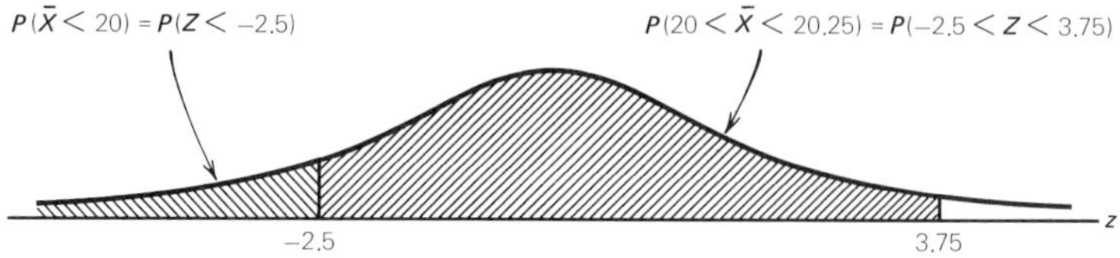

◀

Theorem 8-1 presents a special case of a very general theorem which states "any linear function of normal variables has a normal distribution." The next theorem deals with the case of *independent* normal variables, still a special case of the general theorem.

- *Theorem 8-2: If $Y_1, Y_2, \ldots, Y_k$ are independent with $Y_i \sim \mathrm{N}(\mu_i, \sigma_i^2)$ and $a_1, a_2, \ldots, a_n$ are any set of constants, then*

(8.23) $$L = \sum_{i=1}^{k} a_i Y_i \sim \mathrm{N}(\mu_L, \sigma_L^2),$$

where

(8.24) $$\mu_L = \sum_{i=1}^{k} a_i \mu_i \qquad \text{and} \qquad \sigma_L^2 = \sum_{i=1}^{k} a_i^2 \sigma_i^2$$

▶ ***Example 8-13***

Draw a random sample of size n from $\mathrm{N}(\mu, \sigma^2)$; for $i = 1, 2, \ldots, n$, let $Y_i = X_i$ and $a_i = 1/n$. Then $Y_i \sim \mathrm{N}(\mu, \sigma^2)$ for all i, and Theorem 8-2 gives $L = \bar{X} \sim \mathrm{N}(\mu_L, \sigma_L^2)$ with $\mu_L = \mu$ and $\sigma_L^2 = \sigma^2/n$, thereby giving Theorem 8-1. Also, letting $a_i = 1$ for all i, Theorem 8-2 implies

$$L = X_. = n\bar{X} \sim \mathrm{N}(n\mu, n\sigma^2)$$

Hence, the sample total is also normally distributed if the variable is normal. ◀

▶ ***Example 8-14***

If a sample of size n_1 is drawn from a $\mathrm{N}(\mu_1, \sigma_1^2)$ population and a sample of size n_2 is drawn from a $\mathrm{N}(\mu_2, \sigma_2^2)$ population, Theorems 8-1 and 8-2 may be used to show that

$$\bar{X}_1 - \bar{X}_2 \sim \mathrm{N}\left(\mu_1 - \mu_2, \frac{\sigma_1^2}{n_1} + \frac{\sigma_2^2}{n_2}\right),$$

$$\bar{X}_1 + \bar{X}_2 \sim \mathrm{N}\left(\mu_1 + \mu_2, \frac{\sigma_1^2}{n_1} + \frac{\sigma_2^2}{n_2}\right),$$

$$2\bar{X}_1 - 3\bar{X}_2 \sim \mathrm{N}\left(2\mu_1 - 3\mu_2, \frac{4\sigma_1^2}{n_1} + \frac{9\sigma_2^2}{n_2}\right),$$

and so on. ◀

(Problem 15)

8.7.2 The Student T Family

In Example 8-12 we used the standard normal distribution to calculate probabilities related to $\bar{X}$. For these calculations, we needed to know the value of σ. In cases where σ is unknown to us, we might consider replacing σ in (8.22) by a sample estimate. We no longer have a standard normal variable but a new variable whose distribution is known and will be studied now.

If a random sample of size n is drawn from a $N(\mu, \sigma^2)$ distribution, the quantity $\dfrac{\bar{X} - \mu}{S/\sqrt{n}}$, and similar ones, enters a number of important inference problems. Prior to 1908 this quantity was treated as an approximate Z variable in large sample experiments. William S. Gosset (1876–1937), a statistician working for the Guinness Brewery in Dublin, Ireland, became skeptical of this approximation in small samples and empirically derived the distribution of the modified quantity $\dfrac{\bar{X} - \mu}{S}$, when the sampling was from a $N(\mu, \sigma^2)$ distribution. Since the brewery prohibited the publication of research by its staff members, Gosset submitted his findings to one of his former professors, Karl Pearson (1857–1936). This and Gosset's later research were published under the pseudonym "Student" beginning in 1908 in *Biometrika*, edited by Pearson, see [15].

In 1923, Sir Ronald A. Fisher theoretically derived the distribution of $\dfrac{\bar{X} - \mu}{S/\sqrt{n}}$; this family of distributions has come to be known as the *Student T distributions*. Each member of this family is completely specified by a single parameter, v = degrees of freedom associated with the denominator of the T ratio; for example,

$$T_{n-1} = \frac{\bar{X} - \mu}{S/\sqrt{n}} \tag{8.25}$$

is a T variable with $n - 1$ degrees of freedom. Otherwise, the members of the Student T family have the same general appearances as the standard normal Z distribution; they are symmetric about 0 and have a doubly infinite range, but are flatter for small sample sizes. A T_v variable has mean value 0 if $v > 1$ and variance $v/(v - 2)$ if $v > 2$. In the limit as $v \to \infty$, the distribution of T_v approaches the Z distribution, that is, $T_\infty = Z$. In fact, for $v > 40$ it is virtually impossible to differentiate between the curves for T_v and Z.

We emphasize that the right-hand side of (8.25) is a T variable only when sampling from a normal distribution. A value of the T_v random variable is denoted by $t_{v;\gamma}$ and is that point on the T_v curve for which

$$P(T_v \leqslant t_{v;\gamma}) = \gamma \tag{8.26}$$

Table A.6 gives selected points for the cumulative distribution of T_ν for $\nu = 1(1)40(10)100, 500, 1000, \infty$.

▶ ***Example 8-15***

For $\nu = 13$, Table A.6 gives an entry of 1.7709 in the 0.95 column; that is,

$$P(T_{13} \leqslant 1.7709) = 0.95$$

Entries for $\nu = 22$ with $\gamma = 0.95$ and 0.975 are 1.7171 and 2.0739, respectively. This means:

$$F(-1.7171) = P(T_{22} \leqslant -1.7171) = 0.05$$

$$F(1.7171) = P(T_{22} \leqslant 1.7171) = 0.95$$

$$F(2.0739) = P(T_{22} \leqslant 2.0739) = 0.975$$

$$P(T_{22} > 2.0739) = 0.025$$

$$P(-1.7171 \leqslant T_{22} \leqslant 2.0739) = 0.975 - 0.05 = 0.925$$ ◀

(Problems 29, 32)

8.7.3 The Chi-Square Family

The sum of squares of deviations from the sample mean is the variable portion of the sample variance. We now examine the distribution of SS when sampling from the $N(\mu, \sigma^2)$ distribution.

From mathematical statistics, it is known that a sum of squares of *independent* standard normal variables has a special distribution called a *chi-square distribution.* Specifically,

(8.27)
$$\chi_n^2 = \sum_{i=1}^{n} Z_i^2$$

is a chi-square variable with n degrees of freedom if $Z_1, Z_2, \ldots, Z_n$ are independent $N(0, 1)$ variables.

A special transformation on the sample variance results in a chi-square variable. Recall that the deviations forming the sample SS were not all independent—only $n - 1$ of the deviations were needed for calculating the SS and the variance, this nonindependence has the same impact on the degrees of freedom of our chi-square variable. In normal sampling then the transformed variance is a chi-square variable with $(n - 1)$ degrees of freedom

(8.28)
$$\chi_{n-1}^2 = (n - 1)S^2/\sigma^2 = SS/\sigma^2$$

The chi-square distributions were developed by Pearson [9]. Members of the chi-square family

(i) have nonnegative range,
(ii) depend upon a single parameter, ν = degrees of freedom of SS,
(iii) have mean value equal to the degrees of freedom and variance equal to twice the degrees of freedom, and
(iv) are nonsymmetrical but approach symmetry for large degrees of freedom (actually approach normality in the limit).

A value of the χ_ν^2 random variable is denoted by $\chi_{\nu;\gamma}^2$ and is that point on the χ_ν^2 curve for which

$$P(\chi_\nu^2 \leqslant \chi_{\nu;\gamma}^2) = \gamma \tag{8.29}$$

Note that this represents a deviation from our convention of using capital letters for random variables and corresponding lower-case letters for their values. Table A.7 gives selected points for cumulative distribution of χ_ν^2 for $\nu = 1(1)80$.

▶ ***Example 8-16***

For $\nu = 31$, Table A.7 gives 0.025 and 0.99 entries of 17.539 and 52.213, respectively.

Thus,

$$F(17.539) = P(\chi_{31}^2 \leqslant 17.539) = 0.025$$
$$F(52.213) = P(\chi_{31}^2 \leqslant 52.213) = 0.99$$
$$P(\chi_{31}^2 > 52.213) = 0.01$$
$$P(17.539 \leqslant \chi_{31}^2 \leqslant 52.213) = 0.99 - 0.025 = 0.965$$ ◀

(Problems 30, 33)

8.7.4 The Family of *F* Distributions

If a random sample of size n_1 is drawn from the $N(\mu_1, \sigma_1^2)$ population and an independent random sample of size n_2 is drawn from the $N(\mu_2, \sigma_2^2)$ population, then the transformed ratio of sample variances

$$F_{n_1-1,n_2-1} = \frac{S_1^2/\sigma_1^2}{S_2^2/\sigma_2^2} \tag{8.30}$$

is an F variable with numerator degrees of freedom $n_1 - 1$ and denominator degrees of freedom $n_2 - 1$. Fisher [1] gave the theoretical distribution of $(1/2)\log_e(S_1^2/S_2^2)$, a modified version of (8.30). The importance of the variance ratio (8.30) was discovered by G. W. Snedecor (1881–   ), who subsequently showed that the distribution of (8.30) was merely a transformation of Fisher's result; see [14]. Snedecor named the distribution of (8.30) F in honor of Fisher. As a result, the members of this family are also known as *Snedecor F distributions.*

Likewise, the distribution of the ratio of two independent estimates of a *single* normal variance has the F_{ν_1,ν_2} distribution with parameters ν_1 = degrees of freedom of the numerator estimate of variance and ν_2 = degrees of freedom of the denominator estimate of variance. The members of the F family can also be derived as the distribution of the ratio of two independent chi-square variables, each divided by its degrees of freedom:

$$F_{\nu_1,\nu_2} = \frac{\chi^2_{\nu_1}/\nu_1}{\chi^2_{\nu_2}/\nu_2} \tag{8.31}$$

It is not surprising then that the F curves resemble the χ^2 curves. Each F distribution

(i) has nonnegative range,
(ii) depends upon two parameters, the numerator and denominator degrees of freedom, and
(iii) is nonsymmetrical but approaches symmetry (normality) as both degrees of freedom approach infinity.

An F_{ν_1,ν_2} variable has mean value of $\nu_2/(\nu_2 - 2)$ if $\nu_2 > 2$, and variance of $2\nu_2^2(\nu_1 + \nu_2 - 2)/\nu_1(\nu_2 - 2)^2(\nu_2 - 4)$ if $\nu_2 > 4$ and $\nu_1 > 0$. A useful relationship existing between the F and T variables is that

$$F_{1,\nu_2} = (T_{\nu_2})^2 \tag{8.32}$$

Table A.8 gives selected points of the cumulative distribution of F_{ν_1,ν_2} for various combinations of ν_1, ν_2. These points will be denoted by $F_{\nu_1,\nu_2;\gamma}$.

▶ *Example 8-17*

For $\nu_1 = 15$ and $\nu_2 = 24$, Table A.8 gives 0.05 and 0.995 entries of 0.437 and 3.25, respectively. Therefore,

$$P[F_{15,24} \leqslant 0.437] = 0.05$$

$$P[F_{15,24} \leqslant 3.25] = 0.995;$$

that is,

$$F_{15,24;0.05} = 0.437$$

$$F_{15,24;0.995} = 3.25 \blacktriangleleft$$

(Problems 31, 35)

8.8 A CENTRAL LIMIT THEOREM

In the next three chapters we will study inference about the parameters of normal distributions. This extensive coverage might make it appear that undue emphasis is placed on normality when it has been stated that few, if any, random variables have exact normal distributions. The assumption of normality is *not* critical and may be relaxed when making inference about the mean of any population so long as the sample size is sufficiently large. This important concept is an immediate consequence of the Central Limit Theorem, one of the major theorems in statistical analysis. We state one of the more common versions.

- *Central Limit Theorem: Let the variable X have any distribution with finite mean μ and variance σ^2. Then, in random samples of size n from this distribution, the standardized sample mean $(\bar{X} - \mu)/\sigma_{\bar{X}}$ has a distribution which approaches the* N(0, 1) *distribution as n approaches infinity.*

As a result of the Central Limit Theorem, the Z distribution can be used for inferences about the mean, μ, of any random variable so long as n is large. The degree of approximation in the inference depends upon the population departure from normality and the sample size. For continuous random variables having symmetric distributions, samples of size 20 or more are generally sufficient. For discrete random variables, sample sizes of several hundred may be required. For binomial random variables, for example, good results are obtained for samples as small as $n = 10$ when $p = P(\text{success})$ is near 0.5 (symmetric when $p = 0.5$) whereas comparable results when p is near zero or unity would require considerably larger sample sizes.

If we must estimate the variance σ^2, a Student T distribution would be used instead of the Z distribution. Of course, larger departures from normality can be tolerated with the Z distribution than with the T distributions.

It should be emphasized that the Central Limit Theorem gives the limiting distribution of standardized sample means. When additional information is known about the parent population, stronger results can be stated. If the underlying distribution is normal, for example, Theorem 8-1 gives normality of $\bar{X}$ for *every* sample size.

► *Example 8-18*

Recall for a Poisson distribution with parameter λ that $\mu = \sigma^2 = \lambda$. If 100 units are observed from a Poisson population, the Central Limit Theorem allows us to say that $\bar{X}$ has an approximate N(λ, $\lambda/100$) distribution or that

$$\frac{\bar{X} - \lambda}{\sqrt{\lambda/100}} \doteq Z \sim \mathrm{N}(0, 1) \blacktriangleleft$$

► *Example 8-19*

Even through small sample sizes, the approach of $\bar{X}$ to normality is quite prominent in some cases. Consider the uniform distribution over a range $[a, b]$:

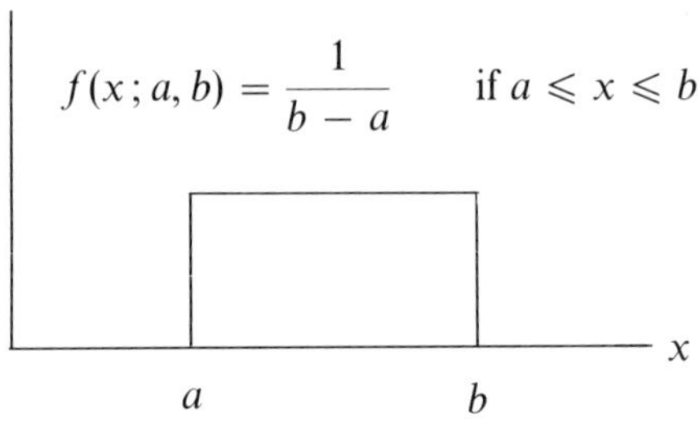

It can be shown for this distribution that $\mu_X = (a + b)/2$ and $\sigma^2 = (b - a)^2/12$. By the Central Limit Theorem, for "large" n, $\bar{X}$ has an approximate normal distribution with mean $\mu_{\bar{X}} = \dfrac{a + b}{2}$ and variance $\sigma_{\bar{X}}^2 = \dfrac{(b - a)^2}{12n}$.

For samples of size $n = 1$, the single observation drawn from the population is the sample mean. Hence, if $n = 1$, $\bar{X}$ has the same distribution as does X.

For sample sizes $n = 2$ and 3 drawn from the uniform distribution, it can be shown that $\bar{X}$ has the following distributions:

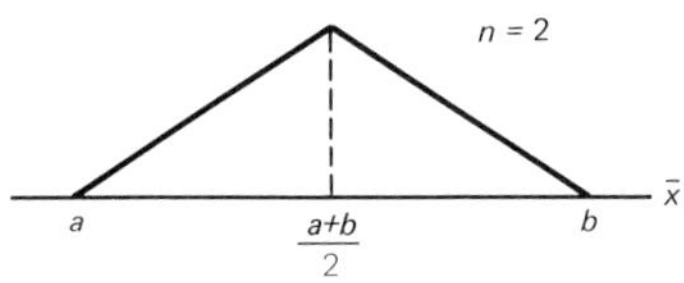

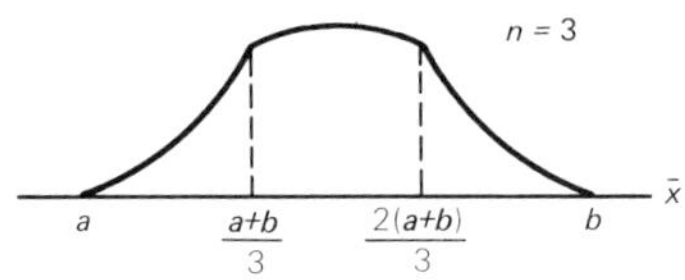

The formulas defining the two curves, $g(\bar{x})$, are of no particular significance; the curve $g(\bar{x})$ will have n branches when X is uniformly distributed.

Whereas a normal distribution would do a poor job of approximating a uniform distribution, it is somewhat amazing that a normal would do a very good job of approximating the distribution of $\bar{X}$ in samples of size 3 or more from a uniform. ◄

8.9 ADDITIONAL TABLES

Over the past few decades computers have aided greatly in the production of numerous tabulations of probabilities and graphs of commonly used statistical distributions. No claim is made that the catalog of tabulations given below is exhaustive. An attempt is made to list the most complete tabulations.

8.9.1 Binomial Tables

The most comprehensive binomial tabulation is the so-called Harvard Tables [4], which gives $1 - F(x - 1; n, p) = P(X \geqslant x; n, p)$ for all combinations of parameters

$$n = 1(1)50(2)100(10)200(20)500(50)1000$$

$$p = .01(.01).50, 1/16, 1/12, 1/8, 1/6, 3/16, 1/3, 3/8, 5/12, 7/16$$

Tables by the Ordinance Corps [17] also give $1 - F(x - 1; n, p)$, to seven decimal places for $n = 1(1)150$ and $p = 0.01(.01).50$. Smaller tabulations include the NBS Tables [7], which give terms for $n = 1(1)50$, and the Romig Tables [12], which give terms for $n = 50(1)100$. For small values of p, the tables by Robertson [11] give terms for $n = 2(1)100(2)200(10)400(20)1000$ when $p = .001(.001).020$, and $n = 2(1)50(2)100(5)200(10)300(20)600(50)1000$ when $p = .021(.001).050$. Last, the tables by Weintraub [18] give terms for $n = 1(1)100$ and $p = .00001, .0001(.0001).001(.001).10$.

8.9.2 Hypergeometric Tables

Fairly extensive tables of the hypergeometric are given in [5]. Terms given are $f(x)$ and $F(x)$ for $N = 1(1)50(10)100$ and for *some* $N = 100(10)200$.

8.9.3 Poisson Tables

Values of $f(x)$ and $1 - F(x)$ for Poisson distributions having

$$\lambda = .001(.001).01(.01).3(.1)15(1)100$$

are given in the Molina Tables [6]. More extensive Poisson tables are the GE Tables [2], which give $f(x)$ and $F(x)$ to eight decimals for numerous λ between 10^{-8} and 205.

8.9.4 Student *T* Tables

Additional percentage points for T distributions may be found in [8]. The coverage is for $\nu = 1(1)100(2)150, 200(100)1000, \infty$, and cumulative percentages of $\gamma = 0.75, 0.90, 0.95, 0.975, 0.99$, and 0.995.

8.9.5 Chi-Square Tables

Percentiles of χ^2 distributions are given in [3] for 23 cumulative percentages and degrees of freedom $\nu = 1(1)150(2)330$.

PROBLEMS

1. If possible, give the family to which the distribution of each of the following variables belongs and specify the parameters of the distribution.

(a) X = number of heads obtained when a fair (well-balanced) coin is tossed 100 times

(b) Steel pins are classified as follows: "usable" if their lengths are between 1.05 and 1.15 inches, "need reworking" if their lengths are over 1.15 inches, and "scrap" if their lengths are under 1.05 inches. The steel pins made by a certain process are such that 86% are "usable," 10% "need reworking," and 4% are "scrap." A random sample of 48 pins is selected from the production. Let

$$X_1 = \text{number that are "usable"}$$
$$X_2 = \text{number that "need reworking."}$$

(c) A card is selected at random from a deck of 52 cards, the result is recorded, the card is laid aside, and a second card is selected from the remaining cards, and so on. From five such selections, let X = number of clubs.

(d) X = time to failure for a certain type of electric light bulb

(e) X = time to failure of a specific kind of electronic tubes

(f) X = vacancies per year in the U.S. Supreme Court

(g) For a random sample of college students,
X_i = number of students in their i-th year, $i = 1, 2, 3, 4, 5$, with 5 denoting any graduate student

2. Suppose that a fair die is tossed five times. Let X denote the number of times 2 or 3 spots show. Give the distribution of X.

3. Suppose a football kicker has an 80% chance of kicking the extra point after any touchdown. In those games where his team scores 5 touchdowns, what is the probability that he kicks the extra point after

(a) exactly 4 touchdowns?

(b) at least two touchdowns?

(c) every touchdown?

If X denotes the number of times this kicker makes the extra point when 5 touchdowns are scored, what is

(d) μ_X? (e) σ_X^2?

4. Suppose a janitor begins his nightly round of buildings with a supply of 8 light bulbs. From past experience, he knows that an average of 5 bulbs are burned out each day. What is the probability on any given night that

(a) he must return for additional bulbs before finishing his rounds?

(b) he has at least five bulbs left upon completion of his rounds?

5. A certain executive makes an average of 60 long-distance telephone calls per month. What proportion of the weeks is he expected to make

(a) 20 or more long-distance calls?

(b) only 10 long-distance calls?

(c) between 10 and 20 long-distance calls?

6. Suppose that bales of cotton are normally distributed with mean $\mu = 100$ pounds and variance $\sigma^2 = 9$. What proportion of cotton bales are
(a) over 105 pounds?
(b) between 90 and 105 pounds?
(c) over 95 pounds?
If six bales of cotton are selected at random from such a lot, what distribution does
(d) their mean weight follow?
(e) their total weight follow?

7. Six electrical fuses in a lot of 36 have faulty connections. If a customer selects four fuses at random from this lot as soon as it is opened, what is the probability that he gets
(a) none with faulty connections?
(b) exactly two with faulty connections?
(c) at least one with a faulty connection?
Let X denote the number of faulty fuses among the 4 drawn and find
(d) μ_X (e) σ_X^2

8. According to genetic theory, the blossom color of second generation sweet peas should be red : white in the 3 : 1 ratio. If a random selection of 20 second generation seeds is planted and all germinate, what is the probability the number of red-blossomed plants will be
(a) exactly twelve?
(b) at least twelve?
(c) less than ten?
If X denotes the number of red-blossomed, second-generation plants among sets of 20, find
(d) μ_X (e) σ_X^2

9. The daily output of silver-producing ore for a certain mine is approximately normal with mean equal to 5,500 pounds and standard deviation of 400 pounds. What proportion of the days does the daily output
(a) exceed 6,000 pounds?
(b) fall below 4,000 pounds?
(c) lie between 4,000 and 6,000 pounds?

10. As an advertising gimmick, many companies put a prize in each package of their product (e.g., cereal, Cracker Jack, and so on). For a certain product, suppose the prizes are put in the packages at random but 4/5 of the prizes are of paper composition, while the remaining are a combination of plastic, wood and metal. If a person buys 10 boxes of the product, what are the chances of getting
(a) exactly 3 nonpaper prizes?
(b) at least 6 paper prizes?
(c) all paper prizes?

11. The reaction times of laboratory animals subjected to a certain stimulus are assumed to be normally distributed with a mean of 1.2 seconds and a variance of 0.16 (second)^2. What proportion of the reaction times are
(a) over 1 second?
(b) under 1.6 seconds?
(c) between 1 and 1.6 seconds?

12. The number of times that a new car must be taken back to the dealer for minor adjustments during the warranty period is assumed to be a Poisson variable with parameter $\lambda = 5$. What is the probability that any new car must be taken back for adjustments
(a) at least once?
(b) never?
(c) at least 10 times?
Let Y denote the number of times the dealer must make minor adjustments on six new cars that he has sold.
(d) What distribution does Y have?
(e) What is the mean of this distribution?
(f) What is the variance of this distribution?

13. A class of 35 students were classified as follows according to nationality:

English	18
French	7
German	3
Spanish	4
African	3

A group of 6 students is to be selected at random to serve on a panel to discuss a current affairs problem. What is the probability the selected group will consist of
(a) 3 English, 1 French, and 2 Spanish students?
(b) 3 English and 1 French student?
(c) at least one from each nationality?

14. It is assumed that the time required, in minutes, for a machine operator to produce a given component follows a normal distribution with a mean $\mu = 16$ and a variance $\sigma^2 = 4$. What is the probability that the time required to produce one component is:
(a) less than 14 minutes?
(b) between 15 and 18 minutes?
If 4 components are produced, what is the probability that the mean of these 4 times is:
(c) between 14 and 18 minutes?
(d) greater than 17 minutes?

15. In the operation of a compression cylinder, the pressures within the cylinder are assumed to be normally distributed with a mean of 355 pounds and a standard deviation of 10 pounds per square inch.
If a single measurement of the pressure is taken, what is the probability it will be
(a) between 342 and 365 pounds?
(b) less than 360 pounds?
(c) over 355 pounds?
(d) If 25 measurements of the pressure are taken, what is the probability that their average will be between 348 and 352 pounds?

16. A motel manager knows from past experience that during any week of June there will be on the average of 10 honeymoon couples among the guests staying in the motel. During any given week of June what is the probability his guests include
(a) exactly 13 honeymoon couples?
(b) at least 6 honeymoon couples?

17. The drying times of a certain lacquer are assumed to be normally distributed with a mean of 90 seconds and a standard deviation of 15 seconds. What proportion of the drying times are expected to be
(a) over 100 seconds?
(b) over 120 seconds?
(c) between 75 and 100 seconds?
(d) equal to 90 seconds?

18. Rolls of paper manufactured by a given paper mill have weights which are normally distributed with a mean $\mu = 200$ pounds and standard deviation of $\sigma = 4$ pounds. What is the probability that an arbitrarily selected roll has weight
(a) between 195 and 205 pounds?
(b) over 210 pounds?
(c) below 198 pounds?
If 4 rolls are selected at random, what is the probability their average weight is
(d) above 203 pounds?
(e) between 196 and 202 pounds?
(f) over 205 pounds?

19. An advisory committtee of 3 persons is to be selected from a group of 8 people: 3 are Democrats, 3 are Republicans, and 2 have no party affiliation. What is the probability that
(a) the number of Democrats on the committee is odd?
(b) the number of Democrats on the committee exceeds the number of non-Democrats?
(c) there is exactly one Democrat and one Republican on the committee?

20. The flight time (airborne to landing) for a given flight between two cities is approximately normal with mean $\mu = 80$ minutes and $\sigma = 5$ minutes. Approximately what proportion of these flights take
(a) over 90 minutes?
(b) between 75 and 88 minutes?
(c) less than 70 minutes?

21. A veterinarian has found that a certain drug is effective in curing 1/3 of the cattle having a given ailment. If, during the next month, he gives the drug to 10 cattle having this ailment, what is the probability that:
(a) at least 4 of these animals are cured?
(b) exactly 5 of these animals are cured?
(c) no more than 4 of these animals are cured?

22. Weight gains of turkeys fed a special ration for five weeks are approximately normal with mean of 6.4 pounds and standard deviation of 0.4 pounds. During the five-week period what proportion of the turkeys are expected to gain
(a) at least 5.6 pounds?

(b) between 6 and 7 pounds?
(c) less than 6 pounds?

23. Three digit random numbers are selected from a random numbers table. Consider 000 as an even number. For a set of 15 such numbers, what is the probability that
(a) more than 8 of them are even?
(b) exactly 7 of them end in 0 or 1?
(c) no more than 9 of them are 200 or greater?

24. In psychological experiments (see [13], for example), a number of subjects each operate a mechanical device which traces a path within a set of boundaries such as the following:

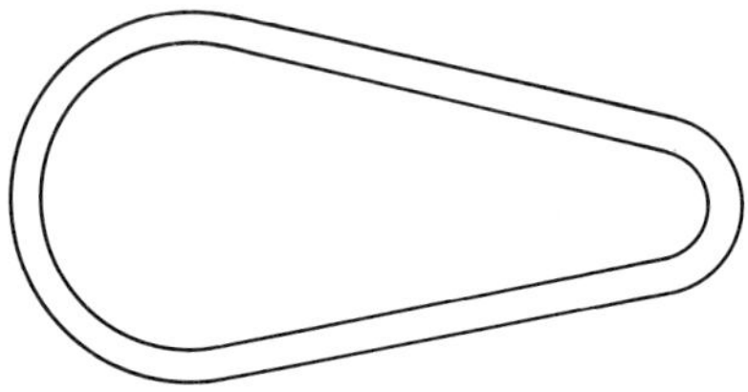

If the traced path goes outside the boundaries, the subject is required to reverse the tracing until the path reenters the boundaries.
Let X = time required to complete the path. Suppose it is reasonable to assume that X is normally distributed with a mean of $\mu = 6$ minutes and a standard deviation of $\sigma = 2$ minutes. What proportion of the subjects are expected to take
(a) no more than 7 minutes to complete the path?
(b) at least 3 minutes to complete the path?
(c) between 2 and 11 minutes to complete the path?

25. Refer to Problem 24 and let

Y = the number of times the traced path goes outside the boundaries for one completion of the path.

On the average, subjects trace outside the boundaries 6 times on the first path completion. What is the probability that any subject traces outside the boundaries during his first path completion
(a) exactly 6 times?
(b) at least twice?
(c) never?

26. Telephone subscribers in a certain large city can expect to receive 8 calls per month from persons dialing wrong numbers. Find the probability that a telephone subscriber in this city
(a) gets at least 10 such calls during the next month.
(b) gets between 5 and 12 such calls during the next month.
(c) gets 5 such calls during the next week.

27. From past experience, 90% of the tulip bulbs that a nursery sells will bloom. If a person buys 15 of these bulbs, what is the probability that
(a) they all bloom?
(b) at least 10 bloom?
(c) no more than a dozen bloom?

28. A manufacturer supplies a certain electrical switch in lots of 50. One of the purchasers of these switches randomly selects 5 switches from each lot, tests them, and if none of these 5 is defective, the lot is accepted. (A rejected lot is returned to the manufacturer.)
(a) If a lot has only 2 defective switches, what is the probability it is accepted?
(b) If a lot has 4 defective switches, what is the probability it is accepted?

29. In each of the following, find the value of W for which
(a) $P[T_{18} > W] = 0.90$
(b) $P[T_{14} > W] + P[T_{14} < -W] = 0.10$
(c) $1 - P[T_{10} < W] = 0.025$
(d) $1 - P[T_{10} > W] = 0.025$
(e) $P[-W < T_{15} < W] = 0.99$
(f) $P[T_W < 1.8331] = 0.95$
(g) $P[T_W > -1.3304] = 0.90$

30. In each of the following, find the value of W for which
(a) $P[\chi_5^2 > W] = 0.10$
(b) $P[\chi_{10}^2 < W] = 0.99$
(c) $P[\chi_{15}^2 > W] = 0.95$
(d) $1 - P[\chi_{10}^2 < W] = 0.975$
(e) $P[\chi_W^2 < 5.578] = 0.10$
(f) $P[\chi_W^2 > 8.672] = 0.95$

31. In each of the following, find the value of W for which
(a) $P[F_{2,10} > W] = 0.05$
(b) $P[F_{3,10} < W] = 0.05$
(c) $1 - P[F_{6,10} > W] = 0.10$
(d) $P[F_{W,10} > 2.66] = 0.05$
(e) $P[F_{5,W} > 0.214] = 0.95$

32. Specify the variance of a Student T variable having degrees of freedom of
(a) 12 (b) 25 (c) 60 (d) 200

33. Specify the mean and variance of a chi-square variable having degrees of freedom of
(a) 15 (b) 27 (c) 48 (d) 127

34. Specify the variance of an F variable having degrees of freedom of
(a) $\nu_1 = 10, \nu_2 = 10$ (b) $\nu_1 = 10, \nu_2 = 20$
(c) $\nu_1 = 6, \nu_2 = 30$ (d) $\nu_1 = 15, \nu_2 = 40$

35. Specify the mean of an F variable having degrees of freedom of
(a) $\nu_1 = 5, \nu_2 = 5$ (b) $\nu_1 = 5, \nu_2 = 12$
(c) $\nu_1 = 12, \nu_2 = 5$ (d) $\nu_1 = 20, \nu_2 = 18$
(e) $\nu_1 = 40, \nu_2 = 1{,}002$ (f) $\nu_1 = \nu_2 = 3$

REFERENCES

[1] Fisher, R. A., "On a Distribution Yielding the Error Functions of Several Well-Known Statistics," *Proceedings of the International Mathematics Congress*, Vol. II, Toronto, 1924.

[2] General Electric Company, *Tables of Terms of Poisson Distribution*, Van Nostrand, Princeton, N.J., 1962.

[3] Harter, H. L., *New Tables of the Incomplete Gamma-Function Ratio and of Percentage Points of the Chi-Square and Beta Distributions*, U.S. Government Printing Office, Washington, D.C., 1964.

[4] Harvard University Computation Laboratory, *Tables of the Cumulative Binomial Probability Distribution*, Harvard University Press, Cambridge, Mass., 1955.

[5] Lieberman, G. J. and D. B. Owen, *Tables of the Hypergeometric Distribution*, Stanford University Press, Stanford, Cal., 1961.

[6] Molina, E. C., *Poisson's Exponential Binomial Limit*, Van Nostrand, Princeton, N.J., 1942.

[7] National Bureau of Standards, *Tables of Binomial Probability Distribution*, Applied Mathematics, Series 6, U.S. Government Printing Office, Washington, D.C., 1950.

[8] Owen, D. B., *Handbook of Statistical Tables*, Addison-Wesley, Reading, Mass., 1962.

[9] Pearson, Karl, "On a Criterion That a Given System of Deviations From the Probable in the Case of a Correlated System of Variables is Such That it Can Reasonably be Supposed to Have Arisen in Random Sampling," *Philosophical Magazine*, **50** (1900).

[10] Poisson, S. D., *Recherches sur la probabilite des jugements en matiere criminelle, et en matiere civile, precedees des regles generales du calcul des probabilities*, 1837.

[11] Robertson, W. H., *Tables of the Binomial Distribution Function for Small Values of p*, Sandia Corporation Monograph SCR-143, Albuquerque, 1960 (available from the Office of Technical Services, Department of Commerce, Washington, D.C.).

[12] Romig, H. G., *50–100 Binomial Tables*, Wiley, New York, 1953.

[13] Sloan, S., "Cognitive Effects of Perceptual-Motor Training," Unpublished M.S. thesis, Kansas State University, Manhattan, Kansas, 1966.

[14] Snedecor, G. W., *Analysis of Variance and Covariance*, Iowa State College Press, Ames, Iowa, 1934.

[15] Student, "On the Probable Error of the Mean," *Biometrika*, 1908.

[16] Tucker, H. G., *An Introduction to Probability and Mathematical Statistics*, Academic Press, New York, 1962.

[17] U.S. Army, Ordinance Corps, *Tables of Cumulative Binomial Probabilities*, Ordinance Corps Pamphlet ORDP20-1, Washington, D.C., September 1952.

[18] Weintraub, S., *Tables of the Cumulative Binomial Probability Distribution for Small Values of p*, The Free Press of Glencoe, New York, 1963.

Chapter 9

STATISTICAL INFERENCE I: ESTIMATION

Estimation of parameters is a central factor in any number of real-life situations. For example, in order to remain competitive, a large electronics firm might ask its engineers to develop a new version of electrical component with a longer lifetime. Before deciding on the feasibility of manufacturing the new components, the administrative officials of the firm would like to know the average lifetime of the new version of components. Thus, if X denotes the time to failure for the new components, administrative officials want to know the value of μ, the mean value of X. This, of course, is impossible unless all new components are operated until failure. As in a number of situations, sampling is necessary because testing the components destroys them.

Consider an agronomist who is testing a new variety of wheat which has been developed specifically for a certain locality. Wheat growers would undoubtedly be interested in the mean yield of this wheat variety. Obtaining information about this specific variety at this one location would involve a simple experiment: plant the new variety of wheat on a number of plots within the given locality. Two comparative experiments of especial interest might be designed to analyze (i) how the new variety compares with other varieties at a single locality, or (ii) how the new variety performs at localities having different soils and/or climatic conditions. Rather than a single mean value, there would be a set of means to be estimated, one for each variety or location.

As another example, suppose a personnel manager is devising an aptitude test and wishes to establish a norm for the test. To get an idea of the norm (whether a mean, median, percentile, or some other form), the proposed test will be given to a large group of employees in the particular field for which the test was devised.

9.1 POINT ESTIMATION IN GENERAL

In the above examples little was said about the underlying distributions. Whether there is interest in specifying these distributions or not, the making of inferences requires a knowledge not only of the underlying distribution but also of the distribution of each statistic entering the analysis. Remember that statistics (estimators) are random variables and have distributions. If for no other reason, the sampling distributions are needed to investigate the various properties of each statistic in order to say which one is "best." Various properties are used to decide which statistic is best, and an estimator which may be best in one particular situation may not even be good in other situations.

Let θ denote any general parameter, and let $\hat{\theta}_1, \hat{\theta}_2, \ldots$ denote a set of statistics which might be used to estimate θ. For example, when sampling from a $N(\mu, \sigma^2)$ population, the estimation of $\theta = \mu$ might be accomplished by using any one of the following statistics:

$$
\begin{aligned}
\hat{\theta}_1 &= \bar{X} = \text{sample mean} \\
\hat{\theta}_2 &= \tilde{X} = \text{sample median} \\
\hat{\theta}_3 &= MR = \text{sample midrange} \\
\hat{\theta}_4 &= X_5 = \text{fifth sample observation}
\end{aligned}
\tag{9.1}
$$

We shall discuss three desirable properties of point estimators which can help us choose among the above as well as any others.

(1) *Unbiasedness*: The average or expected value of the statistic equals the parameter being estimated; that is, $\hat{\theta}$ is unbiased for θ if $E[\hat{\theta}] = \theta$.

An unbiased estimator can also be defined as one whose distribution has mean value equal to the parameter of interest. An estimator, $\hat{\theta}$, that does not have this property is said to be *biased* for θ. A statistic that underestimates the parameter, on the average, is said to be *negatively* biased; one that overestimates is *positively* biased.

(2) *Minimum variance unbiased* (MVU): An estimator possessing this property is unbiased and has a variance that is no greater than any other unbiased estimator.

An estimator with a small variance can be viewed as one whose repeated measurements are close to each other; such an estimator is said to have a high degree of *precision*. On the other hand, the term *accuracy* is related to bias; an unbiased estimator being accurate, biased estimators being "less accurate" than unbiased ones.

(3) *Consistency*: As the sample size increases without bound, the probability of the estimator differing from the parameter approaches 0. Or, with probability 1, the estimator approaches the parameter as the sample size increases indefinitely.

Consistency does not mean that, in any one sample, the value of the estimator must be close to the value of the parameter; we can only say that the probability of this is near 1. Consistency is related to the convergence of a sequence of estimators (the sequence for sample size $n = 1, 2, \ldots$), implying that a consistent estimator "becomes increasingly better" when it is based upon more and more observations. It can be shown that if an estimator is unbiased and has zero variance for the limiting sample size, the estimator is consistent. (The failure of this criterion to hold does not indicate non-consistency; that is, unbiasedness and zero variance is a sufficient but not a necessary condition.)

Determining whether a statistic possesses any of the above properties usually requires considerable analysis, including the determination of the appropriate sampling distributions. Whenever necessary, we will state which properties a particular estimator possesses.

► ***Example 9-1***

In defining the sample variance, the *SS* were divided by degrees of freedom rather than by sample size. This modification makes S^2 an unbiased estimator of σ^2 in infinite populations. ◄

► ***Example 9-2***

For random sampling from any population having a finite mean, μ, the sample mean, $\overline{X}$, is an unbiased estimator of μ. And for most populations, $\overline{X}$ is a consistent and minimum variance estimator of μ. ◄

► ***Example 9-3***

Referring to Example 6-1, the distributions of the sample mean and range are given when sampling from a finite population. It is easy to show that $E[\overline{X}] = 5$; that is, $\overline{X}$ is an unbiased estimator of μ for the sampling scheme described in that example. Also, $E[R] = 5.1 \neq 7 = \eta$, which implies that the sample range is a biased estimator of the population range. The sample range is always a negatively biased estimator of the population range, since r can never exceed the population range.

The estimators $\overline{X}$ and R are consistent, since

$$P(\overline{X} = 5) = P(R = 7) = 1$$

in samples of size 5 (the largest sample size that need be considered, since $N = 5$). ◀

Even though the different point estimators of any parameter can be judged according to various criteria and the "best" one selected, point estimation has an undesirable feature: the probability that an estimator takes on the parameteric value is usually zero. This is clearly so when using a continuous estimator, since continuous variables have zero probabilities at all points; in particular, $P[\hat{\theta} = \theta] = 0$. In the discrete case this probability may be zero for certain sampling schemes and nonzero for others as the following example illustrates.

▶ ***Example 9-4***

When sampling without replacement from the population $\mathscr{P} = \{2, 3, 5, 6, 9\}$, it is fairly easy to establish the probabilities given below for the various sample sizes which might be used in the estimation of $\theta = \mu = 5$ by $\hat{\theta} = \bar{X}$.

n	$P(\bar{X} = 5)$
1	1/5
2	0
3	0
4	1/5
5	1

◀

We will not be concerned with the task of devising new estimation procedures and new estimators of parameters. This is the job of mathematical statisticians, who already have proposed a number of excellent estimation procedures. The least-squares principle (of estimation) will be discussed in Chapter 13, when regression techniques are studied.

(Problems 1(a), (h), (i), (j), (k), (m), 5(a), (c), 7(a))

9.2 *INTERVAL ESTIMATION IN GENERAL*

A measure of reliability (probability) can be included in the inference procedure if, instead of simply a point estimator, an interval of values is given as the estimate of the parameter. As an example, consider the interval

$$\hat{\theta} \pm c \tag{9.2}$$

where c is a judiciously chosen constant. In random sampling from a population, the proportion of intervals (9.2) which contain a parameter θ can be calculated from the distribution of $\hat{\theta}$, with the constant chosen to give any desired proportion. All samples of a given size from a population actually generate a population of similar intervals whose members are now defined.

Definition 9-1

A $100(1 - \alpha)\%$ *confidence interval* (*CI*) for a parameter θ is the set of values encompassed by the interval (L, U), where L and U are *any* functions of the sample observations such that $100(1 - \alpha)\%$ of all similarly constructed intervals contain θ.

Being a function of sample observations, any *CI* is a random variable and involves some statistic $\hat{\theta}$, a point estimator of θ. There is *not* a unique *CI* for any parameter, a fact which follows from the nonuniqueness of point estimators and/or from Definition 9-1 directly.

In actual practice, a small value is generally chosen for α whereby the chance of any *CI* containing θ will be high. The price paid for higher assurance (smaller α) is generally larger intervals: in other words, "the more sure you are, the less you have to be sure of." Point estimation (single-point *CI*) corresponds, in the continuous case at least, to $\alpha = 1$. At the other extreme, $\alpha = 0$ means 100% confidence that the interval contains the parameter, which can be possible only if the interval covers the entire range of the parameter; then we are simply saying that the parameter has some value.

Using the general theory of interval estimation, a $100(1 - \alpha)\%$ *CI* for θ would be constructed according to the following steps:

(9.3)
- (i) Choose a point estimator of θ—call it $\hat{\theta}$.
- (ii) Determine the distribution of $\hat{\theta}$—call it $g(\hat{\theta}; \theta)$. It will depend upon θ.
- (iii) From a $100(1 - \alpha)$ probability with respect to $g(\hat{\theta}; \theta)$, the prescription for setting the *CI* can be determined.

Obviously, the probability in step (iii) could be selected in many ways from central, noncentral, or disjoint components. In an effort to find a "good" *CI*, one could rate the various types of *CI* for a given parameter according to length, ease in calculating, and so on. Also, it seems intuitive that the "best" *CI* should be given when the "best" $\hat{\theta}$ is chosen in step (i).

Instead of using the general theory for setting a *CI* for θ, it is quite often easier to use a function of $\hat{\theta}$ which has a simpler but known distribution. Hopefully, this simpler distribution will be general enough to work for an entire family of distributions and to give the same *CI* as does $g(\hat{\theta}; \theta)$. Any function of the statistic $\hat{\theta}$ may be used provided the following conditions are satisfied:

(9.4)
- (a) The function involves $\hat{\theta}$ and θ but *no* other *unknown* quantities.
- (b) The distribution of the function is known.
- (c) The *distribution* of the function depends upon *no* unknown quantities, not even θ.

If we can successfully find a function of $\hat{\theta}$ which satisfies the conditions (9.4), the distribution of the function becomes a substitute for $g(\hat{\theta}; \theta)$. From this substitute distribution, we merely find an interval in terms of the function of $\hat{\theta}$ and then replace $\hat{\theta}$ and θ to give an expression which is solved for the *CI* for θ. And by judiciously choosing the function of our statistic, we can utilize distributions that are already tabulated.

(Problems 1(b), (c), (d), (e), (f), (g))

9.3 *ESTIMATION FOR THE* N(μ, σ²) *FAMILY*

The remainder of this chapter is devoted to estimation methods for experimental situations where it can safely be assumed that some normal distribution adequately explains the behavior of the random variable, at least over the particular range of interest. Under this assumption of normality, only parametric inference is undertaken. We will examine point and interval estimation for μ and for σ^2.

9.3.1 Point Estimation

Usual estimators for the mean μ and variance σ^2 of a normal distribution are the sample mean, $\bar{X}$, and the sample variance, S^2. Each estimator is unbiased and consistent and has minimum variance for its respective parameter. In addition, $\bar{X}$ is a linear estimator (making it easy to calculate) and has a normal distribution for any sample size (Theorem 8-1). For these reasons, $\bar{X}$ is generally regarded as the best estimator of μ in a normal distribution.

9.3.2 Interval Estimation of μ When σ Is Known

To set a *CI* for μ by the simplified method, we seek a function satisfying conditions (9.4). In view of the remarks made in the preceding subsection about the goodness of $\bar{X}$, we confine our search to functions involving only μ and $\bar{X}$ as unknowns. Using previously stated results, it is easily seen that the random variable

(9.5)
$$Z = \frac{\bar{X} - \mu}{\sigma_{\bar{X}}}$$

is a function which satisfies the conditions (9.4). Thus, the standard normal or Z distribution may be used instead of $g(\bar{x}; \mu)$, to set a *CI* for the mean of a normal distribution if the variance has a known value.

In Figure 9-1 the values z_l and z_u are used to denote lower and upper points of the Z distribution; the area under the Z curve below z_l and above

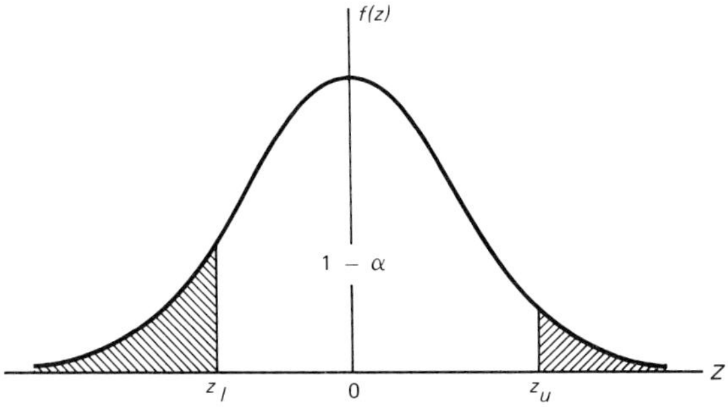

Figure 9-1 Standard Normal Distribution

z_u is α. Whatever the values of z_l and z_u, we have

(9.6)
$$\begin{aligned} 1 - \alpha &= P(z_l < Z < z_u) \\ &= P\left(z_l < \frac{\bar{X} - \mu}{\sigma_{\bar{X}}} < z_u\right) \end{aligned}$$

Algebraic manipulations on the inequality appearing in (9.6) give

(9.7)
$$P(\bar{X} - \sigma_{\bar{X}} z_u < \mu < \bar{X} - \sigma_{\bar{X}} z_l) = 1 - \alpha$$

Now, let

(9.8)
$$L = \bar{X} - \sigma_{\bar{X}} z_u \quad \text{and} \quad U = \bar{X} - \sigma_{\bar{X}} z_l$$

If we assign values to z_l and z_u, the value of α is fixed, and we see from (9.7) that the proportion of all similarly constructed intervals (L, U) which contain $\theta = \mu$ is $1 - \alpha$. Thus, (9.8) specifies $100(1 - \alpha)\%$ confidence limits for the parameter μ of the N(μ, σ^2) population having known variance σ^2. Notice that the upper point, z_u, appears in the lower limit; the lower point, z_l, appears in the upper limit.

If, instead, we fix the value of α, infinitely many choices of z_l and z_u exist. The placement of α (that is, the selection of the area under the Z curve) actually fixes the values of z_l and z_u. Three important cases are

(1) place half of α in each tail of the Z distribution, then $z_u = z_{1-\alpha/2}$ and $z_l = z_{\alpha/2} = -z_{1-\alpha/2}$; the limits (9.8) reduce to

(9.9)
$$L = \bar{X} - \sigma_{\bar{X}} z_{1-\alpha/2} \quad \text{and} \quad U = \bar{X} + \sigma_{\bar{X}} z_{1-\alpha/2}$$

(2) place all of α in the upper tail, then $z_l = -\infty$ and $z_u = z_{1-\alpha}$; the limits (9.8) reduce to

(9.10)
$$L = \bar{X} - \sigma_{\bar{X}} z_{1-\alpha} \quad \text{and} \quad U = +\infty$$

(3) place all of α in the lower tail, then $z_l = z_\alpha = -z_{1-\alpha}$ and $z_u = +\infty$; the limits (9.8) reduce to

$$L = -\infty \quad \text{and} \quad U = \bar{X} + \sigma_{\bar{X}} z_{1-\alpha} \tag{9.11}$$

The limits specified in (9.9) give a *two-sided CI* for the mean, μ, of the $N(\mu, \sigma^2)$ distribution having known variance σ^2. Placing $\alpha/2$ in each tail of the Z distribution gives the shortest† *CI* for μ. The shortest *CI* is desirable simply because it specifies fewer possible values for a parameter than a longer *CI* would.

The limits specified in (9.10) and in (9.11) give, respectively, *one-sided lower-bound* and *one-sided upper-bound CI* for μ. One-sided *CI* are calculated when we want only an upper or only a lower bound on values of a parameter.

The experimental situations where the variance is assumed known are somewhat specialized. The assumption is reasonable when doing extensive experimentation on a given population or when obtaining measurements from equipment having precision specified by the manufacturer. We have given such a detailed discussion of this case because of simplicity of notation involved; *CI* for the remaining cases to be discussed follow the same general format.

► ***Example 9-5***

Using precision equipment, a physicist attempts to measure the velocity of sound in a certain gas. From past experience, the physicist has found that measurements obtained from using this equipment are approximately normal with a standard deviation of $\sigma = 4$ meters per second. One hundred measurements on the velocity gave $\bar{x} = 828.4$ meters per second. The 90% two-sided confidence limits on the true velocity are

$$\begin{aligned} \bar{x} \pm (\sigma/10) z_{0.95} &= 828.4 \pm (4/10)(1.645) \\ &= 828.4 \pm 0.66 \\ &= 827.74;\ 829.06 \end{aligned}$$

The physicist can be 90% confident that the true velocity is in the interval between 827.74 and 829.06 meters per second. ◄

► ***Example 9-6***

If the physicist of the previous example was interested only in what the maximum velocity might be, he would calculate a one-sided upper-bound

† For symmetric distributions, the shortest *CI* is always obtained by placing half of α in each tail of the distribution.

CI. A one-sided 90% upper-bound interval would be

$$L = -\infty \qquad \text{and} \qquad U = \bar{x} + (\sigma/10)z_{0.90}$$
$$= 828.4 + (4/10)1.282$$
$$= 828.4 + 0.51$$
$$= 828.91$$

The physicist can be 90% confident that the true velocity is below 828.91 meters per second.

On the other hand, if the physicist was interested only in what the minimum velocity might be, he would calculate a one-sided lower-bound *CI*. A 90% lower-bound *CI* would be

$$L = 827.89 \text{ meters per second} \qquad \text{and} \qquad U = \infty. ◀$$

(Problem 19)

9.3.3 Interval Estimation of μ When σ² Is Unknown and Estimated by S^2

If the variance, σ^2, is unknown, the quantity $\dfrac{\bar{X} - \mu}{\sigma_{\bar{X}}}$ involves an unknown term $\sigma_{\bar{X}} = \sigma/\sqrt{n}$ and hence does not satisfy conditions (9.4). However, as stated in Section 8.7.2, replacing $\sigma_{\bar{X}}$ by the sample estimator $S_{\bar{X}} = S/\sqrt{n}$ gives a function which has a known distribution, the Student T with $n - 1$ degrees of freedom (Figure 9-2). Thus

(9.12)
$$T_{n-1} = \frac{\bar{X} - \mu}{S/\sqrt{n}}$$

is a quantity that satisfies conditions (9.4) and may be used to set a *CI* for the mean value in the N(μ, σ^2) distribution when S^2 is used as the estimator of σ^2.

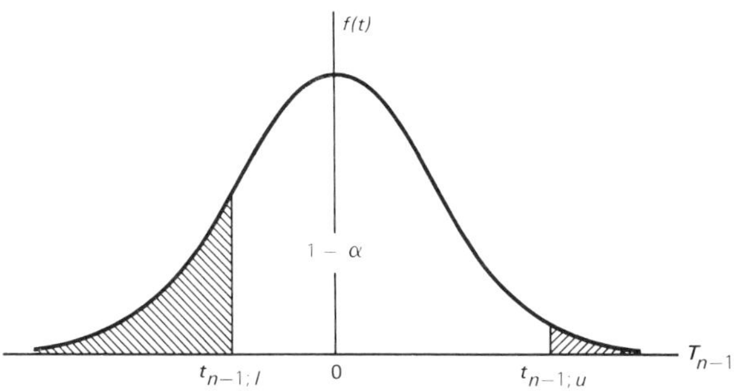

Figure 9-2 Student T_{n-1} Distribution

By similarity of the function in (9.12) to the one in (9.5) and symmetry of the T_{n-1} distribution, the mathematics used in finding the *CI* in this case is completely analogous to those used in finding the *CI* under the *Z* distribution. Consequently, the *CI* for μ under the T_{n-1} quantity can be obtained from (9.9), (9.10), and (9.11) simply by replacing $\sigma_{\bar{X}}$ by $S_{\bar{X}}$ and values of *Z* by values of T_{n-1}. The results are

$$L, U = \bar{X} \pm S_{\bar{X}} t_{n-1;1-\alpha/2} \tag{9.13}$$

for the shortest, 100(1 − α)% two-sided *CI*;

$$L = \bar{X} - S_{\bar{X}} t_{n-1;1-\alpha} \quad \text{and} \quad U = +\infty \tag{9.14}$$

for a 100(1 − α)% one-sided lower-bound *CI*; and

$$L = -\infty \quad \text{and} \quad U = \bar{X} + S_{\bar{X}} t_{n-1;1-\alpha} \tag{9.15}$$

for a 100(1 − α)% one-sided upper-bound *CI*.

▶ ***Example 9-7***

In a nutrition study, 24 calves were fed Ration A exclusively for a six-week period. The gain in weight, *X*, was recorded for each calf, giving the statistics

$$\bar{x} = 22.4 \text{ pounds} \quad \text{and} \quad s = 9.4 \text{ pounds}$$

If it can safely be assumed that the weight gains are distributed approximately normal, the 95% two-sided confidence limits on μ are given by

$$\begin{aligned} \bar{x} \pm (s/\sqrt{24})(t_{23;0.975}) &= 22.4 \pm (9.4/4.9)(2.069) \\ &= 22.4 \pm 4.0 \\ &= 18.4;\ 26.4 \end{aligned}$$

This interval (18.4; 26.4) is one selected from a set of intervals, 95% of which contain the mean six-week weight gain for calves fed Ration A. In other words we can be 95% confident that the mean six-week weight gain for calves fed Ration A is somewhere between 18.4 and 26.4 pounds. ◀

Hereafter the interpretation of *CI* will not be given. The previous examples illustrate the type of statements that are acceptable.

Our final example involves a "large sample" interval estimate of a population mean.

▶ ***Example 9-8***

A vegetable grower wishes to estimate the average weight, in grams (g), per fruit for a particular variety of tomatoes. He records (from the same number of fruit per plant) the average weight per fruit for $n = 50$ tomato plants of this year's crop and obtains

$$x_{.} = 4{,}237 \text{ g} \qquad \text{and} \qquad SS_X = 24{,}876\,(\text{g})^2$$

Applying the Central Limit Theorem, an approximate 99% *CI* for the average weight per fruit is

$$\begin{aligned} L, U &= \bar{x} \pm s_{\bar{X}} t_{49;0.995} \\ &= 84.74 \pm 3.18(2.68) \\ &= 84.74 \pm 8.52 \\ &= 76.22;\ 93.26 \blacktriangleleft \end{aligned}$$

(Problems 3(a), 5(b), 11(a), 15)

*9.3.4 Interval Estimation of μ When σ² Is Unknown and Estimated through the Range

A range estimator of σ was presented in Section 7.8.7. This estimator might be considered when a quick decision is needed (for example, in many manufacturing situations where a few observations are obtained quite often as a check on production). It should be remembered that the time saved in using R rather than S to estimate σ is partially offset by a loss in efficiency.

Replacing $\sigma_{\bar{X}}$ in the denominator of the Z-ratio (9.5) by the sample range R gives

$$T_n^* = \frac{\bar{X} - \mu}{R} \tag{9.16}$$

The random variable T_n^* is known as a *pseudo-T variable*; its distribution when sampling from a normal has been studied and tabulated by Lord [3]. Similar to the Student T distributions, the pseudo-T distributions are symmetric about the mean value of 0, have infinite range, and are indexed by a single parameter, the sample size (not degrees of freedom). Table A.9 gives points $t_{n;\gamma}^*$, where $F(t_{n;\gamma}^*) = \gamma$, for selected γ and n; the lower half of the table is for sampling from two normal populations.

Since the T_n^* variable in (9.16) satisfies conditions (9.4) a *CI* for μ is obtained by considering

$$\begin{aligned} 1 - \alpha &= P[t_{n;l}^* < T_n^* < t_{n;u}^*] \\ &= P\left[t_{n;l}^* < \frac{\bar{X} - \mu}{R} < t_{n;u}^*\right] \end{aligned} \tag{9.17}$$

Again, since (9.17) has the same general form as (9.6), the confidence limits can be obtained from (9.9), (9.10), and (9.11) simply by replacing $\sigma_{\bar{X}}$ by R

and values of Z by values of T_n^*. The results based on the range estimation of variability are

(9.18) $$L, U = \bar{X} \pm Rt^*_{n;1-\alpha/2}$$

for the shortest, 100(1 − α)% two-sided *CI*;

(9.19) $$L = \bar{X} - Rt^*_{n;1-\alpha} \quad \text{and} \quad U = +\infty$$

for a 100(1 − α)% one-sided lower-bound *CI*; and

(9.20) $$L = -\infty \quad \text{and} \quad U = \bar{X} + Rt^*_{n;1-\alpha}$$

for a 100(1 − α)% one-sided upper-bound *CI*.

▶ ***Example 9-9***

The quality-control department weighs 8 boxes of soap taken at random from the assembly line; this process is repeated at 15-minute intervals. For a given sample, they obtain

$$\bar{x} = 83.7 \text{ oz} \quad \text{and} \quad r = 1.2 \text{ oz}$$

Assuming approximate normality of weights, the 95% confidence limits for μ are given by

$$\begin{aligned} \bar{x} \pm r(t^*_{8;0.975}) &= 83.7 \pm 1.2(0.288) \\ &= 83.7 \pm 0.35 \\ &= 83.35; 84.05 \blacktriangleleft \end{aligned}$$

(Problems 9, 13)

9.3.5 Interval Estimation of σ² When μ Is Unknown

Interval estimation of σ^2 will be limited to the case where S^2 is used as the estimator of σ^2. Recalling the material of Section 8.7.3, it is easily seen that

(9.21) $$\chi^2_{n-1} = \frac{(n-1)S^2}{\sigma^2} = \frac{SS}{\sigma^2}$$

is a function that satisfies condition (9.4) and may be used to set a *CI* for the variance of a normal distribution.

Since the χ^2_{n-1} distribution is not symmetric, both the upper and lower percentage points shown in Figure 9-3 will necessarily appear in the final form of the *CI*. Thus,

(9.22) $$\begin{aligned} 1 - \alpha &= P(\chi^2_{n-1;l} < \chi^2_{n-1} < \chi^2_{n-1;u}) \\ &= P(\chi^2_{n-1;l} < SS/\sigma^2 < \chi^2_{n-1;u}) \\ &= P\left[\frac{SS}{\chi^2_{n-1;u}} < \sigma^2 < \frac{SS}{\chi^2_{n-1;l}}\right] \end{aligned}$$

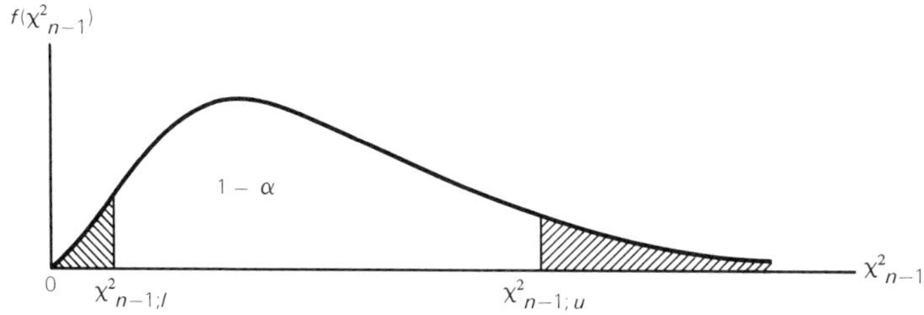

Figure 9-3 The χ^2_{n-1} Distribution

whereby, 100(1 − α)% confidence limits on σ^2 have the following forms:

(9.23) $$L = \frac{SS}{\chi^2_{n-1;1-\alpha/2}} \quad \text{and} \quad U = \frac{SS}{\chi^2_{n-1;\alpha/2}}$$

for equal tail area, 100(1 − α)% two-sided *CI*;

(9.24) $$L = \frac{SS}{\chi^2_{n-1;1-\alpha}} \quad \text{and} \quad U = +\infty$$

for a 100(1 − α)% one-sided upper-bound *CI*; and

(9.25) $$L = 0 \quad \text{and} \quad U = \frac{SS}{\chi^2_{n-1;\alpha}}$$

for a 100(1 − α)% one-sided lower-bound *CI*.

We point out that the two-sided *CI* for σ^2 given by the limits in (9.23) is *not* the shortest *CI* because we are dealing with a nonsymmetric distribution here. But the difference in the length of interval based on (9.23) and the shortest one is usually so slight that it is not worth worrying about.

▶ *Example 9-10*

For the nutrition experiment described in Example 9-7, the 90% two-sided confidence limits for the variance of weight gains of calves fed Ration A are found to be

$$L = \frac{SS}{\chi^2_{23;0.95}} = \frac{23(9.4)^2}{35.2} = 60.3$$

and

$$U = \frac{SS}{\chi^2_{23;0.05}} = \frac{23(9.4)^2}{13.1} = 162.0 \blacktriangleleft$$

(Problems 3(b), (c), 5(d), 7(b))

9.3.6 Interval Estimation of σ^2 When μ Is Known

Since this situation occurs so rarely in practice, it suffices to say that a *CI* for σ^2 may be obtained from the following function, which satisfies conditions (9.4):

$$\chi_n^2 = \sum_{i=1}^{n} (X_i - \mu)^2/\sigma^2 \tag{9.26}$$

That is, the population mean, μ, is used in calculating the sum of squares, and the degrees of freedom on the χ^2 statistic are n.

PROBLEMS

1. In Problem 7 of Chapter 6, the following population was given:

$$\mathscr{P} = \{1, 2, 4, 6, 12\}$$

(a) For each of the 10 random samples of size 3 drawn without replacement, calculate the sample mean, $\bar{x}$, the sample median, $\tilde{x}$, the sample variance, s^2, and the sample standard deviation.
(b) Calculate $\bar{x} \pm s$ for each sample.
(c) What proportion of the intervals in (b) contain the value of μ?
(d) Calculate $\tilde{x} \pm s$ for each sample.
(e) What proportion of the intervals in (d) contain the value of μ?
(f) For each sample, calculate $s \pm 1.8$.
(g) What proportion of the intervals in (f) contain the value of σ^*, where

$$(\sigma^*)^2 = \tfrac{1}{4} \sum_{i=1}^{5} (x_i - \mu)^2 = 19$$

(h) Is $\bar{X}$ an unbiased estimator of μ? Why or why not?
(i) Is $\tilde{X}$ an unbiased estimator of μ? Why or why not?
(j) Is R an unbiased estimator of the population range? Why or why not?
(k) Is S an unbiased estimator of σ^*? Why or why not?
(l) Is S^2 an unbiased estimator of the population variance $(\sigma^*)^2$? Why or why not?

2. (a) For the population $\mathscr{P}$ given in Problem 1, calculate $\bar{x}$ for each sample of size 4 drawn without replacement.
(b) For this sampling scheme, does the sample mean ever equal the value of μ?
(c) Is the sample mean unbiased for μ under this sampling scheme?

3. A metallurgist made 16 measurements on the melting point of a certain manganese alloy obtaining a sample mean of 1,264°C and $s = 9$°C.
(a) Set and interpret a 95% *CI* for the true melting point.
(b) Set and interpret a 99% *CI* for the variance of measurements on the melting point.
(c) What assumptions are necessary for (a) and (b) to be valid?

4. As completed boxes of cereal pass on a conveyor belt to the crating and shipping department, the quality-control personnel randomly select a number of boxes from the belt. The following statistics with respect to the weight of the boxes, in ounces, are recorded:

$$n = 25 \qquad x_{.} = 258 \qquad SS_X = 2{,}160$$

(a) Estimate the true mean weight per box by a 90% *CI*.
(b) Repeat (a) for a 95% *CI*.
(c) Explain the difference in the lengths of these two *CI*. Is this true, in general?
(d) Estimate the variance of weights by a 95% *CI*.
(e) What assumptions does your analysis depend upon?

5. The fourth-year growth, X, of a certain variety of apple trees is assumed to be approximately normal. Fourth-year-growth statistics for 20 apple trees of this variety were, in inches:

$$x_{.} = 286 \qquad \sum x_i^2 = 4{,}272$$

For this particular variety of apple trees
(a) estimate the average fourth-year growth.
(b) estimate the average fourth-year growth by a 95% *CI*.
(c) estimate the variance of fourth-year growths.
(d) estimate σ_X by a 99% *CI*.

6. A machine is set to cut sheets of paper in 14-inch lengths. Forty sheets are selected at random from the machine's output and measured, giving

$$\bar{x} = 14.03 \text{ in.} \qquad SS_X = 0.164\,(\text{in.})^2$$

(a) Set a 99% *CI* for the average length of sheets cut by this machine.
(b) Estimate the variance of sheet lengths by a 90% *CI*.
(c) What assumptions must hold if the results found in (a) and (b) are to be valid statistically?

7. A manufacturer of precision made chemical balances calibrates each balance prior to its sale or use. This is done by repeatedly weighing a standard object so that both the accuracy and the variability of measurements can be checked. Fifteen weighings made by a given balance gave:

$$x_{.} = 750.012 \text{ g} \qquad SS_X = 0.00001708\,(\text{g})^2$$

(a) Estimate the variance of measurements obtained from this balance.
(b) Set an upper 99% *CI*, (0, U), for the true variance of measurements obtained from this balance. (There is no interest in a lower limit, since zero variance would be ideal.)
(c) A particular balance is acceptable from a reliability standpoint if, with 99% assurance, the manufacturer can say that the variance of measurements does not exceed 0.000004 $(\text{g})^2$. Is this balance acceptable by this standard?
(d) Each balance is adjusted until the 95% *CI* for the mean weight of measurements contains the weight of the standard object. Given that the standard object used in this case weighed 50 g, does this balance need further accuracy adjustments?

8. In a time-and-motion study, an industrial engineer times an individual whose task on an assembly line is to solder three wires to a switch. At 20 randomly selected times throughout one day, he found

$$\bar{x} = 23 \text{ sec} \qquad x_{(1)} = 16 \text{ sec}$$

$$x_{(20)} = 35 \text{ sec} \qquad SS_X = 290 \text{ (sec)}^2$$

(a) Estimate the mean assembly time by a 95% *CI*.
(b) Set a 99% *CI* for the standard deviation of assembly times.

***9.** (a) In the previous problem, use the T^* distribution to estimate the mean assembly time by a 95% *CI*.
(b) Explain the difference in lengths of the *CI* obtained from the two distributions.

10. A group of second-generation chickens of a particular parental cross were used in a controlled environment heritability study on egg weight. Ten weeks after laying commenced, a sample of 60 eggs were weighed and revealed:

$$x. = 3{,}580 \text{ g} \qquad SS_X = 1{,}022 \text{ (g)}^2$$

(a) Estimate the average egg weight by a 95% *CI*.
(b) What assumptions are required in this estimation?

11. A chief ingredient in a certain class of insecticides is pyrethrin, obtained from the flowers of pyrethrum plants. Fifteen flower heads are randomly selected from a number of pyrethrum plants under cultivation; their weights, in grams, are:

0.21	0.13	0.14	0.19	0.16
0.16	0.19	0.14	0.18	0.20
0.17	0.12	0.16	0.15	0.24

(a) Set a 95% *CI* for the mean weight of flower heads of pyrethrum plants grown under similar conditions.
(b) What assumptions were used in (a)?
Suppose a reasonable yield of pyrethrin is 0.7% of the flower head weight. Give a 95% interval estimate of pyrethrin yield
(c) per flower head.
(d) per 100 flower heads.

12. Pressure necessary to break (or crack) a certain brand of copper tubing is assumed to have an approximate normal distribution. For a random sample of 16 sections of this particular copper tubing, the following statistics are recorded with respect to the breaking pressure, in pounds per square inch:

$$\bar{x} = 7{,}864 \qquad r = 360 \qquad SS_X = 8{,}280$$

(a) Place 99% limits on the true mean breaking pressure, using the Student T distribution.
(b) Place 95% limits on the true variance of breaking pressures.

***13.** Repeat Problem 12(a) using the T^* distribution.

14. A personnel director wishes to establish norms for a test to be given to prospective key punch operators. After a brief training to familiarize them with the machines, a group of 30 prospective operators were each asked to punch a certain number of cards. The following statistics were obtained:

$$\bar{x} = 13 \text{ min} \qquad SS_X = 3{,}480\,(\text{min})^2$$

Assuming approximate normality for punching times,
(a) set a 90% CI for the mean time to complete the test.
(b) set a 95% CI for the variance of punching times.
(c) set a 95% CI for $\sigma_{\bar{X}}$.

15. Dry weight of food eaten by each of 16 larvae during their larval period (see [5]), in milligrams, was:

57.30	67.00	51.00	64.00
53.82	55.80	51.30	55.50
45.52	65.00	55.20	58.00
58.32	51.00	55.00	64.20

(a) Estimate the average amount of food eaten during the larval period and its standard error.
(b) Give a 90% interval estimate of the average amount of food eaten during the larval period.

16. The heights of 20 mature barley plants selected at random from an experimental plot gave the following statistics, in inches:

$$\bar{x} = 34.7 \qquad SS_X = 14.7 \qquad r = 3.8$$

(a) Give a 95% CI for the mean height of mature barley plants grown under similar conditions.
(b) Estimate the variance of heights of mature barley plants by a 95% CI.

***17.** Repeat Problem 16(a) using the T^* distribution.

18. Twenty specimens (each 1.5 inch in diameter) of half-hard brass were gold plated by a certain technique (see [1]). A nondestructive beta backscatter thickness-measuring instrument was used to determine the thickness of the gold plating at the center of each specimen. The thickness measurements, in microinches, were:

230	225	305	265	260
245	240	330	225	265
235	295	290	215	285
215	295	300	225	330

(a) By a 95% CI, estimate the average thickness of gold plating by this technique.
(b) By a 90% CI, estimate the variance of thickness of gold platings made by this technique.

19. A physicist makes 12 determinations of the focal length, X, of a certain optical lens and finds $\bar{x} = 24.3$ cm. His technique and equipment have, in the past, exhibited a variance of $\sigma^2 = 3\,(\text{cm})^2$. Assuming approximate normality of determinations,
(a) estimate the true focal length and give its standard error.
(b) set a 99% CI for the true focal length.
(c) set a 98% CI for the true focal length.
(d) explain the difference in the lengths of the two CI found in parts (b) and (c). Is this true in general?

20. The amount of glutathione present in lemon peels was measured for 30 lemons of a certain variety, giving (in mg/100 g):

$$\bar{x} = 41.6 \qquad SS_X = 2{,}966$$

(a) Set a 95% *CI* for the average glutathione per 100 g of lemon peels.
(b) Give a 90% *CI* for $\sigma_{\bar{X}}$.

21. Using the data in Problem 13 of Chapter 7,
(a) set a 95% *CI* for the true mean time required to trace the object.
(b) set a 90% *CI* for the standard deviation of tracing times.

22. Refer to the data given in Problem 17 of Chapter 7.
(a) Set a 90% *CI* for the monthly average spent on electricity.
(b) Set a 95% *CI* for the variance of monthly electricity costs.
If you were told that the data given represent eighteen consecutive months,
(c) does normality of Y seem reasonable? Why or why not?
(d) does normality of X seem reasonable? Why or why not?
(e) Discuss the assumptions used in (a) and (b).

23. Assume that the clean fleece weights in Problem 26 of Chapter 7 represent those of 18 randomly selected sheep.
(a) Set a 95% *CI* for the average clean fleece weight for the flock from which these sheep were selected.
(b) Set a 90% *CI* for the variance of clean fleece weights.

24. Using the data given in Problem 28 of Chapter 7,
(a) set a 99% *CI* for the true temperature at which the gas liquefies (in Absolute degrees).
(b) set a 99% *CI* for the variance of temperatures that might be observed under his experimental procedure.
(c) Discuss assumptions used in this inference.

REFERENCES

[1] Brewer, D. H., "Solder for Thick Gold Plating," *Welding Journal, Research Supplement*, **49**, 10 (1970).
[2] Fisher, R. A., "On the Mathematical Foundations of Theoretical Statistics," *Philosophical Transactions of the Royal Society of London*, Series A, **222** (1922).
[3] Fisher, R. A., "Theory of Statistical Estimation," *Proceedings of the Cambridge Philosophical Society*, **22**, 5 (1925).
[4] Lord, E., "The Use of Range in Place of Standard Deviation in the t-Test," *Biometrika*, **34** (1947).
[5] Fraenkel, G. and M. Blewett, "The Utilization of Metabolic Water in Insects," *Bulletin of Entomological Research*, **35** (1944–45).

Chapter 10

STATISTICAL INFERENCE II: TESTS OF HYPOTHESES

Inference under the general heading of *estimation* may be defined as a process of specifying all distributions which might reasonably explain the variable being investigated. Of a slightly different nature are the problems of determining whether one set of distributions is more reasonable than a second set. Statistical analyses falling within this latter framework form the second general area of statistical inference, known as *statistical hypothesis testing*.

10.1 TESTING PROBLEMS

Among the many factors influencing the weight of a candy bar, one is the amount of chocolate covering applied at the final step. Suppose the machine depositing the chocolate requires cleaning and adjusting, at irregular intervals, whenever the average amount deposited falls below 1 g. Obviously, the candy maker does not want to stop the process to clean and adjust the machine unless it is necessary. Nor does he want the average amount of chocolate deposited to be below 1 g—customers will be unhappy, he may be charged with false advertising if he sells underweight candy bars, and so on. Suppose that, from past experience, he knows that the random variable

$$X = \text{amount of chocolate deposited on a candy bar, in grams,}$$

has an approximate $N(\mu, 0.04)$ distribution. At 15-minute intervals he checks the amount of chocolate deposited on $n = 9$ candy bars (obtained by weighing each before and after the chocolate is deposited). On the basis of these results, he must decide whether the distribution of X has a mean value above or below 1 g. We will illustrate various aspects of this decision-making process in the next few sections.

Let us take another look at the experimental situations presented at the beginning of Chapter 9, and see how different objectives might lead to testing problems rather than the estimation problems considered earlier. Suppose the administrative officials of the electronic firm are not particularly interested in knowing a specific value of the mean lifetime. Rather, they would like to state with reasonably high confidence that the mean time to failure of their new type of components exceeds the mean of their old components by more than 100 hours. Maybe they feel that the changes in the manufacturing process, advertising, and so on, demand such an increase in the mean time to failure to warrant adoption and subsequent manufacture. Here the decision is whether or not the failure times of new and old components have different distributions—different to the extent that the mean time to failure of new components exceeds the mean time to failure of old components by 100 or more hours.

For the wheat example, the agronomist may wish to know if the new variety gives an average yield at least as great as some standard variety for the locality. Knowing the exact mean yield for the new variety may not be important to him. In fact, knowing the mean yield for the standard variety may not really be important either, so long as the new variety can be demonstrated as being "better" or "no better" than the standard variety. Again, he must decide whether or not he is dealing with two different distributions.

10.2 DEFINITIONS AND GENERAL CONCEPTS

A *statistical hypothesis* is a statement about the distribution of a variable under investigation. One single statement may specify (i) only the family, (ii) only parameters for a fixed family, or (iii) both family and parameters. In a large percentage of the situations calling for a test of hypothesis, the underlying form of the distribution is assumed, whereby the hypothesis simply specifies values of one or more parameters. It is not necessary to specify all parameters; those of no particular interest may be left unspecified.

The statement subjected to the test procedure is called the *null hypothesis* and is denoted by H_0. The *alternative hypothesis*, denoted by H_1, is the statement of some or all possibilities not found in H_0. Note that H_0 and H_1 are mutually exclusive; they may or may not form an exhaustive set.

▶ ***Example 10-1***

The following are typical hypotheses:

(a) H_0: The weights of all students at your school are N(130,900).
H_1: The weights of all students at your school are N(140,900).

(b) H_0: The number of hailstorms per year in Rains County, Texas, is a Poisson variable.
H_1: The number of hailstorms per year in Rains County, Texas, is not a Poisson variable.

(c) For weight gains assumed to be normally distributed,
$H_0: \mu \leqslant 20$
$H_1: \mu > 20$

(d) For the candy example of Section 10.1, the company would probably consider
$H_0: \mu = 1$ g
$H_1: \mu < 1$ g
(This particular choice, and others that the candy company might consider, are discussed in Section 10.2.2.)

Each hypothesis in (a) completely specifies a distribution, but they are not exhaustive; those in (b) specify only the form of the distribution and are exhaustive; those in (c) leave the parameter σ^2 unspecified but are exhaustive; those in (d) specify all unknown parameters but are not exhaustive. ◀

It should be pointed out that for all tests we will consider: (1) the null hypothesis *always* contains at least "an equality" which specifies a "reference distribution," (2) hypotheses *never* specify values of statistics, and (3) the alternative hypothesis usually contains the statement of most interest to the person making the test. The specification at the equality of the null hypothesis is called the *null value*, that value indicating either no difference or a difference of no consequence to the tester.

Definition 10-1

A *statistical test of hypothesis* is a rule used for deciding whether or not H_0 should be rejected.

That function of the sample observations used in making the decision about H_0 is called the *test statistic*. The set of all experimental values of the test statistic is called the *sample space* of the test. Some points of the sample space will be quite unlikely *if* H_0 *is true*. That subset of the sample space which leads to the decision "reject H_0" is the *critical region* (CR), the *region of rejection*.

Since there is a chance that the results of any random sampling scheme might be misleading in defining the parent population, the same must be said of a statistical test. No matter which decision is reached about H_0, there is a possibility of an error; we may reject H_0 when it is true, or we may fail

Table 10-1 Errors in Hypotheses Testing

Decision	True Situation	
	H_0 true	H_0 false
Reject H_0	Type I error	No error
Fail to reject H_0	No error	Type II error

to reject H_0 when it is false. Table 10-1 depicts the four cases and the errors in hypotheses testing. The use of "error" or "erroneous decision" should not be construed as a mistake; instead, "error" refers to unusual, extreme samples drawn by chance.

For any statistical test, each of the errors occurs with a certain probability. Customary notations for these are

$$\alpha = P(\text{Type I error}) = P(\text{reject } H_0 | H_0 \text{ true}) \quad (10.1)$$

and

$$\beta = P(\text{Type II error}) = P(\text{Fail to reject } H_0 | H_0 \text{ false}) \quad (10.2)$$

The quantity α is called the size of the *CR* or the level of significance of the test.

▶ ***Example 10-2***

For the candy problem discussed in Section 10.1, $X \sim N(\mu, 0.04)$, and Example 10-1 stated appropriate hypotheses:

$$H_0: \mu = 1 \text{ g} \qquad \text{versus} \qquad H_1: \mu < 1 \text{ g}$$

The null value is 1 g. Consider the following test procedure:

(1) Use $\bar{X}$ as a test statistic.
(2) Reject H_0 if $\bar{X} \leqslant 0.85$ g
(3) Do not reject H_0 if $\bar{X} > 0.85$ g

If H_0 is true, $\bar{X}$ has a normal distribution with $\mu = 1$ and $\sigma_{\bar{X}}^2 = 0.04/9$. A Type I error occurs when H_0 is true (the average amount deposited is really 1 g), but because we observe $\bar{X} \leqslant 0.85$, we reject H_0 and conclude that the average amount deposited is below 1 g. The probability of such an error would be

$$\begin{aligned}\alpha &= P(\bar{X} < 0.85|\mu = 1)\\ &= P\left(Z < \frac{0.85 - 1}{0.2/3}\right)\\ &= P(Z < -2.25)\\ &= F_Z(-2.25) = 0.01222\end{aligned}$$

When H_0 is true, a Type II error cannot be made.

If H_0 is false, $\bar{X}$ has a normal distribution with mean μ below 1 g and $\sigma_{\bar{X}} = 0.2/3$. We make a Type II error by failing to detect a mean below 1 g; this error occurs when we observe $\bar{X} > 0.85$ and fail to reject H_0. The probability of a Type II error can be specified by

$$\beta = P[\bar{X} > 0.85|\mu],$$

where μ is a value specified by H_1, a value below 1 g. Each value of μ below 1 g gives a different value of β. Thus, for $\mu = 0.95$

$$\begin{aligned}\beta &= P[\bar{X} > 0.85|\mu = 0.95]\\ &= P\left[Z > \frac{0.85 - 0.95}{0.2/3}\right]\\ &= P[Z > -1.50]\\ &= 1 - F_Z(-1.50)\\ &= 1 - 0.06681\\ &= 0.93319\end{aligned}$$

and for $\mu = 0.8$

$$\begin{aligned}\beta &= P[\bar{X} > 0.85|\mu = 0.8]\\ &= P[Z > 0.75]\\ &= 0.22663\end{aligned}$$

When H_0 is false, we cannot make a Type I error. ◀

(Problem 7)

10.2.1 Power of a Test

If the candy machine is really depositing an average of 0.95 g of chocolate, the probability is 0.93319 (from Example 10-2) of failing to arrive at the conclusion $\mu < 1$. In other words, when the mean is $\mu = 0.95$ g, the probability of arriving at the conclusion $\mu < 1$ is $1 - 0.93319 = 0.06681$. Our test procedure certainly has little chance of detecting a mean only 0.05 g below the null value. We note, however, that the probability of detecting a mean 0.2 g below the null value ($\mu = 0.8$ g) is $1 - 0.22667 = 0.77337$. We would certainly hope, and it is indeed true for our test procedure, that means below 0.8 g are detected with even higher probabilities.

Definition 10-2

The probability that a test procedure detects a false null hypothesis is the *power of the test.*

The power of a test may be written in the following ways:

$$
\begin{aligned}
\text{Power} &= 1 - \beta \\
&= 1 - P(\text{Type II error}) \\
&= 1 - P(\text{fail to reject } H_0 | H_1 \text{ is true}) \qquad (10.3) \\
&= P(\text{reject } H_0 | H_1 \text{ is true}) \\
&= P(\text{test statistic is in the } CR | H_1 \text{ is true})
\end{aligned}
$$

It is impossible to calculate a unique value for the power of a statistical test unless the alternative hypothesis specifies a unique distribution. If the power is plotted for each distribution of H_1, the result is called the *power curve* for that test. An ideal power curve would be zero for all distributions under H_0 and would have a value of one for all distributions under H_1.

For most statistical tests, the calculation of the power curve is very difficult. Under H_0 the test statistic has a known distribution, but under H_1 the test statistic has either an unknown or a complex distribution. Fortunately, tables and graphs exist for calculating power of many common tests.

A power curve is given in the next example. Usually four to six points are adequate for drawing a reasonably good curve. Additional material on power appears in starred sections at the end of each remaining chapter.

▶ ***Example 10-3***

The power curve will be drawn for the candy problem. For our test procedure, the power is, by (10.3), $P(\bar{X} < 0.85|\mu)$. Evaluating this probability at the following values of μ reveals:

μ	Power
0.95	0.06681
0.9	0.22663
0.85	0.5
0.8	0.77337
0.75	0.93319
0.7	0.98778

From this we obtain the power curve:

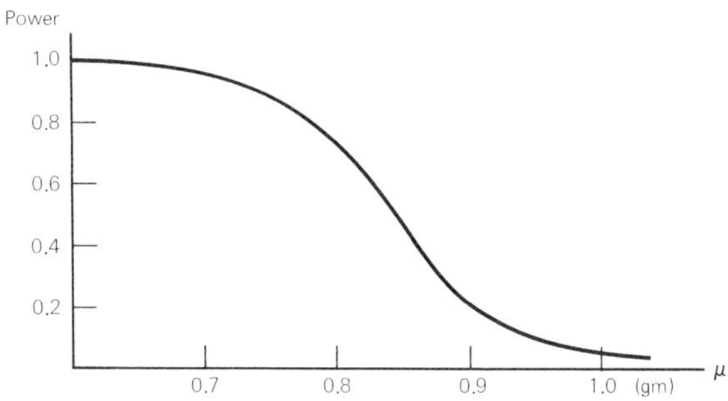

From the power curve, we can read the power of our test for any value of μ below 1 g. Thus, the power is 0.4 when $\mu = 0.86$ and 0.6 when $\mu = 0.84$, 0.8 when $\mu = 0.79$ and 0.95 when $\mu = 0.74$. We see that the power of the above test increases as values specified by the alternative deviate from the null value by a larger amount. The power is greater when $\mu = 0.8$ than when $\mu = 0.9$.

Suppose μ is really 0.9; then the power of the above test is only 0.22663. If $\mu = 0.9$ represents a mean value that we would like to detect, we can increase the power of the test (the chance of detection) by increasing the sample size. Increasing the sample size reduces the standard error of the mean—the distribution of $\bar{X}$ is more concentrated about the mean—and smaller discrepancies from the null value can be detected. ◀

The power curve given in Example 10-3 is for one particular hypothesis test. Different types of power curves are obtained for other tests.

10.2.2 One- and Two-Sided Alternatives

In the candy example, we considered the null hypothesis $H_0: \mu = 1$ g against a "composite" alternative $H_1: \mu < 1$ g. This choice of hypotheses was made so that, on our first encounter of hypothesis testing, we could anchor certain ideas in the simplest manner. By using this pair of hypotheses, we were asking whether the underlying distribution had a mean value *equal to* 1 g or a mean value *below* 1 g. But what about the mean values *above* 1 g?

If the machine depositing the chocolate has a chamber capacity only "slightly above" 1 g, the average amount deposited can be only "slightly above" 1 g. For this type of situation, we would be justified in testing as we did earlier—ignoring values of μ above 1 g.

If the chocolate chamber does not have a small capacity—because it holds a supply for several hours of operation or because there is a constant supply—the composite null hypothesis $H_0: \mu \geqslant 1$ g would be used. Even so, 1 g is the null value, the mean value which divides the set of interest ($\mu < 1$) from the uninteresting set ($\mu \geqslant 1$).

For economy reasons, the set $\mu > 1$ might be of interest to the candy maker. He may not mind putting a little more than 1 g of chocolate on each candy bar but a large amount over 1 g may be unprofitable. Now the candy company wishes to detect mean values above 1 g as well as mean values below 1 g. That is, the relevant hypotheses would be the single-valued null, $H_0: \mu = 1$ g, against the composite alternative, $H_1: \mu \neq 1$ g.

Alternative hypotheses of the form $H_1: \theta \neq \theta_0$ are termed *two-sided* while either $H_1: \theta < \theta_0$ or $H_1: \theta > \theta_0$ would be termed *one-sided.* If the tester is interested in a single parametric value against all other values, a two-sided alternative would be appropriate. If values of the parameter either less than or greater than a certain point are the only concern, a one-sided alternative would be used. As stated earlier, the directional of most interest to the tester would be placed in H_1.

With respect to the electronic problem discussed earlier, the engineers should conclude that $\mu > 2{,}100$ hours or that $\mu \leqslant 2{,}100$ hours. The conclusion $\mu > 2{,}100$ hours would lead to adoption of the new components, while the conclusion $\mu \leqslant 2{,}100$ hours would imply that the new components are not sufficiently better than the components presently manufactured. When the new components are adopted, the costs and problems of changing the manufacturing process would be incurred; presumably, this should be avoided unless there is substantial evidence that the new components are better (as measured by longer lifetime). Rejecting the new components simply means that further research must be undertaken to develop better components. Not marketing a better component at the present time may or may not be crucial, depending upon whether or not competitors will be introducing better components shortly. If introducing better components is not immediately important, the engineers would test $H_0: \mu \leqslant 2{,}100$ versus $H_1: \mu > 2{,}100$.

In general, the tester attempts to discredit any claim made by any other person or agency. For example, a drug company claims less than 3%

impurities will be found in a certain drug it sells. If the Pure Food and Drug Administration of the federal government attempts to show there are more than 3% impurities, it must be fairly certain of this statement if and when they make it. Consequently, the agency chemists would use $H_1: p > 0.03$.

(Problem 2)

10.2.3 Critical Regions

The critical region of a hypothesis test consists of those values of the test statistic which are most unlikely under H_0 but quite likely under H_1. Critical values must have this feature because H_0 is rejected in favor of H_1.

Consider a test of hypothesis of the form $H_0: \mu = \mu_0$ against an alternative $H_1: \mu < \mu_0$, as in the candy example. When sampling from any normal distribution specified by H_1, the probability of obtaining a value of $\overline{X}$ below μ_0 must be greater than 0.5; see the shaded area in Figure 10-1. Standardizing a value of $\overline{X}$ below μ_0 results in a negative Z value. Hence, negative values of Z are more likely (probability greater than 0.5) for every distribution under H_1. If μ is considerably less than μ_0, the probability of a negative Z value is near unity. Therefore, the critical values for a normal test having an alternative $H_1: \mu < \mu_0$ would be negative values of Z; that is, $Z < z_\alpha$, where z_α is the *critical point* corresponding to a test of size α.

Similarly, a normal test having an alternative of the form $H_1: \mu > \mu_0$ would have a critical region in the opposite tail of the Z distribution; namely, $Z > z_{1-\alpha}$.

And, as you might guess, a normal test involving a two-sided alternative $H_1: \mu \neq \mu_0$ would lead to a critical region consisting of both tails of the Z distribution; namely, $|Z| \geqslant z_{1-\alpha/2}$. Relative to μ_0, "large" or "small" values of $\overline{X}$ would support H_1; such values would standardize into extreme values of Z, either positive or negative.

A rationalization as given above can be used to determine the critical region of other statistical tests. Hereafter we will merely state the results when they are comparable to the above. We will analyze new situations as they occur.

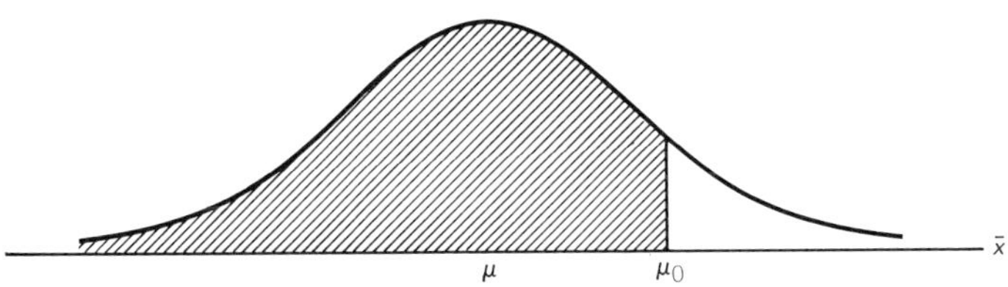

Figure 10-1 $P(\overline{X} < \mu_0 | \mu$ has a value less than $\mu_0)$

10.2.4 A Five-Step Procedure in Making a Statistical Test of Hypothesis

For the sake of uniformity, the following is suggested as a guide for making a test of hypothesis.

(10.4)

① Set up H_0 and H_1. (Both hypotheses are actually determined as soon as the experiment is fully designed.)
② Give the test statistic and its distribution. (The test statistic will usually be the function used in setting a *CI* for θ but with θ replaced by the null value, say θ_0.)
③ Choose α and give the critical region. (The critical region is completely determined by H_1, the test statistic, and α.)
④ Conduct the experiment, obtain the data, and make the computations. Make the decision about the rejection of H_0.
⑤ Interpret the results in practical terms.

Note that the first three steps can, and should, be set up *before* any data have been obtained. We will adhere to the steps in (10.4) in all examples. Without data, we will simply give the first three steps.

10.2.5 Some Philosophy of Hypothesis Testing

When the underlying distribution, the two hypotheses, and the test statistics are specified, it can be shown that the three quantities n, α, and β of a statistical test are functionally related so that only two of them can be chosen arbitrarily. In practice, n and α are usually chosen, since β depends upon values of *unknown* parameters in H_1 (except when H_1 specifies a unique distribution). In any test the ideal would be a minimum value for both error probabilities, α and β. It is quite unfortunate, however, that a *decrease* in either α or β is accompanied by an *increase* in the other when n remains constant. Thus, since the distribution and hypotheses are fixed, the only solution is to choose that test statistic which gives the minimum value for β when α and n are specified.

In the candy example, our test of $H_0: \mu = 1$ versus $H_1: \mu < 1$ was based on random samples of size $n = 9$ from a normal distribution. The value of α was calculated from $P(\overline{X} \leqslant 0.85|\mu = 1)$ and had a value of 0.01222; and when $\mu = 0.85$, $\beta = P(\overline{X} \geqslant 0.85|\mu = 0.85) = 0.50$. To get a *smaller* value for α, we need to choose a "critical value" for $\overline{X}$ below 0.85. But then the probability of a Type II error would be

$$\beta = P(\overline{X} > \text{critical point below } 0.85|\mu),$$

a larger value than $P(\overline{X} > 0.85|\mu = 1)$. For example, if we use a critical point of 0.8,

$$\alpha = P(\overline{X} \leqslant 0.8|\mu = 1) = P(Z \leqslant -3) = 0.00135,$$

and when $\mu = 0.85$,

$$\beta = P(\overline{X} > 0.8 | \mu = 0.85) = P(Z > -0.75) = 0.77337$$

A statistical test of hypothesis is made in the following manner:

(a) The test statistic is set up as though H_0 were true. This is why the null value is used in the test function—see step ② of (10.4)—and why we require an equality in H_0.
(b) The critical region is chosen in such a way that if H_1 were true, values in the critical region are more likely to occur.

If H_0 is *not* true, the test procedure should be more likely to reject H_0, since the test statistic measures sample results relative to the null value.

Only trivial hypotheses can be *proved*; it is generally much easier to disprove hypotheses. Consider, for example, H_0: There are no faulty flash bulbs in this stockpile. Without testing the entire stockpile, this hypothesis cannot be proved; finding one faulty flash bulb would disprove H_0.

In nontrivial cases, either decision in a hypothesis test carries with it the possibility of an error; hence it will never be known whether H_0 is proved or disproved. Rejecting or not rejecting H_0 does not disprove or prove the hypothesis. When H_0 is rejected, there is evidence against H_0; in this case, we proceed *as though* H_1 is probably true—we placed the odds in favor of H_0 but came up with strong evidence against H_0. Failure to reject or "acceptance of" H_0 merely means that the experiment provided no evidence against H_0; in this case we proceed *as though* H_0 could still be true.

By fixing the value of n and α, the experimenter is actually controlling the probability of a Type I error. This, coupled with the fact that $\beta = P(\text{Type II error})$ is an unknown quantity, means that the stronger interpretation results if and when H_0 can be rejected. For the cases when H_0 is rejected, the chances of an error, namely α, can be included in the interpretation. When H_0 is not rejected, the only possible error is a Type II error which occurs with an unknown frequency of β; thus, very little is known about making an erroneous decision in this case.

In very broad terms, the interpretations would appear as follows:

(i) When H_0 is rejected: "the statement given in H_1 is true *unless* a $100\alpha\%$ error has occurred."
(ii) When H_0 is not rejected: "there is no sample evidence against H_0" or, equivalently, "the statement given in H_0 is *true* unless a Type II error of *unknown* value has occurred."

Clearly, the interpretation is very weak when H_0 is not rejected. Failure to reject H_0 implies only that the sample could reasonably have come from distributions hypothesized by H_0; that is, the difference between the hypothesized and sampled populations was not great enough to be detected by the given experiment. On the other hand, rejection of H_0 has stronger implications: only $100\alpha\%$ of the samples drawn from the distribution hypothesized by H_0 will indicate the falsity of the statement in H_0.

10.3 TESTS FOR THE NORMAL FAMILY

Tests of parametric hypotheses about normal distributions are discussed under four general settings. Each case presented in the next four subsections actually has three subdivisions corresponding to the different types of alternative hypotheses. The first three steps will be given for each test using a general size, α. Specific examples will be given but power of these tests is deferred to Section 10.5.

10.3.1 Tests about μ of N(μ, σ²) When σ² Is Known

A conclusion about the mean of a normal distribution would be based upon $\bar{X}$ whenever feasible. From previous discussion, we can say that the standardized sample mean is appropriate for tests about μ when the variance σ^2 is known. Let μ_0 denote the null value of μ.

When σ^2 is known, tests about the mean of a normal have as the first three steps

† ① ⓐ $H_0: \mu \leqslant \mu_0$, $H_1: \mu > \mu_0$ ① ⓑ $H_0: \mu \geqslant \mu_0$, $H_1: \mu < \mu_0$ ① ⓒ $H_0: \mu = \mu_0$, $H_1: \mu \neq \mu_0$

(10.5) ② Under H_0, $\dfrac{\bar{X} - \mu_0}{\sigma_{\bar{X}}}$ is a Z variable; that is, if H_0 is true,

$$Z = \frac{\bar{X} - \mu_0}{\sigma_{\bar{X}}}$$

③ Reject H_0 if
ⓐ $Z \geqslant z_{1-\alpha}$ ⓑ $Z \leqslant -z_{1-\alpha}$ ⓒ $|Z| \geqslant z_{1-\alpha/2}$

In case ⓒ the alternative can be written as $H_1: \{\mu < \mu_0 \text{ and } \mu > \mu_0\}$, while the critical region is equivalent to $\{Z \leqslant -z_{1-\alpha/2} \text{ and } Z \geqslant z_{1-\alpha/2}\}$. Note that the inequalities in H_1 specify the position of the critical regions for this test procedure.

▶ ***Example 10-4***

Suppose the physicist in Example 9-5 was measuring the velocity of sound in an effort to identify the particular gas. He is fairly certain of its composition

† In a given problem only one pair of hypotheses would be used, of course. The null hypotheses in cases ⓐ and ⓑ may contain only the equality. The critical regions given in step ③ correspond to the hypotheses of the same letter. This system of specifying tests will be adhered to throughout the remainder of this book.

which should give a velocity of 830 meters per second (m/sec). He decides to make a 10% test to check his hypothesis. Assuming normality, his test is

① $H_0: \mu = 830$ m/sec
$H_1: \mu \neq 830$ m/sec

② If H_0 is true, $\dfrac{\bar{X} - 830}{\sigma_{\bar{X}}}$ is a Z variable.

③ For $\alpha = 0.10$, reject H_0 if $|Z| \geqslant z_{0.95} = 1.645$.

④ $n = 100$; $\bar{x} = 828.4$; $\sigma = 4$

$$Z = \frac{828.4 - 830}{4/10} = \frac{-1.6}{0.4} = -4.0$$

Thus, H_0 is rejected.

⑤ The true velocity of sound in this gas is *not* 830 m/sec *unless* a 10% error has occurred. ◀

10.3.2 Tests about μ of N(μ, σ²) When σ² Is Unknown and Estimated by S^2

A majority of tests about the mean value of normal distribution fall in this category. Only in rather specialized cases is the variance σ^2 known. When μ and σ^2 are estimated by $\bar{X}$ and S^2, the usual test procedure is the one based on Student T distribution:

① ⓐ $H_0: \mu \leqslant \mu_0$
$H_1: \mu > \mu_0$

① ⓑ $H_0: \mu \geqslant \mu_0$
$H_1: \mu < \mu_0$

① ⓒ $H_0: \mu = \mu_0$
$H_1: \mu \neq \mu_0$

(10.6) ② Under H_0, $\dfrac{\bar{X} - \mu_0}{S/\sqrt{n}}$ is a T_{n-1} variable.

③ Reject H_0 if
ⓐ $T_{n-1} \geqslant t_{n-1;1-\alpha}$ ⓑ $T_{n-1} \leqslant -t_{n-1;1-\alpha}$
ⓒ $|T_{n-1}| \geqslant t_{n-1;1-\alpha/2}$

Again, the position of the critical region corresponds to the directional indicated in the alternative hypothesis.

▶ ***Example 10-5***

Suppose the following question was to be answered by the nutrition study of weight gains of calves in Example 9-7: "Is there any evidence that the 6-week weight gains of calves fed Ration A average over 20 pounds?" A

1% test of hypothesis for this experiment would have the following form, under the assumption of normality:

① $H_0: \mu \leqslant 20$
$H_1: \mu > 20$

② If H_0 is true, $\dfrac{\overline{X} - 20}{S_{\overline{X}}}$ is a T_{23} variable.

③ H_0 is rejected if $T_{23} \geqslant t_{23;0.99} = 2.50$

④ $n = 24$; $\bar{x} = 22.4$; $s = 9.4$

$$T_{23} = \frac{22.4 - 20}{9.4/4.9} = \frac{2.4}{1.92} = 1.25$$

Therefore, we fail to reject H_0.

⑤ There is no sample evidence that the 6-week average weight gain for calves fed Ration A exceeds 20 pounds. ◀

▶ *Example 10-6*

Bottles of catsup are filled automatically by a machine which must be adjusted periodically to increase or decrease the average content per bottle. The amount placed in the bottle is assumed to be a normal variable, and each bottle is supposed to contain 18 oz. It is important to detect an average content significantly above or below 18 oz; too much catsup per bottle would obviously be unprofitable, while too little would be a poor business practice. A random sample of 24 bottles filled by a given machine were each weighed, giving

$$\bar{x} = 18.34 \text{ oz} \qquad \text{and} \qquad SS_X = 12.37\,(\text{oz})^2$$

Is there any evidence at the 5% level that the machine needs adjusting now?

① $H_0: \mu = 18$ versus $H_1: \mu \neq 18$

② Under H_0, $T_{23} = \dfrac{\overline{X} - 18}{S/\sqrt{24}}$

③ Reject H_0 if $|T_{23}| \geqslant 2.069$

④ $s_{\bar{X}}^2 = \dfrac{12.37}{24(23)} = 0.0224$

$$T_{23} = \frac{18.34 - 18}{0.15} = 2.27$$

Therefore, H_0 is rejected.

⑤ The mean content per catsup bottle differs from 18 oz unless a 5% error has been made: that is, evidence indicates that the machine needs adjusting. ◀

(Problems 1(a), 3(a), (b), 5(a))

*10.3.3 Tests about μ of N(μ, σ²) When σ² Is Unknown and Estimated through the Range

When testing hypotheses about the mean of a normal, a substitute for the Student T test is a procedure based on the range. Called the pseudo-T test, it is easier and quicker to perform, but it is less powerful than the Student T test. For small sample sizes, say less than 15, the sacrifice of power is not too great.

The pseudo-T test procedure is as follows:

① ⓐ $H_0: \mu \leqslant \mu_0$ $\quad H_1: \mu > \mu_0$ ① ⓑ $H_0: \mu \geqslant \mu_0$ $\quad H_1: \mu < \mu_0$ ① ⓒ $H_0: \mu = \mu_0$ $\quad H_1: \mu \neq \mu_0$

② Under H_0, $T_n^* = \dfrac{\bar{X} - \mu_0}{R}$

(10.7)

③ Reject H_0 if

ⓐ $T_n^* \geqslant t^*_{n;1-\alpha}$ ⓑ $T_n^* \leqslant -t_{n;1-\alpha}$ ⓒ $|T_n^*| \geqslant t^*_{n;1-\alpha/2}$

As with the two previous tests, the position of the critical region is dictated by alternative values.

▶ ***Example 10-7***

If all boxes of soap in Example 9-9 are to have an average weight of 84 oz, is there any evidence that the boxes are underweight? Assuming normality, a 5% test of this hypothesis would be

① $H_0: \mu \geqslant 84$
$H_1: \mu < 84$

② If H_0 is true, $\dfrac{\bar{X} - 84}{R}$ is a T_8^* variable.

③ For $\alpha = 0.05$, reject H_0 if $T_8^* \leqslant -t^*_{8;0.95} = -0.230$

④ $\bar{x} = 83.7$; $r = 1.2$

$$T_8^* = \frac{83.7 - 84}{1.2} = \frac{-0.3}{1.2} = -0.25$$

Therefore, H_0 is rejected.

⑤ The average weight per box of soap is less than 84 oz unless a 5% error has been made. ◀

(Problem 6)

10.3.4 Tests about σ^2 of $N(\mu, \sigma^2)$ When μ Is Unknown

The customary test about the variance of a normal distribution involves the chi-square distribution. The first three steps of the test procedure are:

(10.8)

① ⓐ $H_0: \sigma^2 \leqslant \sigma_0^2$, $H_1: \sigma^2 > \sigma_0^2$ ① ⓑ $H_0: \sigma^2 \geqslant \sigma_0^2$, $H_1: \sigma^2 < \sigma_0^2$ ① ⓒ $H_0: \sigma^2 = \sigma_0^2$, $H_1: \sigma^2 \neq \sigma_0^2$

② Under H_0, $\chi^2_{n-1} = \dfrac{(n-1)S^2}{\sigma_0^2}$

③ Reject H_0 if

ⓐ $\chi^2_{n-1} \geqslant \chi^2_{n-1;1-\alpha}$ ⓑ $\chi^2_{n-1} \leqslant \chi^2_{n-1;\alpha}$ ⓒ $\chi^2_{n-1} \leqslant \chi^2_{n-1;\alpha/2}$ or $\chi^2_{n-1} \geqslant \chi^2_{n-1;1-\alpha/2}$

► ***Example 10-8***

A manufacturer of chemical balances calibrates each instrument by repeatedly weighing a given object. Each balance is to have a certain precision, the acceptable standard being a variance of 5 (mg)^2 or less. To check a given balance, 20 weighings are made, giving $s^2 = 6.1$ (mg)^2. A 5% test to check the precision of this balance is:

① $H_0: \sigma^2 \leqslant 5$ versus $H_1: \sigma^2 > 5$
② Under H_0 $\chi^2_{19} = (n-1)S^2/5$
③ Reject H_0 if $\chi^2_{19} \geqslant 30.14$

④ $\chi^2_{19} = \dfrac{19(6.1)}{5} = 23.18$

Therefore, H_0 is not rejected.
⑤ For the balance tested, there is no evidence that the variability of measurements is greater than the acceptable standard. ◄

► ***Example 10-9***

Suppose in the process of filling catsup bottles (Example 10-6) it is of interest to know if the standard deviation of the amount per bottle is significantly different from 1/2 oz. A 10% test of this hypothesis would be:

① $H_0: \sigma^2 = (1/2)^2 = 1/4$ versus $H_1: \sigma^2 \neq 1/4$

② Under H_0, $\chi^2_{23} = \dfrac{SS_X}{1/4} = 4SS_X$

③ Reject H_0, if $\chi^2_{23} \leqslant 13.09$ or if $\chi^2_{23} \geqslant 35.17$

④ $\chi^2_{23} = 4(12.37) = 49.48.$
Therefore, H_0 is rejected.

⑤ The variance of the amount of catsup per bottle differs from 1/4 unless a 10% error has occurred. ◀

(Problems 3(c), 13(b))

10.3.5 Tests about σ^2 of $N(\mu, \sigma^2)$ When μ Is Known

Should a problem be encountered in this category, the χ^2 test of the previous section would be modified as discussed in Section 9.3.6. That is, the *SS* would be calculated using the known value of μ, and the χ^2 test statistic would have n degrees of freedom. A likely encounter of this case occurs when we study the differences of variables having the same means; for this situation, $\mu = 0$.

10.4 UNDERLYING ASSUMPTIONS

The same comments apply here as for the interval estimations discussed in the last chapter. All procedures of Section 10.3 require random sampling and normally distributed observations. Other assumptions, such as known variance, determine which particular test procedure is appropriate.

Violation of necessary assumptions results in approximate tests. In cases where the Central Limit Theorem may be applied, the assumption of normality may be relaxed when testing hypotheses about the mean of a distribution.

**10.5 POWER OF TESTS FOR THE NORMAL FAMILY*

A basic discussion of the power of a test may be found in Section 10.2.1. This section contains a catalog of power calculations for the normal tests of Section 10.3.

Recall that the power of a test can be written as the conditional probabilities

$$\text{Power} = P(\text{reject } H_0 | H_1 \text{ is true})$$

(10.9)

$$= P(\text{test statistic is in the critical region} | H_1 \text{ is true})$$

It will be convenient to let θ_a denote the assumed (or actual) value of θ when the alternative hypothesis is true. Then, the difference $\Delta_a = \theta_a - \theta_0$ will enter power calculations. For notational convenience, let P_a denote the power when $\theta = \theta_a$.

*10.5.1 The *Z* Test

The power for any *Z* test is simply a probability from the *Z* distribution. For a one-sided upper test, it follows from (10.9) that

$$P_a = P[Z \geqslant z_{1-\alpha} | \mu = \mu_a]$$

$$= P\left[\frac{\overline{X} - \mu_0}{\sigma_{\overline{X}}} \geqslant z_{1-\alpha} | \mu = \mu_a\right]$$

$$(10.10) \qquad = P\left[\frac{\overline{X} - \mu_a + \mu_a - \mu_0}{\sigma_{\overline{X}}} \geqslant z_{1-\alpha} | \mu = \mu_a\right]$$

$$= P\left[\frac{\overline{X} - \mu_a}{\sigma_{\overline{X}}} \geqslant z_{1-\alpha} - \Delta_a/\sigma_{\overline{X}} | \mu = \mu_a\right]$$

$$= P[Z \geqslant z_{1-\alpha} - \Delta_a/\sigma_{\overline{X}}] = P[Z \geqslant -z_\alpha - \Delta_a/\sigma_{\overline{X}}]$$

$$= F(z_\alpha + \Delta_a/\sigma_{\overline{X}})$$

Similar calculations for a one-sided lower Z test show that

$$(10.11) \qquad P_a = F(z_\alpha - \Delta_a/\sigma_{\overline{X}})$$

and for a two-sided Z test

$$(10.12) \qquad P_a = F(z_{\alpha/2} - \Delta_a/\sigma_{\overline{X}}) + F(z_{\alpha/2} + \Delta_a/\sigma_{\overline{X}})$$

▶ ***Example 10-10***

If the unknown gas in Example 10-4 actually has a velocity of 830.8 m/sec, the chance of this being detected by the Z test is found by the following calculation:

$$\Delta_a = 830.8 - 830 = 0.8$$

$$P_a = F_z(-1.645 + 0.8/0.4) + F_z(-1.645 - 0.8/0.4)$$

$$= F_z(-3.645) + F_z(0.355)$$

$$= 0.0005 + 0.6388 = 0.6393$$

There is about a 64% chance of rejecting $H_0: \mu = 830$ m/sec when the true velocity is really 830.8 m/sec. ◀

*10.5.2 The *T* Test

The power of the T test is more difficult to calculate than the power of a Z test. For a one-sided upper test,

$$P_a = P[T_{n-1} \geqslant t_{1-\alpha;n-1} | \mu = \mu_a]$$

$$(10.13) \qquad = P\left[\frac{\overline{X} - \mu_a + \mu_a - \mu_0}{S_{\overline{X}}} \geqslant t_{1-\alpha;n-1} | \mu = \mu_a\right],$$

which is a probability from a *noncentral* T distribution. Fortunately, graphs exist which can be used to obtain the power when Δ_a is taken to be a constant times σ. Letting $\delta_a = \sqrt{n}|\Delta_a|/\sigma$, the power of any *one-sided* T test can be read from Appendix B.1.

The graphs can also be used for the power of a two-sided T test of size $\alpha^* = 2\alpha$ for the four values of α given. However, as δ_a approaches zero, the error in the values of power read from these graphs approaches a maximum of $\alpha^*/2$. For $\delta_a > 1$, the error vanishes. The curves for infinite degrees of freedom, $\nu = \infty$, can be used to obtain the power of Z tests.

▶ ***Example 10-11***

For the analysis performed in Example 10-5, what is the power of the T test if the true mean gain is actually one-fourth standard deviation above the hypothesized value? one-half standard deviation above? one standard deviation above? The following is obtained from Appendix B.1 with $\alpha = 0.01$ (using the curve $\nu = 24$):

Δ_a	δ_a	P_a
$\sigma/4$	$\sqrt{1.5} = 1.22$	0.13
$\sigma/2$	$\sqrt{6} = 2.45$	0.50
σ	$\sqrt{24} = 4.90$	0.988

◀

▶ ***Example 10-12***

The power curve for the catsup experiment considered in Example 10-6 can be drawn from the following values obtained with the aid of Appendix B.1, using $\alpha = 0.025$ and the curve for $\nu = 24$:

Δ_a	δ_a	P_a
$\pm\sigma/4$	$\sqrt{1.5} = 1.22$	0.22
$\pm\sigma/3$	$\sqrt{2.67} = 1.63$	0.34
$\pm\sigma/2$	$\sqrt{6} = 2.45$	0.65
$\pm\sigma$	$\sqrt{24} = 4.90$	0.996

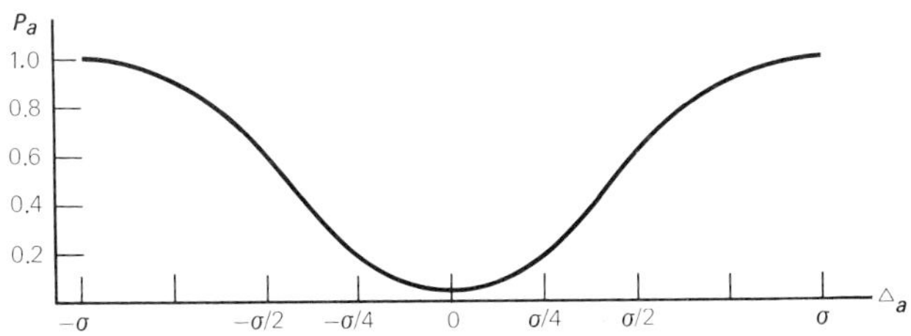

◀

*10.5.3 The Pseudo-T Test

The power of a pseudo-T test is almost equal to that of the corresponding T test if $n \leqslant 10$; see Daly [1] and Lord [2]. Thus, Appendix B.1 may be used to give approximate power (on the high side) of the T^* test. Remember that the pseudo-T test is a poor substitute for the Student T test when $n > 10$.

*10.5.4 The Chi-Square Test

The power function for a chi-square test on the variance of a normal can be easily found. For a one-sided upper test,

$$\begin{aligned} P_a &= P[\chi^2_{n-1} \geqslant \chi^2_{n-1;1-\alpha} | \sigma^2 = \sigma_a^2] \\ &= P[SS/\sigma_0^2 \geqslant \chi^2_{n-1;1-\alpha} | \sigma^2 = \sigma_a^2] \\ &= P[SS/\sigma_a^2 \geqslant (\sigma_0^2/\sigma_a^2)\chi^2_{n-1;1-\alpha} | \sigma^2 = \sigma_a^2] \\ &= P[\chi^2_{n-1} \geqslant (\sigma_0^2/\sigma_a^2)\chi^2_{n-1;1-\alpha}] \\ &= 1 - F(R_a^2 \chi^2_{n-1;1-\alpha}), \end{aligned} \tag{10.14}$$

where $R_a^2 = \sigma_0^2/\sigma_a^2$. Similar calculations for a one-sided lower test give

$$P_a = F(R_a^2 \chi^2_{n-1;\alpha}), \tag{10.15}$$

and for a two-sided chi-square test

$$P_a = 1 - F(R_a^2 \chi^2_{n-1;1-\alpha/2}) + F(R_a^2 \chi^2_{n-1;\alpha/2}) \tag{10.16}$$

Since the point $R_a^2 \chi^2_{n-1;\gamma}$ rarely appears in a χ^2 table, interpolation would have to be done. Curves are available for finding the power of a chi-square test; these appear in Appendix B.2.

► ***Example 10-13***

The power curve for the test in Example 10-8 passes through the following points obtained from Appendix B.2, with $\alpha = 0.05$ and $\nu = 19$ (interpolating between the $\nu = 15$ and 20 curves):

σ_a^2	R_a	P_a
5.5	$\sqrt{10/11} = 0.953$	0.08
6.0	$\sqrt{5/6} = 0.913$	0.13
7.5	$\sqrt{2/3} = 0.817$	0.28
10.0	$\sqrt{1/2} = 0.707$	0.60
15.0	$\sqrt{1/3} = 0.577$	0.95

◄

PROBLEMS

1. A paper manufacturer measures 60 randomly selected sheets of paper and obtains the following with respect to X, their width:

$$\bar{x} = 8.55 \text{ in.} \qquad s^2 = 0.054 \text{ (in.)}^2$$

(a) At the 5% level, test $H_0: \mu = 8.5$ in. against $H_1: \mu \neq 8.5$ in.

(b) What assumptions are required in (a) for statistical validity? Are these assumptions realistic?

2. Packages of frozen peas are advertised as weighing 10 oz.

(a) Formulate suitable hypotheses, H_0 and H_1, in which the food processor would be interested. Explain.

(b) Formulate suitable hypotheses, H_0 and H_1, in which a consumer would be interested. Explain.

A random sample of 18 packages of the frozen peas gave the following statistics:

$$x. = 175 \text{ oz} \qquad r = 1.6 \text{ oz} \qquad s^2 = 0.18 \text{ (oz)}^2$$

(c) As a consumer, test the appropriate hypothesis at the 5% level. Use the Student T procedure.

(d) Test $H_0: \sigma^2 = 0.15 \text{ (oz)}^2$ versus $H_1: \sigma^2 \neq 0.15 \text{ (oz)}^2$ at the 2% level.

3. The Pure Food and Drug Commission is investigating a charge that a certain drug manufacturer is supplying a reagent with more than the advertised impurities. The advertised amount is 15 parts per million (ppm). A research chemist for the Commission tests the contents of 25 bottles and obtains the following in his analysis:

$$\bar{x} = 17.8 \text{ ppm} \qquad SS_X = 1{,}250 \text{ (ppm)}^2$$

(a) Formulate a suitable hypothesis and test it at the 5% level.

(b) What assumptions were made in (a)?

(c) At the 2.5% level, test that the standard deviation of impurities does not exceed 6 ppm.

4. In a manufacturing process, a machine is supposed to turn out ball bearings with a mean diameter of 0.200 in. It is suspected that the mean diameter has increased because of machine wear. A random sample of 100 bearings is selected from the process and measured, giving

$$\bar{x} = 0.2120 \qquad s^2 = 0.0016 \text{ (in.)}^2$$

At the 1% level, is there any evidence that the mean has increased?

5. A rechargeable flashlight is powered by nickel-cadmium batteries. The manufacturer advertises that a "fully charged flashlight will give light brighter than an ordinary flashlight for at least 3 hours." An independent testing laboratory obtains 12 of the flashlights, has them fully charged, and records the times at which the light intensity diminishes (as measured by a photometer) below that of an ordinary flashlight. These times, in minutes, were:

174	194	177	190	210	174
188	172	183	170	185	168

(a) As a member of the testing laboratory, test the claim (from a consumer's point of view) at the 5% level.
(b) Test that the variance of such times is at least 200 $(\text{min})^2$. Let $\alpha = 0.01$.

***6.** Perform the test in Problem 5(a) using the T^* distribution.

7. You are making a 10% test of $H_0: \theta = 40$ versus $H_1: \theta \neq 40$. Give the values of each of the following, if possible:
(a) P (fail to reject $H_0 | \theta = 30$)
(b) P (fail to reject $H_0 | \theta = 40$)
(c) P (reject $H_0 | \theta = 40$)
(d) P (reject $H_0 | \theta = 30$)
(e) P (reject $H_0 | \theta = 40.1$)

8. At a vegetable stand, a gardener is selling tomatoes in bulk and by the box. The boxes were designed to hold one bushel (60 pounds) when full. The gardener claims his tomatoes are of a quality that the boxes contain at least 61 pounds when full. Ten full boxes are selected at random, their contents weighed, giving in pounds:

60.8	59.7	61.2	60.6	61.5
59.0	59.6	60.4	60.4	60.2

(a) Is there evidence, at the 5% level, against his claim?
(b) Is there evidence, at the 5% level, that his boxes of tomatoes contain less than one bushel?

9. A group of secondary education student teachers were given two and one half days' training in interpersonal communication group work (McFarland [3]). The effect of such a training session on the dogmatic (general authoritarianism) nature of the student teachers was measured by the difference of scores on the Rokeach Dogmatism test given before and after the training session. For

$$X = \text{post score minus pre score on the Rokeach Dogmatic Scale}$$

the following differences were recorded:

−16	−5	+4	+19	−40
−16	−29	+15	−2	0
+5	−23	−3	+16	−8
+9	−14	−33	−64	−33

Higher scores mean the person is more dogmatic. Assuming that differences arising from such training sessions are approximately normal, can it be concluded that the training session makes student teachers (in secondary education) less dogmatic? Do the appropriate analysis with $\alpha = 5\%$.

10. Yields, X, of sweet potatoes, in pounds, from 15 experimental plots (0.1 acre each) were:

1,420	1,410	1,433	1,434	1,427
1,441	1,446	1,436	1,428	1,448
1,423	1,440	1,429	1,442	1,444

(a) Test $H_0 : \mu_X \leqslant 1{,}430$ lb versus $H_1 : \mu_X > 1{,}430$ lb at the 5% level.
(b) Test $H_0 : \sigma_X \leqslant 10$ lb versus $H_1 : \sigma_X > 10$ lb at the 1% level.
(c) What assumptions were used in (a) and (b)? Are these reasonable?

11. A new variety of wheat is supposed to grow to a height of 38 or more inches. Thirty-six randomly selected full grown plants were measured and the following obtained (in inches):

$$\bar{x} = 36.9 \qquad s^2 = 20.78$$

Is there evidence that the new variety does not grow to an average height of 38 inches? Make the appropriate test with $\alpha = 0.05$.

12. Fifteen liters of water from a tributary of a particular river had the following amounts of dissolved oxygen (in mg/liter):

4.13	5.38	5.14	4.41	5.03
4.07	4.92	5.08	4.60	4.16
5.61	4.67	4.52	4.72	4.27

Is there evidence, at the 5% level, to conclude that the mean oxygen per liter is below 5 mg for this tributary?

13. Newly made tires are subjected to a number of tests. A particular tread wear test is made by mounting tires on a special machine which rotates tires at a fixed speed and keeps their tread at a certain pressure in contact with a rotating drum. Twenty tires of a given type are tested and for their tread wear, X, measured in millimeters (mm), the following was recorded:

$$x. = 28.62 \qquad SS_X = 2.43$$

(a) Do tires of this particular type have mean tread wear in excess of 1.25 mm when subjected to this tread wear test? Let $\alpha = 0.01$.
(b) Do tires of this particular type have tread wear (as measured by the tread wear test) whose variance exceeds 0.1 $(\text{mm})^2$? Let $\alpha = 0.05$.

***14.** For the tread wear of the 20 tires in Problem 13, it is given that $r = 0.58$ mm. Repeat Problem 13(a), using the T^* distribution.

15. Refer to Problem 13 of Chapter 7, and assume that the time required by a subject to trace the object is an approximate normal variable. Using the data given,
(a) test, at the 5% level, that the mean tracing time is 90 sec
(b) test, at the 10% level, $H_0 : \sigma \leqslant 5$ sec

16. Assuming that the clean fleece weights given in Problem 26 of Chapter 7 are from 18 sheep randomly selected from his flock, can the rancher conclude that his flock has clean fleece weights which average more than 4.5 lb? Let $\alpha = 0.05$.

***17.** What are the chances of rejecting H_0 in Problem 1 if the mean width, μ, of paper is actually
(a) $8.5 - \sigma/3$? (b) $8.5 + \sigma/2$? (c) $8.5 + 3\sigma/4$?

***18.** Determine the power of the test in Problem 2(c) if the mean weight of packages is actually
(a) $10 - \sigma/2$ (b) $10 - \sigma$ (c) $10 - 3\sigma/4$

***19.** Determine the power of the test in Problem 2(d) if
(a) $\sigma^2 = 0.12$ (b) $\sigma^2 = 0.18$ (c) $\sigma^2 = 0.30$

***20.** (a) Draw the power curve for the test in Problem 3(a); that is, determine the power at three or four points and draw the curve.
(b) Draw the power curve for the test in Problem 3(c).

***21.** What are the chances of rejecting H_0 in Problem 5(a) if the mean time of bright light is actually
(a) $180 - \sigma$? (b) $180 - 2\sigma/3$? (c) $180 - \sigma/4$?

***22.** What are the chances of rejecting H_0 in Problem 5(b) if
(a) $\sigma^2 = 150$? (b) $\sigma^2 = 100$? (c) $\sigma^2 = 80$? (d) $\sigma^2 = 50$?

***23.** (a) Draw the power curve for the test in Problem 8(a).
(b) Draw the power curve for the test in Problem 8(b).

***24.** (a) Draw the power curve for the test in Problem 10(a).
(b) Draw the power curve for the test in Problem 10(b).

REFERENCES

[1] Daly, J. F., "The Use of the Sample Range in an Analogue of Student's *t*-Test," *Annals of Mathematical Statistics*, **17** (1946).

[2] Lord, E., "Power of the Modified *t*-Test (u-Test) Based on Range," *Biometrika*, **37** (1950).

[3] McFarland, B. N., "An Analysis of the Effect of Interpersonal Communication Group Work on Dogmatism and Self Concept of Student Teachers," unpublished Ed.D. dissertation, New Mexico State University, 1970.

Chapter 11

INFERENCE ABOUT TWO NORMAL POPULATIONS

As discussed so far, inference has been limited to one normal population. Clearly, the outcome of any single-population experiment can be compared only with preassigned specifications. But experiments dealing with two or more populations are frequently necessary in comparisons of different methods, treatments, processes, and so on. For example, an educator might wish to compare the average grades received by two groups of students enrolled in an elementary statistics course, one group taught exclusively by lecturing, the other by visual-aids presentations. In the electronic problem discussed earlier, officials of the firm might wish to compare the mean time to failure of new components with the mean time to failure of standard components. And the candy maker might test the equality of variances for the amounts of chocolate deposited by two different machines.

This chapter deals exclusively with various analyses of two normal populations. Later chapters deal with more than two populations and with populations having special structure.

11.1 SAMPLE DATA AND POINT ESTIMATORS

The experimental data consist of a random sample of n_1 observations drawn from the "first" population and a random sample of n_2 observations drawn from the "second" population. Observations will be denoted by

$$x_{11}, x_{12}, \ldots, x_{1n_1} \qquad (11.1)$$
$$x_{21}, x_{22}, \ldots, x_{2n_2}$$

where x_{ij} refers to the j-th observation from the i-th population. The two sample totals have values given by

(11.2) $$x_{1\cdot} = \sum_{j=1}^{n_1} x_{1j} \quad \text{and} \quad x_{2\cdot} = \sum_{j=1}^{n_2} x_{2j}$$

and the two sample means by

(11.3) $$\bar{x}_{1\cdot} = x_{1\cdot}/n_1 \quad \text{and} \quad \bar{x}_{2\cdot} = x_{2\cdot}/n_2$$

The sums of squares for the two sets of data are calculated by

(11.4) $$SS_1 = \sum_{j=1}^{n_1} (x_{1j} - \bar{x}_{1\cdot})^2 = \sum_{j=1}^{n_1} x_{1j}^2 - x_{1\cdot}^2/n_1$$
$$SS_2 = \sum_{j=1}^{n_2} (x_{2j} - \bar{x}_{2\cdot})^2 = \sum_{j=1}^{n_2} x_{2j}^2 - x_{2\cdot}^2/n_2$$

whereby the two sample variances have values

(11.5) $$s_1^2 = SS_1/(n_1 - 1) \quad \text{and} \quad s_2^2 = SS_2/(n_2 - 1)$$

▶ ***Example 11-1***

A person can choose between two routes of traveling to work. One, a city route, is much shorter but is congested with traffic. The major portion of the second (longer) route is freeway. Over a two-month period, the person selects 10 days at random to travel each route. He keeps track of the time to travel to work on these days and obtains (to the nearest tenth of a minute):

City route: x_{1j}	Freeway: x_{2j}
14.1	13.3
15.6	14.7
18.3	14.1
13.7	16.6
16.4	12.9
14.4	15.5
17.8	14.3
15.9	16.8
16.1	14.0
17.0	17.5

$x_{1\cdot} = 159.3$	$x_{2\cdot} = 149.7$
$\bar{x}_{1\cdot} = 15.93$	$\bar{x}_{2\cdot} = 14.97$
$\sum x_{1j}^2 = 2558.93$	$\sum x_{2j}^2 = 2262.99$
$x_{1\cdot}^2/10 = 2537.65$	$x_{2\cdot}^2/10 = 2241.01$
$SS_1 = 21.28$	$SS_2 = 21.98$
$s_1^2 = 2.36$	$s_2^2 = 2.44$ ◀

(Problems 5(a), (b))

11.2 A LOOK AT ASSUMPTIONS

The following assumptions are basic to two-population experiments:

(11.6)
(i) random sampling
(ii) independence of the samples
(iii) each of the sampled populations is normal: $X_{ij} \sim \mathrm{N}(\mu_i, \sigma_i^2)$, for $i = 1, 2$
(iv) $\sigma_1^2 = \sigma_2^2$

Randomness is necessary if we wish to use tabulated distributions to attach probability statements to our inferences. Analysis may be possible with nonrandom data—provided, however, that the nature of the nonrandomness is very well specified—but the resulting techniques exceed the scope of this presentation.

Independence is generally assured when separate experiments are conducted at two different locations, on two different sets of material, and so on. Independence does not exist when observations are related because of some logical pairing, such as before-after readings on the same animal or measurements from identical twins. Section 11.6 treats experiments where the independence condition does not hold.

As might be expected, the assumption of normality can be relaxed when making inference about the means. This is a consequence of the Central Limit Theorem. As the two sample sizes increase, larger departures from normality can be tolerated without seriously affecting the probabilities.

If the variances are known, the fourth assumption need not hold. When the variances are unknown, exact inferences require the homogeneity of variances. Various cases, including that of nonhomogeneous variances, are treated in Section 11.5.

11.3 TESTING EQUALITY OF TWO NORMAL VARIANCES

If the assumption of equal variances is questionable or seems unreasonable, but all other assumptions in (11.6) hold, a test of $H_0: \sigma_1^2 = \sigma_2^2$ can be made.

By (8.27), it follows that

(11.7) $$\chi^2_{n_1-1} = SS_1/\sigma_1^2 \qquad \text{and} \qquad \chi^2_{n_2-1} = SS_2/\sigma_2^2$$

The two chi-squares in (11.7) can be shown to be independent under the assumption of independent samples. Then, by (8.30), the variance-ratio stated in (8.29) is obtained:

(11.8) $$F_{n_1-1,n_2-1} = \frac{S_1^2/\sigma_1^2}{S_2^2/\sigma_2^2}$$

Thus, under the hypothesis of equal variances, the right-hand side of (11.8) reduces to a ratio of sample variances, a ratio that becomes the test statistic. A 100α% test of equality of the two normal variances has as the first three steps:

(11.9)

① ⓐ $H_0: \sigma_1^2 \leqslant \sigma_2^2$ ① ⓑ $H_0: \sigma_1^2 \geqslant \sigma_2^2$ ① ⓒ $H_0: \sigma_1^2 = \sigma_2^2$
$H_1: \sigma_1^2 > \sigma_2^2$ $H_1: \sigma_1^2 < \sigma_2^2$ $H_1: \sigma_1^2 \neq \sigma_2^2$

② Under H_0, $F_{n_1-1,n_2-1} = S_1^2/S_2^2$

③ Reject H_0 if

ⓐ $F_{n_1-1,n_2-1} \geqslant F_{n_1-1,n_2-1;1-\alpha}$

ⓑ $F_{n_1-1,n_2-1} \leqslant F_{n_1-1,n_2-1;\alpha}$

ⓒ $F_{n_1-1,n_2-1} \geqslant F_{n_1-1,n_2-1;1-\alpha/2}$
or if $F_{n_1-1,n_2-1} \leqslant F_{n_1-1,n_2-1;\alpha/2}$

► ***Example 11-2***

The drying times of two different brands of paint were investigated. Eleven measurements were obtained from each brand, giving (in coded form):

$\bar{x}_{1\cdot} = 57$ $SS_1 = 920$ $r_1 = 22$ $n_1 = 11$

$\bar{x}_{2\cdot} = 64$ $SS_2 = 670$ $r_2 = 16$ $n_2 = 11$

Assuming normality, a 5% test of equal variances is

① $H_0: \sigma_1^2 = \sigma_2^2$
$H_1: \sigma_1^2 \neq \sigma_2^2$

② If H_0 is true, $F_{10,10} = S_1^2/S_2^2$

③ Reject H_0 if $F_{10,10} \geqslant F_{10,10;0.975} = 3.72$
or if $F_{10,10} \leqslant F_{10,10;0.025} = 0.269$

④ $$F_{10,10} = \frac{920/10}{670/10} = 1.37$$

Therefore, H_0 is not rejected.

⑤ There is no evidence at the 5% level of significance that the two variances differ. ◄

(Problems 1(a), (b), (c), 3(a), (b), 9(a))

11.4 INFERENCE ABOUT THE MEANS OF TWO INDEPENDENT NORMAL DISTRIBUTIONS

We shall consider inference about μ_1 and μ_2 individually and in terms of their difference. Most two-population experiments are designed to compare μ_1 and μ_2; hence, the difference $\theta = \mu_1 - \mu_2$ represents a parametric function of utmost importance. Point estimation of θ is accomplished by using the corresponding difference of sample means,

$$\hat{\theta} = \bar{X}_{1\cdot} - \bar{X}_{2\cdot}. \tag{11.10}$$

In random sampling from independent $N(\mu_1, \sigma_1^2)$ and $N(\mu_2, \sigma_2^2)$ distributions, it is well known (see Section 8.7.1) that

$$\begin{aligned} &\bar{X}_{1\cdot} \sim N(\mu_1, \sigma_1^2/n_1) \\ &\bar{X}_{2\cdot} \sim N(\mu_2, \sigma_2^2/n_2) \\ &\bar{X}_{1\cdot} - \bar{X}_{2\cdot} \sim N(\mu_1 - \mu_2, \sigma_1^2/n_1 + \sigma_2^2/n_2) \end{aligned} \tag{11.11}$$

where n_1 and n_2 are the respective sample sizes.

11.4.1 Inference When σ_1^2 and σ_2^2 Are Known, Equal or Unequal

By standardizing the appropriate variable in (11.11), inference about μ_1, μ_2, or $\theta = \mu_1 - \mu_2$ may be made in the usual manner with the standard normal distribution. Thus

$$\begin{aligned} Z &= \frac{\bar{X}_{1\cdot} - \mu_1}{\sigma_1/\sqrt{n_1}} = \frac{\bar{X}_{1\cdot} - \mu_1}{\sigma_{\bar{X}_{1\cdot}}}, \\ Z &= \frac{\bar{X}_{2\cdot} - \mu_2}{\sigma_2/\sqrt{n_2}} = \frac{\bar{X}_{2\cdot} - \mu_2}{\sigma_{\bar{X}_{2\cdot}}}, \text{ and} \\ Z &= \frac{\bar{X}_{1\cdot} - \bar{X}_{2\cdot} - (\mu_1 - \mu_2)}{\sqrt{\dfrac{\sigma_1^2}{n_1} + \dfrac{\sigma_2^2}{n_2}}} = \frac{\hat{\theta} - \theta}{\sigma_{\hat{\theta}}} \end{aligned} \tag{11.12}$$

are functions that may be used for inference about μ_1, μ_2, and $\theta = \mu_1 - \mu_2$, respectively.

Experiments dealing with one population having known variance are rare; those involving two or more populations having known variances are even rarer. For this reason, we shall not pursue this case any further.

11.4.2 Inference When σ_1^2 and σ_2^2 Are Unknown but Equal

If the variances are unknown but equal, each of the quantities in (11.12) contains two unknowns: the parameter of interest in the numerator and a standard error in the denominator. Because of the unknown standard errors, these quantities cannot be used for inference about μ_1, μ_2, and $\theta = \mu_1 - \mu_2$. Replacing the standard errors in (11.12) by sample estimators gives Student T variables rather than Z variables.

Let the common variance be denoted simply by σ^2. An unbiased estimator of σ^2 based on $n_1 - 1$ degrees of freedom is given by the "first" sample variance. Likewise, the variance of the "second" sample is an unbiased estimator of σ^2 based upon $n_2 - 1$ degrees of freedom. Since these two estimators are independent, we may combine this information to give a single, better estimator of σ^2. Called the *pooled estimator* of σ^2 and denoted by S_P^2, its value is

$$s_P^2 = \frac{\text{Pooled } SS}{\text{Pooled } df} = \frac{SS_1 + SS_2}{(n_1 - 1) + (n_2 - 1)} \tag{11.13}$$

Note that the pooled estimator of σ^2 is based upon $n_1 + n_2 - 2$ degrees of freedom. Under the assumption of a common variance and independent distributions, each sample contains valuable information about σ^2 which is not ignored by the pooled variance. Pooling of information to calculate an estimate of the common variance is accomplished as illustrated in (11.13), by adding sums of squares not variances. This is a general concept that we will encounter again in future work.

It can be shown that unbiased estimators of three standard errors appearing in (11.12) are:

$$\begin{aligned} S_{\bar{X}_{1\cdot}} &= S_P/\sqrt{n_1} \\ S_{\bar{X}_{2\cdot}} &= S_P/\sqrt{n_2} \\ S_{\hat{\theta}} = S_{\bar{X}_{1\cdot} - \bar{X}_{2\cdot}} &= S_P\sqrt{\frac{1}{n_1} + \frac{1}{n_2}} \end{aligned} \tag{11.14}$$

Using these three standard errors, we can form the following Student T variables:

$$\begin{aligned} T_{n_1+n_2-2} &= \frac{\bar{X}_{1\cdot} - \mu_1}{S_P/\sqrt{n_1}} \\ T_{n_1+n_2-2} &= \frac{\bar{X}_{2\cdot} - \mu_2}{S_P/\sqrt{n_2}} \\ T_{n_1+n_2-2} &= \frac{\bar{X}_{1\cdot} - \bar{X}_{2\cdot} - (\mu_1 - \mu_2)}{S_P\sqrt{\dfrac{1}{n_1} + \dfrac{1}{n_2}}} = \frac{\hat{\theta} - \theta}{S_{\hat{\theta}}} \end{aligned} \tag{11.15}$$

We emphasize that each T variable in (11.15) has $n_1 + n_2 - 2$ degrees of freedom because the pooled estimator of variance has these degrees of freedom.

Tests and CI about μ_1, μ_2, and $\theta = \mu_1 - \mu_2$ have the usual form afforded by the Student T theory. Thus, $100(1 - \alpha)\%$ confidence limits for μ_1, μ_2, and $\mu_1 - \mu_2$, respectively, are:

(11.16)

$$\bar{x}_{1\cdot} \pm (s_P/\sqrt{n_1})t_{n_1+n_2-2;1-\alpha/2}$$

$$\bar{x}_{2\cdot} \pm (s_P/\sqrt{n_2})t_{n_1+n_2-2;1-\alpha/2}$$

$$\bar{x}_{1\cdot} - \bar{x}_{2\cdot} \pm \left(s_P\sqrt{\frac{1}{n_1} + \frac{1}{n_2}}\right)t_{n_1+n_2-2;1-\alpha/2}$$

We shall give only the structure of a test about the difference of two means. To test that $\mu_1 - \mu_2$ has a specified difference, say δ_0, the first three steps of a $100\alpha\%$ test would be:

① ⓐ $H_0: \mu_1 - \mu_2 \leqslant \delta_0$ $\quad H_1: \mu_1 - \mu_2 > \delta_0$

① ⓑ $H_0: \mu_1 - \mu_2 \geqslant \delta_0$ $\quad H_1: \mu_1 - \mu_2 < \delta_0$

① ⓒ $H_0: \mu_1 - \mu_2 = \delta_0$ $\quad H_1: \mu_1 - \mu_2 \neq \delta_0$

(11.17) ② Under H_0, $$T_{n_1+n_2-2} = \frac{\bar{X}_{1\cdot} - \bar{X}_{2\cdot} - \delta_0}{S_P\sqrt{\frac{1}{n_1} + \frac{1}{n_2}}}$$

③ H_0 is rejected if

ⓐ $T_{n_1+n_2-2} \geqslant t_{n_1+n_2-2;1-\alpha}$

ⓑ $T_{n_1+n_2-2} \leqslant t_{n_1+n_2-2;\alpha}$

ⓒ $|T_{n_1+n_2-2}| \geqslant t_{n_1+n_2-2;1-\alpha/2}$

Tests about a single mean follow the structure given in Section 10.3.2 except that the T variables are now defined as in (11.15), that is, using S_P and having "full" degrees of freedom.

▶ *Example 11-3*

Suppose the two brands of paint in Example 11-2 have composition which leads one to believe that the drying time for the second brand is at least 10 minutes longer than the drying time for the first brand. The equality of variances was not refuted by a preliminary test, see Example 11-2. Assuming approximate normality of drying time, a 1% test of this hypothesis would be

① $H_0: \mu_2 - \mu_1 \geqslant 10$ versus $H_1: \mu_2 - \mu_1 < 10$

② If H_0 is true, $$T_{20} = \frac{\bar{X}_{2\cdot} - \bar{X}_{1\cdot} - 10}{S_{\bar{X}_{1\cdot} - \bar{X}_{2\cdot}}}$$

③ Reject H_0 if $T_{20} \leqslant -t_{20;0.99} = -2.528$

④ $$s_P^2 = \frac{920 + 670}{10 + 10} = 79.5$$

$$s_{\bar{X}_1. - \bar{X}_2.} = \sqrt{2(79.5)/11} = \sqrt{14.45} = 3.8$$

$$T_{20} = \frac{64 - 57 - 10}{3.8} = -0.76$$

Therefore, we fail to reject H_0.

⑤ There is no evidence at the 1% level that the mean drying time for the first brand is shorter than the mean drying time for the second brand by more than 10 minutes. ◀

► ***Example 11-4***

Using the data of Example 11-1, is there any evidence at the 5% level that the average driving time differs for the two routes? Assuming normality and independence of the two sets of times, a preliminary test of $H_0: \sigma_1^2 = \sigma_2^2$ reveals no significant difference of σ_1^2 and σ_2^2. Thus,

① $H_0: \mu_1 = \mu_2$ versus $H_1: \mu_1 \neq \mu_2$

② Under H_0, $T_{18} = \dfrac{\bar{X}_{1\cdot} - \bar{X}_{2\cdot}}{S_P\sqrt{1/5}}$

③ Reject H_0 if $|T_{18}| \geqslant 2.101$

④ $$s_P^2 = \frac{43.26}{18} = 2.40$$

$$s^2_{\bar{X}_1. - \bar{X}_2.} = 2.40(2/10) = 0.48$$

$$T_{18} = \frac{15.93 - 14.97}{\sqrt{0.48}} = 1.39$$

Therefore, we fail to reject H_0.

⑤ There is no evidence at the 5% level that the average driving time differs for the two routes. ◀

(Problems 1(d), (e), 3(c), 9)

*11.4.3 Tests about the Difference of Means When σ_1^2 and σ_2^2 Are Equal, Unknown, and Estimated through the Range

As in the one-sample case, the calculation of the sample variance may be too tedious or time-consuming. For example, in quality-control situations on an assembly line, the quickest decision is mandatory. Estimating the common variance σ^2 through the sample ranges facilitates a quicker decision (than using the pooled estimator) but with a slight loss in power. Let

$$R_1 = \text{range of "first" sample}$$
$$R_2 = \text{range of "second" sample} \tag{11.18}$$
$$R. = R_1 + R_2$$

Then the two-sample pseudo-T variable is

$$T_m^* = \frac{\bar{X}_{1\cdot} - \bar{X}_{2\cdot} - (\mu_1 - \mu_2)}{R.} \tag{11.19}$$

where $m = n_1 = n_2$. For unequal sample sizes, it is recommended that the average sample size be used.

The lower half of Table A.9 gives two-sample T^* points for equal sample sizes. Then, the first three steps of the pseudo-T test are:

(11.20)

① ⓐ $H_0: \mu_1 - \mu_2 \leqslant \delta_0$ ① ⓑ $H_0: \mu_1 - \mu_2 \geqslant \delta_0$
$H_1: \mu_1 - \mu_2 > \delta_0$ $H_1: \mu_1 - \mu_2 < \delta_0$

① ⓒ $H_0: \mu_1 - \mu_2 = \delta_0$
$H_1: \mu_1 - \mu_2 \neq \delta_0$

② Under H_0, $T_m^* = \dfrac{\bar{X}_{1\cdot} - \bar{X}_{2\cdot} - \delta_0}{R.}$

③ Reject H_0 if

ⓐ $T_m^* \geqslant t^*_{m;1-\alpha}$ ⓑ $T_m^* \leqslant -t^*_{m;1-\alpha}$ ⓒ $|T_m^*| \geqslant t^*_{m;1-\alpha/2}$

► ***Example 11-5***

The hypothesis stated in Example 11-3 will now be tested by the pseudo-T distribution.

① $H_0: \mu_2 - \mu_1 \geqslant 10$ versus $H_1: \mu_2 - \mu_1 < 10$

② Under H_0, $T_{11}^* = \dfrac{\bar{X}_{2\cdot} - \bar{X}_{1\cdot} - 10}{R.}$

③ Reject H_0 if $T_{11}^* \leqslant -t^*_{11;0.99} = -0.170$

④ $T_{11}^* = \dfrac{64 - 57 - 10}{16 + 22} = -3/38 = -0.08$

Therefore, we fail to reject H_0.

⑤ Same as in Example 11-3. ◄

(Problem 17)

11.4.4 Testing the Equality of Means When Both Variances Are Unknown and Unequal (The Behrens-Fisher Problem)

This is one of the classical problems in statistics for which no exact solution has yet been found. Several approximate procedures have been proposed; for a discussion of some of these, refer to [1], [2], [3], [6], [9], and [10]. It is stated in [10] and has been shown experimentally that mild departures from equality of variances can be tolerated without greatly affecting probabilities of Student T tests so long as the sample sizes are equal or nearly so.

A widely used testing criterion proposed for the Behrens-Fisher problem arises from a straightforward modification of the regular two-sample T variable given in (11.15). Denoted by T', it is

$$T' = \frac{\bar{X}_{1\cdot} - \bar{X}_{2\cdot} - (\mu_1 - \mu_2)}{\sqrt{\dfrac{S_1^2}{n_1} + \dfrac{S_2^2}{n_2}}} \tag{11.21}$$

Welch [9] has shown that T' has an approximate Student T distribution, but its degrees of freedom must be calculated from a formula. Depending upon the degree of approximation proposed, several formulas are given for calculating the degrees of freedom (see [1], [9], and [10]). Rather than calculating the degrees of freedom, which usually does not result in an integer, we shall adopt the procedure suggested in [3]: calculating the critical point directly by a formula.

In the framework of our testing procedure, the approximate $100\alpha\%$ test is:

(11.22)

① ⓐ $H_0: \mu_1 - \mu_2 \leqslant \delta_0$ ① ⓑ $H_0: \mu_1 - \mu_2 \geqslant \delta_0$

$H_1: \mu_1 - \mu_2 > \delta_0$ $H_1: \mu_1 - \mu_2 < \delta_0$

① ⓒ $H_0: \mu_1 - \mu_2 = \delta_0$

$H_1: \mu_1 - \mu_2 \neq \delta_0$

② Under H_0, $T' = \dfrac{\bar{X}_{1\cdot} - \bar{X}_{2\cdot} - \delta_0}{\sqrt{\dfrac{S_1^2}{n_1} + \dfrac{S_2^2}{n_2}}}$

③ Reject H_0 if

ⓐ $T' \geqslant t'_{1-\alpha}$ ⓑ $T' \leqslant t'_{\alpha}$ ⓒ $|T'| \geqslant t'_{1-\alpha/2}$,

where the critical points in step ③ are weighted means of regular Student T points, and defined by

$$t'_{\gamma} = (w_1 t_{n_1-1;\gamma} + w_2 t_{n_2-1;\gamma})/(w_1 + w_2)$$
$$w_i = s_i^2/n_i \qquad \text{for } i = 1, 2 \tag{11.23}$$

For $n_1 = n_2 = n$, equal sample sizes,

$$t'_{\gamma} = t_{n-1;\gamma}, \tag{11.24}$$

that is, the critical point is a regular T point with only one-half the degrees of freedom (this loss in df being the price we must pay for unequal variances). It should be pointed out that the test in (11.22) is on the conservative side; that is, the significance level is lower than the value of α used. This is especially true when $n_1 = n_2 = n$, since T' defined in (11.21) reduces to the regular two-sample T variable given in (11.15).

► ***Example 11-6***

A toy manufacturer produces a given toy at two different plant locations. A large proportion of the toys produced and sold by the one plant are returned as defective. Because of the nature of the defect and the fact that the two plants have different sources of raw material, it was suspected that the trouble lies in the quality of wooden pegs used. The company decided to test 40 wooden pegs from each source to see if they have the same mean breaking strength. Data obtained (in coded form) gave the following statistics:

$$\bar{x}_{1\cdot} = 99.62 \qquad \bar{x}_{2\cdot} = 77.70$$

$$SS_1 = 1{,}385 \qquad SS_2 = 15{,}950$$

$$s_1^2 = 35.51 \qquad s_2^2 = 408.97$$

Since there was no *a priori* information about the two variances, a preliminary test of $H_0: \sigma_1^2 = \sigma_2^2$ was made and indicated a highly significant difference between σ_1^2 and σ_2^2.

Independence is present because of different sources of supply. An approximate 10% test of equal mean breaking strength is

① $H_0: \mu_1 = \mu_2$ versus $H_1: \mu_1 \neq \mu_2$

② Under H_0, $T' = \dfrac{\bar{X}_{1\cdot} - \bar{X}_{2\cdot}}{\sqrt{\dfrac{S_1^2}{n_1} + \dfrac{S_2^2}{n_2}}}$

③ Reject H_0 if $|T'| \geqslant t'_{0.95}$, where $t'_{0.95} = t_{39;0.95} = 1.685$ since $n_1 = n_2$.

④ $T' = \dfrac{99.62 - 77.70}{\sqrt{\dfrac{35.51 + 408.97}{40}}}$

Therefore, H_0 is rejected.

⑤ The mean breaking strength differs for the two sources of wooden pegs unless a 10% error has been made. ◄

If there is either a priori knowledge or evidence from a preliminary test that the variances differ, we might seriously question whether a test about the means should be made. It depends on our objectives. Different variances define different distributions; if a conclusion of different distributions is sufficient, a test about the means is unnecessary. We would make a test about the means, however, if inference about the means is important (as in Example 11-6).

11.5 DEPENDENT SAMPLES: PAIRED DATA

This section considers tests about the difference of means when the two populations are not independently distributed. The sample observations have a logical pairing; members of each pair have a common factor. Examples of paired data are:

(a) the medical readings for a patient before and after administering a drug,
(b) observations on pairs of identical twins subjected to two different treatments,
(c) measurements made on the same object by two different tools or devices, and
(d) drying times of two paints (see Example 11-2) when both are painted at n different atmospheric conditions, both on n different boards, and so on.

The measurements so obtained can be separated into two groups, but the two samples are *not* independent; they are related. Because of this lack of independence, the previous analysis for two populations fails to be applicable.

The data are as displayed in (11.1) with $n_1 = n_2 = n$. Then, for $j = 1, 2, \ldots, n$, a difference is calculated for the j-th pair:

(11.25) $$d_j = x_{1j} - x_{2j}$$

The mean difference is

(11.26) $$\bar{d} = \frac{1}{n}\sum d_j = d./n = \bar{x}_1. - \bar{x}_2.$$

which is an unbiased estimator of $\mu_D = \mu_1 - \mu_2$. Measuring the variability is also accomplished from the n differences:

(11.27) $$s_D^2 = \frac{1}{n-1}\left[\sum d_j^2 - d.^2/n\right]$$

Consequently, the variance of the mean difference is given by

(11.28) $$s_{\bar{D}}^2 = s_D^2/n$$

Under the assumptions of related measurements whose differences are randomly selected from a $N(\mu_D, \sigma_D^2)$ population, a $100\alpha\%$ test about μ_D would have as the first three steps:

① ⓐ $H_0: \mu_D \leqslant \delta_0$ ① ⓑ $H_0: \mu_D \geqslant \delta_0$

$H_1: \mu_D > \delta_0$ $H_1: \mu_D < \delta_0$

① ⓒ $H_0: \mu_D = \delta_0$

$H_1: \mu_D \neq \delta_0$

(11.29) ② Under H_0, $T_{n-1} = \dfrac{\bar{D} - \delta_0}{S_{\bar{D}}}$

③ Reject H_0 if

ⓐ $T_{n-1} \geqslant t_{n-1;1-\alpha}$ ⓑ $T_{n-1} \leqslant t_{n-1;\alpha}$

ⓒ $|T_{n-1}| \geqslant t_{n-1;1-\alpha/2}$

Note that the degrees of freedom for the T variable of this test are only $n - 1$, whereas the regular two-sample T test has $2(n - 1)$ degrees of freedom when $n_1 = n_2 = n$. This loss in degrees of freedom can be compensated for only by a strong relationship between the pairs. It can be shown that the differences of paired measurements, say $D_j = X_{1j} - X_{2j}$, have variance given by

(11.30) $$\sigma_D^2 = \sigma_1^2 + \sigma_2^2 - 2\rho\sigma_1\sigma_2$$

where σ_1^2 and σ_2^2 are the variances of the variables X_1 and X_2, respectively, and ρ is a parameter measuring the strength of the relationship. (This parameter will be discussed in detail in Chapter 13.) For independent samples (random pairings), the measurements are unrelated, and the last term in (11.30) vanishes. If $\sigma_1 = \sigma_2 = \sigma$, (11.30) reduces to

(11.31) $$\sigma_D^2 = 2\sigma^2(1 - \rho)$$

As ρ approaches unity (meaning the variables are strongly related), we see from either (11.30) or (11.31) that σ_D^2 gets smaller. Smaller variance generally means more powerful tests and shorter *CI*. Equality of variances σ_1^2 and σ_2^2 is not necessary though for (11.29) to be valid.

► ***Example 11-7***

A drug company advertises that its new diet pills result in an average loss of no more than 12 pounds during the first week. A government agency, in checking the claim, obtains 30 volunteers who wish to lose weight. Each volunteer is weighed at the start and after taking the diet pills for one week, giving the following losses:

14.5, 10.6, 11.7, 12.0, 12.4, 13.6, 13.3, 12.8, 12.7, 13.1
13.0, 12.8, 12.6, 11.9, 11.8, 12.2, 13.2, 12.7, 13.2, 12.8
13.9, 13.3, 13.0, 12.5, 12.9, 10.9, 12.4, 12.1, 11.6, 12.8

Calculations reveal:

$$d. = 378.3, \qquad \text{and so} \qquad \bar{d} = 12.61$$

$$\sum d_j^2 = 4{,}789.89$$

$$s_D^2 = \tfrac{1}{29}[4{,}789.89 - (378.3)^2/30]$$

$$= \tfrac{1}{29}[19.53] = 0.673$$

$$s_{\bar{D}} = \sqrt{0.673/30} = 0.15$$

A 1% test of the claim by the government agency would be:

① $H_0: \mu_D \leqslant 12$ versus $H_1: \mu_D > 12$

② Under H_0, $T_{29} = \dfrac{\bar{D} - 12}{S_{\bar{D}}}$

③ Reject H_0 if $T_{29} \geqslant 2.462$

④ $T_{29} = \dfrac{12.61 - 12}{0.15} = 4.07$

⑤ The mean weight loss exceeds 12 pounds unless a 1% error has occurred. People taking these diet pills can expect to lose more than 12 pounds during the first week. ◀

(Problems 6, 7, 8)

*11.6 POWER OF TESTS

Power of Z tests using statistics in (11.12) will not be discussed because of their rarity in practice. Should one of these tests be encountered, formulas (10.10) through (10.12) with the appropriate modification of Δ_a and $\sigma_{\bar{X}}$ give the power.

Power of the approximate T' test for the Behrens–Fisher problem and the pseudo-T test will not be treated.

*11.6.1 Power of the F Test of Variances

For the test of equality of two normal variances (11.9), the standard F tables may be used to determine the power. This is easily seen by rewriting the statistic in (11.8) as follows:

$$F_{n_1-1,n_2-1} = (S_1^2/S_2^2)(\sigma_2^2/\sigma_1^2) \tag{11.32}$$

Under H_0, $R^2 = \sigma_1^2/\sigma_2^2$ is unity, whereby S_1^2/S_2^2 is an F variable. However, under H_1, $R^2 \neq 1$, and the expression in (11.32) does not simplify. Letting R_a denote a particular ratio of σ_1 to σ_2 under the alternative, it follows that the power related to the ratio R_a is

$$P_{R_a} = P[\text{reject } H_0|R_a] = P[S_1^2/S_2^2 > F_{n_1-1,n_2-1;1-\alpha}|R_a] \quad (11.33)$$

when $H_1: \sigma_1^2 > \sigma_2^2$ is the alternative. Thus,

$$P_{R_a} = P[S_1^2/S_2^2R_a^2 > (1/R_a^2)F_{n_1-1,n_2-1;1-\alpha}] = P[F_{n_1-1,n_2-1} > (1/R_a^2)F_{n_1-1,n_2-1;1-\alpha}], \quad (11.34)$$

which may be read (approximately in most cases) from Table A.10. For a one-sided lower test, a similar argument would reveal

$$P_{R_a} = P[F_{n_1-1,n_2-1} < (1/R_a^2)F_{n_1-1,n_2-1;\alpha}], \quad (11.35)$$

and for a two-sided test

$$P_{R_a} = P[F_{n_1-1,n_2-1} < (1/R_a^2)F_{n_1-1,n_2-1;\alpha/2}] + P[F_{n_1-1,n_2-1} > (1/R_a^2)F_{n_1-1,n_2-1;1-\alpha/2}] \quad (11.36)$$

For equal sample sizes, the power may be read directly from Appendix B.6. Entries are $\nu = n_1 - 1$, $R_a = \sigma_1/\sigma_2$ or $R_a^* = \sigma_2/\sigma_1$ (whichever is less than unity), and α for one-sided tests, $\alpha/2$ for two-sided tests. For two-sided tests, an error approaches a maximum of $\alpha/2$ as R_a approaches 1 but vanishes as P_{R_a} becomes greater than 0.30.

► ***Example 11-8***

For the test of equality of variances in Example 11-2, the power will be determined when $R_a = 0.4$, 0.8, 1.25, and 1.50. Since $n_1 = n_2 = 11$, we can use Appendix B.6 with $\nu = 10$, $\alpha/2 = 0.025$, $R_a = 0.4$, 0.8 and $R_a^* = 0.8$, 0.67. The values of power are read from Appendix B.6 and obtained from using (11.36); they are

	Power	
R_a	Appendix B.6	Formula (11.36)
0.4	0.78	0.78
0.8	0.095	0.095
1.25	0.095	0.100
1.50	0.21	0.231 ◄

*11.6.2 Power of the Student T Tests

For the T tests using statistics given in (11.15), Appendix B.1 gives the power. For a single mean, say μ_1, the appropriate quantities are α ($\alpha/2$ for two-sided tests), $\nu = n_1 + n_2 - 2$, and

$$\delta = \sqrt{n_1}|\mu_1' - \mu_1^*|/\sigma \tag{11.37}$$

where μ_1^* is the null value of μ_1 and μ_1' is a value of μ_1 under the alternative. A similar δ is obtained for a test about μ_2 by replacing the subscript 1 by 2 on the μ's and n.

In the case of testing two means, the appropriate quantities are α ($\alpha/2$ for two-sided tests), $\nu = n_1 + n_2 - 2$, and

$$\delta = \frac{|\delta_1 - \delta_0|}{\sigma\sqrt{1/n_1 + 1/n_2}} \tag{11.38}$$

where $\delta_1 = \mu_1' - \mu_2'$ is a difference of the means under the alternative and $\delta_0 = \mu_1^* - \mu_2^*$ is the difference of the means under the null.

For a paired data test (11.29), power is obtained from Appendix B.1 with α ($\alpha/2$ for two-sided tests), $\nu = n - 1$, and

$$\delta = \sqrt{n}|\delta_1 - \delta_0|/\sigma_D \tag{11.39}$$

where δ_0 and δ_1 are the same as in (11.38).

► ***Example 11-9***

For the test about the difference of means in Example 11-3, $\delta_0 = 10$. The power of this test will be calculated for $\delta_1 = \mu_2' - \mu_1' = 10 - 0.5\sigma$, $10 - \sigma$, and $10 - 2\sigma$. With $\alpha = 0.01$, $\nu = 20$, and

$$\delta = 0.5\sqrt{11/2} = 1.17$$
$$\delta = \sqrt{11/2} = 2.34$$
$$\delta = 2\sqrt{11/2} = 4.69$$

the values of power are found from Appendix B.1 to be

δ_1	power
$10 - 0.5\sigma$	0.075
$10 - \sigma$	0.32
$10 - 2\sigma$	0.95 ◄

▶ ***Example 11-10***

Using the test in Example 11-4, we find the power for $\delta_1 = \pm 0.5\sigma, \pm\sigma, \pm 1.5\sigma, \pm 2\sigma$. Here we use Appendix B.1 with $\alpha/2 = 0.025$, $\nu = 18$, and $\delta = \sqrt{5}|\delta_1|/\sigma$, since $\delta_0 = 0$. Thus,

δ_1	power
$\pm 0.5\sigma$	0.18
$\pm \sigma$	0.55
$\pm 1.5\sigma$	0.88
$\pm 2\sigma$	0.988 ◀

PROBLEMS

1. Statistics calculated in a two-population experiment were:

$$n_1 = 12 \qquad \bar{x}_{1\cdot} = 42.7 \qquad SS_1 = 900$$

$$n_2 = 12 \qquad \bar{x}_{2\cdot} = 54.3 \qquad SS_2 = 1212$$

(a) Test $H_0: \sigma_1^2 = \sigma_2^2$ versus $H_1: \sigma_1^2 \neq \sigma_2^2$ at the 5% level.
Based on the results of (a),
(b) give the standard error of each mean estimate.
(c) give the standard error of $\bar{X}_{1\cdot} - \bar{X}_{2\cdot}$.
(d) test, at the 1% level, $H_0: \mu_1 \geqslant \mu_2$ versus $H_1: \mu_1 < \mu_2$.
(e) What assumptions were required in (a)? in (b) and (c)?

2. Repeat Problem 1, using all statistics given except that $SS_2 = 3612$.

3. A two-population experiment gave the following information:

$$n_1 = 15 \qquad x_{1\cdot} = 312.9 \qquad SS_1 = 446.05$$

$$n_2 = 18 \qquad x_{2\cdot} = 437.1 \qquad SS_2 = 568.5$$

(a) Test $H_0: \sigma_1^2 = \sigma_2^2$ versus $H_1: \sigma_1^2 \neq \sigma_2^2$ at the 1% level.
Based on the results of (a),
(b) give the standard error of $\bar{X}_{1\cdot} - \bar{X}_{2\cdot}$.
(c) test $H_0: \mu_1 \geqslant \mu_2$ versus $H_1: \mu_1 < \mu_2$ at the 5% level.

4. In the comparison of two processes, the following information was recorded:

$$n_1 = 25 \qquad x_{1\cdot} = 420 \qquad \sum x_{1i}^2 = 7738$$

$$n_2 = 25 \qquad x_{2\cdot} = 495 \qquad \sum x_{2i}^2 = 10559$$

(a) Estimate the two population means.
(b) Estimate the two population variances.
(c) Do the population variances differ significantly? Let $\alpha = 0.05$.

(d) Estimate $\mu_2 - \mu_1$ and give the standard error of this estimator.
(e) Set a 90% *CI* for $\mu_2 - \mu_1$.
(f) Test the equality of means at the 1% level.
(g) Set a 95% *CI* for μ_1.

5. The following are calculations from a two-population experiment:

sample size	total	raw *SS*
$n_1 = 18$	$x_{1\cdot} = 540$	$\sum x_{1j}^2 = 18330$
$n_2 = 20$	$x_{2\cdot} = 500$	$\sum x_{2j}^2 = 14400$

(a) Estimate the two population means.
(b) Estimate the two population variances.
(c) Test $H_0: \sigma_1^2 = \sigma_2^2$ (against a two-sided alternative) at the 5% level.
(d) Give the standard error of each sample mean.
(e) Estimate $\mu_1 - \mu_2$ and give the standard error of this estimator.
(f) Set a 99% *CI* for $\mu_1 - \mu_2$.
(g) At the 5% level, is there evidence that the means are significantly different?
(h) Set a 95% *CI* for μ_2.
(i) Discuss assumptions required in the above.

6. In a human engineering study, two groups of ten people each were timed in performing a series of tasks. Group I was provided with music while group II was not. The results (in seconds) were:

person	with music (x_{1j})	without music (x_{2j})	$d_j = x_{1j} - x_{2j}$
1	22	24	2
2	26	23	−3
3	15	19	4
4	20	27	7
5	26	32	6
6	21	17	−4
7	17	21	4
8	15	18	3
9	18	22	4
10	23	26	3
Totals	203	229	26

$SS_1 = 148 \qquad \sum d_j^2 = 180$

$SS_2 = 188 \qquad d_{\cdot}^2/10 = 68$

(a) Is this a paired data experiment? Why or why not?
(b) Is there any reason to believe that the average performance time is greater when no music is provided? Let $\alpha = 0.05$.

7. A well-known television commercial claims that cars get more mileage (on the average) from gasoline with an additive than from gasoline without. When fifteen cars were selected at random, the mileages obtained for each car from 2 gallons without the additive and then from 2 gallons with the additive were as follows:

Car	with (x_{2j})	without (x_{1j})	$d_j = x_{2j} - x_{1j}$
Buick	19.6	19.1	0.5
Cadillac	18.3	18.5	−0.2
Chevrolet (C)	26.1	25.8	0.3
Chevrolet (I)	24.4	21.2	3.2
Chrysler	20.7	19.6	1.1
Dodge	23.3	22.4	0.9
Ford (G)	25.6	26.0	−0.4
Ford (M)	28.2	27.3	0.9
Hillman	32.7	34.1	−1.4
Mercury	27.7	25.5	2.2
Oldsmobile	21.1	20.8	0.4
Plymouth	25.6	24.4	1.2
Pontiac	20.9	21.2	−0.3
Rambler	28.6	26.8	1.8
Volkswagen	30.5	30.1	0.4
Totals	373.4	362.8	10.6

$SS_1 = 250 \qquad \sum d_j^2 = 25.50$

$SS_2 = 270 \qquad d_.^2/15 = 7.49$

(a) Is this a paired data experiment? Why or why not?
(b) Does the additive significantly increase gas mileage? Make the appropriate test at the 5% level.

8. The cellulose content was determined for each of 20 samples for each of two varieties of alfalfa hay. The results (in mg/g) were:

variety	average cellulose	raw SS
1	146.7	430,832
2	135.2	365,760

(a) Is there any evidence that these two varieties have different variance of measurements on the cellulose content? Let $\alpha = 0.01$.
(b) Is there any evidence that these two varieties have mean cellulose contents that differ? Let $\alpha = 0.05$.
(c) Set a 99% CI for $\mu_1 - \mu_2$.
(d) Discuss assumptions required.

9. To study the effect of relative humidity (RH) on the weight of newly formed pupae, Fraenkel and Blewett [4] used environments of 70% RH and 20% RH. A number of first-stage larvae were placed on the same diet at each of these RH with the following weights (in mg) recorded for newly formed pupae:

70% RH			20% RH		
22.5	26.5	25.5	19.0	21.3	21.2
24.2	23.5	31.0	18.0	18.3	20.6
24.5	25.5	24.0	14.5	18.0	17.1
29.2	27.0	27.8	18.5	18.5	19.5
27.5	23.5	23.5	19.0	18.8	
24.5	23.5		19.5	17.9	

(a) Test that the variance of weights is the same for the two RH. Let $\alpha = 0.05$.
(b) Estimate the difference in average weight of newly formed pupae under the two RH and give the standard error of this estimate.
(c) Is the average weight under 70% RH over 5 mg higher than the average weight under 20% RH? Do the appropriate analysis with $\alpha = 5\%$.

10. Two methods of calculating the area under cardiac output curves were examined by Steadham and Blackwell [8]. One was a standard method, the Stewart–Hamilton (S–H) method, while the second was a speedier new method. The areas, in centimeters, under 10 cardiac output curves were given as follows:

	Area	
Curve	S–H method	New method
1	38.9	41.9
2	33.4	33.0
3	25.5	25.0
4	28.4	27.8
5	26.8	26.1
6	24.9	24.8
7	29.0	25.0
8	20.3	19.4
9	32.6	30.6
10	27.5	28.2

Do the two methods give the same area, on the average? Do the appropriate analysis with $\alpha = 0.01$.

11. An experiment was conducted to study the effect, if any, of potassium on root growth of bean plants. Thirty seedlings were subjected to each of two media, one containing potassium, the other containing no potassium. The root growths, in millimeters, gave:

with potassium	without potassium
$n_1 = 30$	$n_2 = 30$
$x_{1\cdot} = 48.1$	$x_{2\cdot} = 41.7$
$SS_1 = 1.24$	$SS_2 = 0.93$

(a) Test $H_0: \sigma_1^2 = \sigma_2^2$ at the 1% level.
(b) Is there any evidence that potassium increases average root growth on beans? Make the appropriate test with $\alpha = 5\%$.

12. Two methods for determining the density of solid solutions are (1) by direct measuring and (2) calculated from an X-ray technique. Prior research with various solids has indicated that densities obtained by the X-ray method tend to be significantly lower than the densities obtained by the direct method. Ridley [5] determined the density of 19 idium–cadmium specimens by each method obtaining (in gm/cm^3):

Specimen	X-ray	Direct	Specimen	X-ray	Direct
1	7.287	7.285	11	7.304	7.304
2	7.292	7.291	12	7.304	7.302
3	7.295	7.295	13	7.306	7.306
4	7.301	7.299	14	7.305	7.304
5	7.301	7.300	15	7.306	7.305
6	7.303	7.301	16	7.309	7.310
7	7.302	7.301	17	7.312	7.311
8	7.304	7.303	18	7.313	7.313
9	7.303	7.304	19	7.317	7.318
10	7.303	7.300			

(a) Do the results of this experiment substantiate the findings of the prior research; that is, do the earlier statements apply to idium–cadmium solids? Do the appropriate analysis with $\alpha = 0.05$.
(b) Discuss the assumptions required in your analysis.

13. Two procedures for determination of amylase in human body fluids were studied by Smith and Roe [7]. The "original" method is considered to be an acceptable standard method. The "new" method uses a smaller volume of water making it more convenient as well as more economical. For 15 subjects, the following data resulted, in units per milliliter:

Subject	Original	New	Subject	Original	New
1	38	46	9	69	85
2	48	57	10	59	74
3	58	73	11	81	96
4	53	60	12	44	55
5	75	86	13	56	71
6	58	67	14	50	63
7	59	65	15	60	74
8	46	58			

It is claimed that amylase values obtained by the new method average at least 12 units greater than the corresponding values from the original method. Test the claim at the 1% level.

14. The amount of thiamine present in cooked pork roasts was determined for 15 roasts cooked at each of two temperatures. The results were summarized as follows, in micrograms per gram (μg/g):

Temperature	Thiamine present	Variance
1	$\bar{x}_{1\cdot} = 13.41$	$s_1^2 = 0.288$
2	$\bar{x}_{2\cdot} = 12.93$	$s_2^2 = 0.177$

(a) Do the populations have significantly different variances? Let $\alpha = 0.01$.
(b) Estimate $\mu_1 - \mu_2$ and give the standard error of this estimate.
(c) Set a 95% *CI* for $\mu_1 - \mu_2$.
(d) Test that the average thiamine content is the same for both temperatures. Let $\alpha = 0.02$.

15. Serum amylase values were determined for two groups of people (see [7]). Group I fasted overnight, while group II did not. Experimental results, in units per milliliter, gave:

Group I	Group II
$n_1 = 122$	$n_2 = 62$
$\bar{x}_{1\cdot} = 62.2$	$\bar{x}_{2\cdot} = 74.3$
$s_1 = 22.99$	$s_2 = 22.39$

Does fasting significantly lower the serum amylase value? Do the appropriate analysis with $\alpha = 0.01$.

***16.** Repeat Problem 6(b) using the T^* distribution.

***17.** Repeat Problem 9 using the T^* distribution.

***18.** Find the power of the test in Problem 5(g), if the difference of the means is actually
(a) 0.1σ (b) 0.2σ (c) 0.5σ (d) σ

***19.** Draw the power curve for the test in Problem 6(b).

***20.** Draw the power curve for the test in Problem 8(b).

***21.** Find the power of the test in Problem 9(c), if the difference of the means is actually
(a) $5 + 0.15\sigma$ (b) $5 + 0.25\sigma$ (c) $5 + 0.33\sigma$ (d) $5 + 0.50\sigma$

***22.** Draw the power curve for the test in Problem 10.

***23.** Draw the power curve for the test in Problem 1(a).

***24.** Find the power of the test in Problem 5(c) if
(a) $\sigma_1 = 1.2\sigma_2$ (b) $\sigma_1 = 1.5\sigma_2$ (c) $\sigma_1 = 2\sigma_2$ (d) $\sigma_1 = 3\sigma_2$

***25.** Find the power of the test in Problem 4(c) if
(a) $\sigma_1 = 1.25\sigma_2$ (b) $\sigma_1 = 0.5\sigma_2$ (c) $\sigma_1 = 1.5\sigma_2$ (d) $\sigma_1 = 2\sigma_2$

***26.** Draw the power curve for the test in Problem 8(a).

REFERENCES

[1] Aspin, A. A., "An Examination and Further Development of a Formula Arising in the Problem of Comparing Two Mean Values," *Biometrika*, **35** (1948).

[2] Aspin, A. A., "Tables for Use in Comparison Whose Accuracy Involves Two Variances, Separately Estimated," *Biometrika*, **36** (1949).

[3] Cochran, W. G., and G. E. Cox, *Experimental Designs*, 2nd ed., Wiley, New York, 1957.

[4] Fraenkel, G., and M. Blewett, "The Utilisation of Metabolic Water in Insects," *Bulletin of Entomological Research*, **35** (1944–45).

[5] Ridley, N., "Densities of Idium-cadmium Solid Solutions," *British Journal of App. Physics*, Series 2, **1** (1968).

[6] Scheffe, H., *The Analysis of Variance*, Wiley, New York, 1959.

[7] Smith, B. W., and J. H. Roe, "A Micromodification of the Smith and Roe Method for the Determination of Amylase in Body Fluids," *Journal of Biological Chemistry*, **227** (1957).

[8] Steadham, R. E., and L. H. Blackwell, "A New Method for Determination of the Area under a Cardiac Output Curve," *IEEE Transactions on Bio-Medical Engineering*, **BME-17**, 4 (1970).

[9] Welch, B. L., "The Significance of the Difference between Two Means When the Population Variances Are Unequal," *Biometrika*, **29** (1938).

[10] Welch, B. L., "The Generalization of 'Student's' Problem When Several Different Population Variances Are Involved," *Biometrika*, **34** (1947).

Chapter 12

REGRESSION ANALYSIS

A frequent problem for research workers is investigating relationships of two or more variables in order to predict certain responses when the values of all but one variable are available. In more specific terms, the problem is determining a model and finding a functional relation, so that the behavior of a dependent variable, say Y, can be described from values of one or more independent variables, say $X_1, X_2, \ldots$.

To gain some insight into the diversity of this subject, we consider some examples. A well-known law in physics says that an object having mass m and moving with a velocity v has kinetic energy (KE) given by $KE = \frac{1}{2}mv^2$. Thus, kinetic energy depends on the mass of the object and the square of its velocity. The interest, I, which is realized on an investment depends on the principle invested, P, the rate of return, R, and the duration or time, T. Simple interest is given by the formula $I = PRT$. An educator may wish to know how a student's grade in college English, Y, is related to X, his high school grade average. A chemist may be interested in predicting the yield, Y, of a chemical process given the amount, X, of catalyst, or the reaction temperature, or both.

Other examples of related variates are

(a) Y = resale value of car
X = age of the car

(b) Y = wear on a machine
X = hours of operation

(c) Y = number of neutrons emitted by a radioactive material
X = time elapsed

(d) $Y =$ student's grade in a statistics course
$X =$ student's grade in a prerequisite mathematics course

(e) $Y =$ resale value of a car
$X_1 =$ age of car
$X_2 =$ number of owners
$X_3 =$ number of cars of the same type available for resale.

12.1 GENERAL CONCEPTS

Independent variables, the X variables, are also called secondary, auxiliary, or concomitant variables; we will use *auxiliary* in this book. Putting forth additional time and effort to measure auxiliary variables is hopefully compensated for by better inference. An adjustment is made for the information contained by the auxiliary variable; that is, the influence which X tends to produce is incorporated in the inferential process.

While it may be easy to decide which variables might be related, it is more difficult to arrive at the exact relationship existing among them. In some cases the relationship is obtained from theoretical considerations or prior experience. For many situations, especially in areas where little research has been done, plotting the data might reveal a trend or relationship. As might be expected, cases involving only one independent X variable are the simplest in terms of both computation and interpretation.

Relationships are classified as *linear* or *nonlinear*, no matter how many independent variables are involved. The use of functional relationships for prediction purposes is termed *regression analysis.* In the remainder of this chapter, only *linear* relationships involving one independent X variable will be considered. This is the simplest type of functional relationship and is called *simple linear regression* or merely *the regression of Y on X*.

A linear regression analysis might be used even if it is known that the relationship between Y and X is definitely not linear. Should the experiment be restricted to a small range of X, a linear function may serve as a good approximation to the exact function; see Figure 12-1.

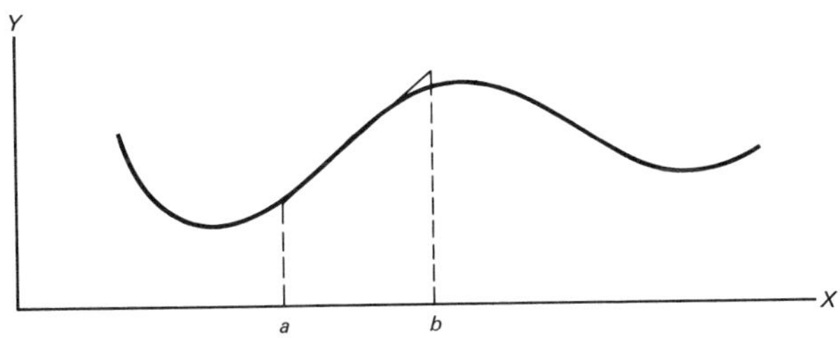

Figure 12-1 A Linear Approximation of a Nonlinear Relationship

Regression techniques were introduced by Karl Pearson in the early 1900's in his work of checking Galton's "law of universal regression" for inheritance. Galton stated that tall fathers tended to have tall sons and short fathers short sons, but Pearson's analysis gave evidence of just the opposite: fathers had sons whose heights "regressed" toward a norm.

A significant functional relation among a set of variables should not be interpreted as a "cause-and-effect" relationship. All that can be said is that predictions will be better (more precise) if the auxiliary information is taken into account. One reason for being so cautious is that the dependent and independent variables used in the investigation might be highly related to other variables which are in fact causing the experimental variable to change in the fashion exhibited by the data.

12.2 THE LINEAR REGRESSION MODEL AND ASSUMPTIONS

In simple linear regression, the points of the sample space are paired values (x_i, y_i) of the variables (X, Y). All points having a fixed X coordinate represent a population; the Y values of each such set have a distribution, a mean, a variance, and so on; see Figure 12-2. The population regression curve is the locus of points giving the Y population means for the various X values. In symbols, a linear regression curve would be represented by

(12.1) $$\mu_{Y|X} = \beta_0 + \beta_1 X$$

and is the straight line on which the Y population means, $\mu_{Y|X}$, are assumed to lie. Points on this line are denoted by

(12.2) $$\mu_{Y|x} = \beta_0 + \beta_1 x$$

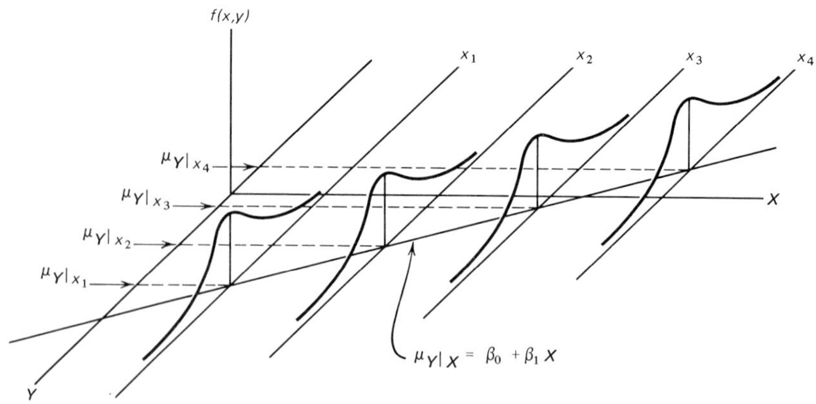

Figure 12-2 The Linear Regression of Y on X

where $\mu_{Y|x}$ denotes the mean value of the Y population specified by $X = x$; see Figure 12-2, where four Y populations and their means are displayed. On the right-hand side of (12.2), β_0 denotes the Y intercept, the mean value of Y when $X = 0$, and β_1 denotes the slope of the line, the change in the mean of Y per unit change in X. Under the linearity assumption, the points of the sample space can be represented by the *linear regression model*

$$y_i = \beta_0 + \beta_1 x_i + \varepsilon_i \tag{12.3}$$

where ε_i measures the vertical deviation of the point (x_i, y_i) from the population mean given by (12.2). Figure 12-3 gives an illustration of these deviations.

As seen from (12.1) or (12.3), there are two parameters associated with any linear regression function. Thus inference about the population regression lines reduces to inference about β_0 and β_1, or some function thereof. Various phases of inference in regression analyses depend upon the following assumptions:

(i) The linear model adequately describes the data.
(ii) The values of X are measured without error.
(iii) For $X = x$, there is a population of Y values which are normally and independently distributed about a mean of $\mu_{Y|x} = \beta_0 + \beta_1 x$ and with a variance of $\sigma^2_{Y|x}$; that is, $Y \sim \text{NID}(\beta_0 + \beta_1 x, \sigma^2_{Y|x})$ for all x.
(iv) $\sigma^2_{Y|x} = \sigma^2_\varepsilon$ for all x—all populations have the same variance.

Note that $\mu_{Y|x}$ and $\sigma^2_{Y|x}$ are really *conditional means* and *variances*; they are, respectively, the mean and variance of the Y population "given that X has the value x." It should be emphasized that only the first two assumptions and independence of the Y's are necessary in calculating the estimated regression

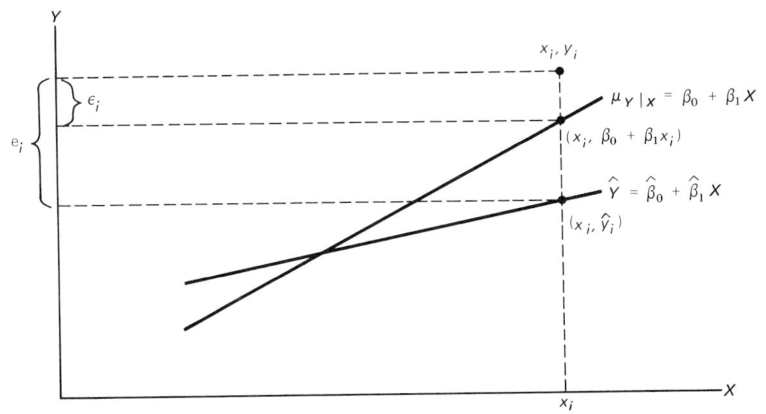

Figure 12-3 Deviations from Population and Sample Regressions

line. To obtain valid inferences from the usual sampling distributions, all assumptions will be required.

We might make another point about the regression assumptions. It is not necessary to have fixed X values; that is, regression may be used for predictive purposes even when X and Y have a joint distribution. All that is required is that the stated assumptions hold for the specific set of X values under investigation. Thus, in a study involving ages and weights of persons, age could be reasonably taken as a predictive variable. When dealing with heights and weights of people, however, one may wish to use X = height as a predictor of weight even though X and Y most likely have a joint distribution.

12.3 LEAST-SQUARES ESTIMATION

One of the most important problems in a regression study is the estimation of mean values of various Y populations. This is accomplished through a sample regression line which would be estimated from a random sample of paired observations: (x_i, y_i), for $i = 1, 2, \ldots, n$. If the sample points do not lie on a straight line, there are many straight lines that can be chosen to estimate (12.1). Merely drawing a line through a set of plotted data has the drawback that two attempts are almost certain to give two different lines.

Of all the lines that could be drawn through a set of sample points, there is only one for which "the sum of squares of vertical deviations of the y_i from the fitted line" is a minimum. The linear equation having this minimum sum of squares is called the *Least-Squares* (LS) equation and is denoted by

$$\hat{Y} = \hat{\beta}_0 + \hat{\beta}_1 X \tag{12.4}$$

Using $\hat{Y}$ in place of $\hat{\mu}_{Y|X}$ for the point estimator of $\mu_{Y|X}$ is a well-established practice. When $X = x$, the point on the line would be labeled as

$$\hat{y} = \hat{\beta}_0 + \hat{\beta}_1 x \tag{12.5}$$

Most, if not all, sample points will deviate from the LS line; therefore, the points (x_i, y_i) of the sample can be represented by

$$y_i = \hat{\beta}_0 + \hat{\beta}_1 x_i + e_i = \hat{y}_i + e_i \tag{12.6}$$

where e_i is the vertical deviation of the point from the *fitted* line; these deviations also are illustrated in Figure 12-3. From the definition of the LS line, it follows that the quantity to be minimized is

$$\sum_{i=1}^{n} e_i^2 = \sum_{i=1}^{n} (y_i - \hat{\beta}_0 - \hat{\beta}_1 x_i)^2 \tag{12.7}$$

and is called the *residual sum of squares*. Minimizing the expression in (12.7) is a problem in the calculus, performed by taking derivatives and equating to 0. The system of equations obtained from this minimization process is referred to as the *normal equations* (though they have nothing whatsoever to do with the normal family):

(12.8)
$$n\hat{\beta}_0 + \hat{\beta}_1 x. = y.$$
$$\hat{\beta}_0 x. + \hat{\beta}_1 \sum x_i^2 = \sum x_i y_i$$

The solution to system (12.8) is

(12.9)
$$\hat{\beta}_0 = \bar{y} - \hat{\beta}_1 \bar{x}$$
$$\hat{\beta}_1 = \frac{\sum (x_i - \bar{x})(y_i - \bar{y})}{\sum (x_i - \bar{x})^2} = \frac{CP_{XY}}{SS_X}$$

where CP stands for *cross-products*. Expanding the terms in the second equation of (12.9), we obtain a form which is easier to use computationally†:

(12.10)
$$\hat{\beta}_1 = \frac{\sum x_i y_i - x.y./n}{\sum x_i^2 - x.^2/n}$$

It can be shown that the least squares estimators of β_0 and β_1 are minimum variance, unbiased, and consistent under the assumptions given.

► ***Example 12-1***

For a simple illustration of some regression calculations, consider the following data:

x_i	y_i	x_i^2	$x_i y_i$	y_i^2
1	3	1	3	9
2	4	4	8	16
3	4	9	12	16
4	5	16	20	25
5	6	25	30	36
$x. = 15$	$y. = 22$	$\sum x_i^2 = 55$	$\sum x_i y_i = 73$	$\sum y_i^2 = 102$
$\bar{x} = 3$	$\bar{y} = 4.4$			

$$\hat{\beta}_1 = \frac{73 - 15(22)/5}{55 - 15(15)/5} = 7/10 \quad \text{and} \quad \hat{\beta}_0 = 4.4 - 0.7(3) = 2.3$$

The LS regression line is: $\hat{Y} = 2.3 + 0.7\,X$ ◄

† Some authors use lower-case letters in regression settings to represent deviations from the mean. Thus, other forms of (12.9) and (12.10) will be encountered in the statistical literature.

▶ ***Example 12-2***

In a study of the regression of $Y =$ college English grades, on the variable $X =$ high school grade average, a sample of 100 persons who had completed freshmen college English revealed:

$$x. = 272 \qquad \sum x_i^2 = 843.20$$

$$y. = 256 \qquad \sum y_i^2 = 726.37$$

$$\sum x_i y_i = 760.40$$

$$SS_X = 843.20 - (272)^2/100 = 103.36$$

$$SS_Y = 726.37 - (256)^2/100 = 71.01$$

$$CP_{XY} = 760.40 - 256(272)/100 = 64.08$$

$$\hat{\beta}_1 = 64.08/103.36 = 0.62$$

$$\hat{\beta}_0 = 2.56 - 0.62(2.72) = 0.87$$

The LS equation is: $\hat{Y} = 0.87 + 0.62\,X$ ◀

(Problems 1(a), 3(a), 5(a), (b), 7(a), (b))

12.4 INFERENCE PROBLEMS IN REGRESSION

We have already discussed one method of obtaining point estimates of the two parameters in a linear regression model. A predicting equation, the LS regression line, is formed from these estimates. As indicated earlier, the mean values of Y populations are estimated through the LS line.

In addition to the inference problems just mentioned, some of the other important phases of a regression study are:

(a) prediction of individual performances,
(b) estimation of the regression variance and standard errors of various estimators,
(c) tests and confidence intervals about the β's, mean values, and the like,
(d) adjustment of Y values, and
(e) allocation of X values.

Additional problems are encountered when certain of the regression assumptions seem questionable. The experimenter might formulate hypotheses to test (i) the linearity of the model, (ii) equality of variances, (iii) equality of slopes for similar experiments, and so on. Most of these are specialized problems in that they require a sufficient amount of data; refer to [2], [5], and [7].

12.5 THE REGRESSION VARIANCE

Referring to Figure 12-2 and the regression assumptions, we see that the common population variance, σ_ε^2, is actually a measure of the dispersion of Y values about the true regression line. Thus σ_ε^2 is an internal or within population variability. The closer the points are to the line, the smaller is σ_ε^2 and the stronger is the relationship.

Note that the variability about the regression line, σ_ε^2, is not the same thing as σ_Y^2, which measures the dispersion of all Y values about their mean, μ_Y. We might call the quantity σ_Y^2 the *total variability* in Y, since it reflects the dispersion if the regression function is ignored. In light of a significant regression, the variability measure σ_ε^2 is smaller than the variability σ_Y^2.

Figure 12-4 illustrates the composition of $y_i - \bar{y}$, the total deviation of y_i from the sample mean, $\bar{y}$. It is seen that $y_i - \bar{y}$ can be partitioned into two meaningful components; namely, $\hat{y}_i - \bar{y}$, which represents that portion attributable to the regression function, and $y_i - \hat{y}_i = e_i$, which represents the deviation of y_i from the regression line. The sum of squares of the residuals e_i is used in estimating σ_ε^2. An unbiased estimator of σ_ε^2 is the *regression variance*, denoted by S_e^2 and having values

(12.11) $$s_e^2 = \frac{1}{n-2}\sum (y_i - \hat{y}_i)^2 = \frac{1}{n-2}\sum e_i^2$$

where $(n - 2)$ are the degrees of freedom of the residual SS.

A calculating form for $\sum e_i^2$ can be obtained by adjusting the total SS for Y. Since

(12.12) $$y_i - \bar{y} = (y_i - \hat{y}_i) + (\hat{y}_i - \bar{y}),$$

it follows that

(12.13) $$SS_Y = \sum (y_i - \bar{y})^2 = \sum [(y_i - \hat{y}_i) + (\hat{y}_i - \bar{y})]^2,$$

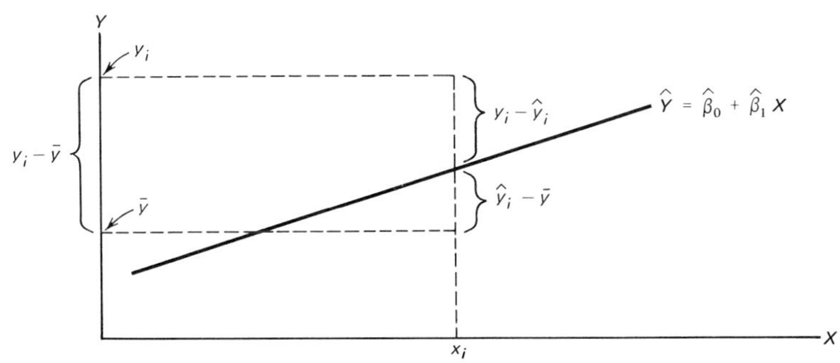

Figure 12-4 Partitioning the Total Deviation of Y

which, upon squaring and simplifying, becomes

(12.14) $$SS_Y = \sum (y_i - \hat{y}_i)^2 + \sum (\hat{y}_i - \bar{y})^2$$

The first term on the right-hand side of (12.14) is $\sum e_i^2$, while the remaining term, called the *SS due to regression*, can be written in the following ways:

(12.15)
$$\begin{aligned}\sum (\hat{y}_i - \bar{y})^2 &= \hat{\beta}_1 CP_{XY} \\ &= \hat{\beta}_1 \sum (x_i - \bar{x})(y_i - \bar{y}) \\ &= [\sum (x_i - \bar{x})(y_i - \bar{y})]^2 / \sum (x_i - \bar{x})^2 \\ &= (CP_{XY})^2/SS_X\end{aligned}$$

Thus the residual *SS* has a simple calculating form

(12.16) $$\sum e_i^2 = SS_Y - (CP_{XY})^2/SS_X$$

As indicated earlier, the total variability in *Y*, as measured by SS_Y, is adjusted by removing the variability due to regression. Equation (12.16) reflects this adjustment.

► ***Example 12-3***

Using the data in Example 12-1, we find

$$SS_Y = 102 - (22)^2/5 = 5.2$$

$$SS_X = 55 - (15)^2/5 = 10$$

$$CP_{XY} = 73 - (22)(15) = 7$$

whereby the residual *SS* is

$$\sum e_i^2 = 5.2 - (7)^2/10 = 0.3$$

and the regression variance is

$$s_e^2 = 0.3/3 = 0.1$$

In the absence of regression, the variability in *Y* would be measured by

$$s_Y^2 = 5.2/4 = 1.3$$ ◄

► ***Example 12-4***

Referring to the high school and college English grades of Example 12-2, the residual *SS* is

$$\sum e_i^2 = 71.01 - (64.08)^2/103.36$$

$$= 71.01 - 39.73 = 31.28$$

which gives a regression variance of

$$s_e^2 = 31.28/98 = 0.32 \blacktriangleleft$$

(Problems 1(b), 3(b), 5(c), 7(c))

12.6 *INFERENCE ABOUT* $\boldsymbol{\beta_0}$ *AND* $\boldsymbol{\beta_1}$

Unbiased and minimum variance properties of $\hat{\beta}_0$ and $\hat{\beta}_1$ have already been stated. From the regression assumptions stated earlier, it can be shown that $\hat{\beta}_0$ and $\hat{\beta}_1$ have the following variances:

$$\text{(12.17)} \qquad \begin{aligned} \sigma_{\hat{\beta}_0}^2 &= \sigma_\varepsilon^2[1/n + \bar{x}^2/SS_X] \\ \sigma_{\hat{\beta}_1}^2 &= \sigma_\varepsilon^2/SS_X \end{aligned}$$

In view of the unbiased estimation of σ_ε^2, these have the following unbiased estimators:

$$\text{(12.18)} \qquad \begin{aligned} S_{\hat{\beta}_0}^2 &= S_e^2[1/n + \bar{x}^2/SS_X] \\ S_{\hat{\beta}_1}^2 &= S_e^2/SS_X \end{aligned}$$

It is fairly easy to show that $\hat{\beta}_0$ and $\hat{\beta}_1$ are each linear functions of the Y values. Specifically

$$\text{(12.19)} \qquad \begin{aligned} \hat{\beta}_0 &= \sum d_i y_i \\ \hat{\beta}_1 &= \sum c_i y_i \end{aligned}$$

where $c_i = (x_i - \bar{x})/SS_X$ and $d_i = 1/n - \bar{x}c_i$. Combining the unbiasedness with (12.17) and Theorem 8-2, we have the following:

$$\text{(12.20)} \qquad \begin{aligned} \hat{\beta}_0 &\sim \text{N}(\beta_0, \sigma_{\hat{\beta}_0}^2) \\ \hat{\beta}_1 &\sim \text{N}(\beta_1, \sigma_{\hat{\beta}_1}^2) \end{aligned}$$

Inference in the form of tests or CI can therefore be made by the Student T distribution in the usual manner. With respect to β_1, tests would have structure as follows:

(12.21)

① ⓐ $H_0: \beta_1 \leqslant \beta_1^*$; $H_1: \beta_1 > \beta_1^*$ ① ⓑ $H_0: \beta_1 \geqslant \beta_1^*$; $H_1: \beta_1 < \beta_1^*$ ① ⓒ $H_0: \beta_1 = \beta_1^*$; $H_1: \beta_1 \neq \beta_1^*$

② Under H_0, $\dfrac{\hat{\beta}_1 - \beta_1^*}{S_{\hat{\beta}_1}} = T_{n-2}$

③ Reject H_0 if
ⓐ $T_{n-2} \geqslant t_{n-2;1-\alpha}$ ⓑ $T_{n-2} \leqslant t_{n-2;\alpha}$
ⓒ $|T_{n-2}| \geqslant t_{n-2;1-\alpha/2}$

The asterisk denotes a null value here, to avoid confusion with β_0.

A 100(1 − α)% *CI* for β_1 would be specified by the limits

$$L, U = \hat{\beta}_1 \pm s_{\hat{\beta}_1} t_{n-2;1-\alpha/2} \tag{12.22}$$

Replacing the subscripts throughout (12.21) and (12.22) would give a similar set of formulas and procedures for inference about β_0.

► ***Example 12-5***

Using the data on high school and college English grades from Examples 12-2 and 12-4, we obtain

$$s^2_{\hat{\beta}_1} = 0.32/103.36 = 0.0031;\ s_{\hat{\beta}_1} = 0.056$$

$$s^2_{\hat{\beta}_0} = 0.32[1/100 + (2.72)^2/103.36)]$$

$$= 0.32[0.01 + 0.07] = 0.0256;\ s_{\hat{\beta}_0} = 0.16$$

A 99% *CI* for β_0 is specified by

$$L, U = 0.87 \pm 0.16(2.364)$$

$$= 0.87 \pm 0.38 = 0.49;\ 1.25$$

A 5% test that the regression slope is no more than 0.5 is:

① $H_0: \beta_1 \leqslant 0.5$ versus $H_1: \beta_1 > 0.5$

② Under H_0, $T_{98} = \dfrac{\hat{\beta}_1 - 0.5}{S_{\hat{\beta}_1}}$

③ Reject H_0 if $T_{98} \geqslant 1.6602$

④ Since $\hat{\beta}_1 = 0.62$,

$$T_{98} = \frac{0.62 - 0.5}{0.056} = 2.14$$

Therefore, H_0 is rejected.

⑤ The regression slope β_1 exceeds 0.5 unless a 5% error has been made. ◄

(Problems 1(c), 3(c), 7(e), 9(c))

12.7 INFERENCE ABOUT MEAN VALUES AND INDIVIDUAL PERFORMANCES

While the most important use of the LS equation is estimating mean values for certain *Y* populations, a secondary application is predicting what an individual *Y* value might be. For example, in the grade-point data of Example 12-2, one adviser may wish to estimate the mean or average performance in college English of students having a high school average equal to 2.80, while another adviser may not be interested in the average for all such students but, rather, the performance of *one* advisee who has a high school average of 2.80.

In general terms, the two problems can be stated as follows:

(i) Estimating $\mu_{Y|x}$, the mean value of Y when $X = x$—these estimates will be denoted by $\hat{y}$, or by $\hat{y}(x)$ if there is a need for more specific identification.

(ii) Predicting a single value of Y when $X = x$—these estimates will be denoted by y, or by $y(x)$ if there is a need for more specific identification.

The two point estimates are obtained from the LS equation and coincide for a fixed value of X; that is,

$$\hat{y}(x) = y(x) = \hat{\beta}_0 + \hat{\beta}_1 x, \tag{12.23}$$

which is not surprising since any individual observation is *expected* to occur at the mean value. Of course, a single value is less stable (more variable) than a mean value. The appropriate variances are:

$$\begin{aligned} \sigma^2_{\hat{y}(x)} &= \sigma^2_\varepsilon[1/n + (x - \bar{x})^2/SS_X] \\ \sigma^2_{y(x)} &= \sigma^2_\varepsilon[1 + 1/n + (x - \bar{x})^2/SS_X] = \sigma^2_{\hat{y}(x)} + \sigma^2_\varepsilon \end{aligned} \tag{12.24}$$

Unbiased estimators of these variances are, respectively:

$$\begin{aligned} S^2_{\hat{y}(x)} &= S^2_e[1/n + (x - \bar{x})^2/SS_X] \\ S^2_{y(x)} &= S^2_e[1 + 1/n + (x - \bar{x})^2/SS_X] \end{aligned} \tag{12.25}$$

Having established that $\hat{\beta}_0$ and $\hat{\beta}_1$ are linear functions of the y_i, we see from (12.23) that the same is true of $\hat{Y}(x)$ and $Y(x)$. Consequently,

$$T_{n-2} = \frac{\hat{Y}(x) - (\beta_0 - \beta_1 x)}{S_{\hat{y}(x)}} \tag{12.26}$$

and

$$T_{n-2} = \frac{Y(x) - (\beta_0 - \beta_1 x)}{S_{y(x)}} \tag{12.27}$$

under the regression assumptions. Tests and *CI* for means follow in the usual manner via the Student T distribution with $n - 2$ degrees of freedom. Tests about individual performances are not made but "prediction intervals" can be made from (12.27). Note that *CI* and prediction intervals are intervals on points, not intervals on the regression line.

► ***Example 12-6***

For the grade-point data in Examples 12-2, 4, and 5, the mean English grade for all students having a 2.0 high school average would be estimated as

$$\hat{y}(2.0) = 0.87 + 0.62(2.0) = 2.11$$

with the following variance and standard error:

$$s^2_{\hat{y}(2.0)} = 0.32[1/100 + (2 - 2.72)^2/103.36]$$
$$= 0.32(0.0105) = 0.0034$$
$$s_{\hat{y}(2.0)} = 0.058$$

The English grade of one student having a 3.0 high school average would be predicted as

$$y(3.0) = 0.87 + 0.62(3) = 2.73$$

with the following variance and standard error:

$$s^2_{y(3)} = 0.32[1 + 1/100 + (3 - 2.72)^2/103.36]$$
$$= 0.32(1.0103) = 0.3233$$
$$s_{y(3)} = 0.57$$

A 95% *CI* for the mean English grade for all students having a 2.0 high school average is

$$\text{L, U} = 2.11 \pm 0.058(1.984)$$
$$= 2.11 \pm 0.12$$
$$= 1.99; 2.23$$

A 95% prediction interval for the English grade of one student having a 3.0 high school average is

$$\text{L, U} = 2.73 \pm 0.57(1.984)$$
$$= 2.73 \pm 1.13$$
$$= 1.60; 3.86$$ ◄

(Problems 9(d), (e), 11(d), (e))

12.8 THE UNCERTAINTIES OF EXTRAPOLATION

Let I denote the interval of X values with endpoints equal to the smallest and largest x appearing in the regression study. For any Y populations defined by an x belonging to I, inference about the mean and individual performances would be accomplished as outlined in the preceding section.

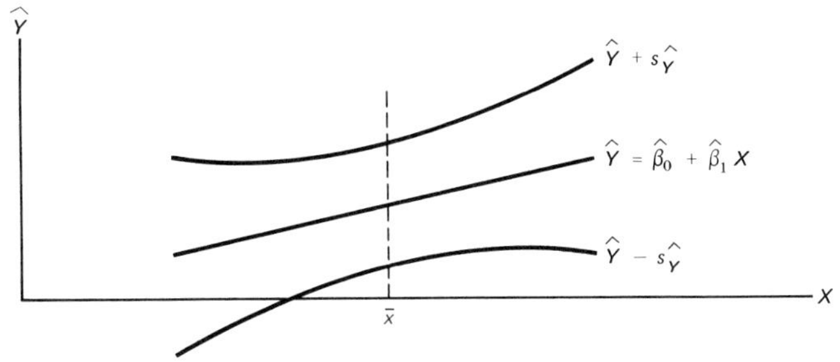

Figure 12-5 Least-Squares Line and Standard Error Bands

On the other hand, for an x outside I, predictions about Y require *extrapolation*, that is, extending the relationship outside the realm of the study. This practice is generally hazardous, since the same relationship cannot be guaranteed outside the experimental range, most obviously when using a linear model to approximate a curvilinear relation.

Now let us look at results of extrapolation with reference to reliability. Restricting discussion to the linear case, suppose we are willing to assume that the line extends linearly outside the experimental range. It is seen from (12.24) or (12.25) that predictions become more variable as x deviates more from $\bar{x}$. Consequently, predictions at the extremes of the experimental range become highly uncertain; those outside get progressively worse. Figure 12-5 shows the standard error bands about an LS line—this is not a *CI* about the line. We see that the least variable predictions occur at $X = \bar{x}$.

12.9 OTHER PHASES OF REGRESSION

The subdivisions of this section will treat topics of adjustment, allocation, and the coefficient of determination.

12.9.1 A Coefficient of Determination

Section 12.4 indicated that a small value of σ_ε^2 or its estimator, s_e^2, corresponds to a strong linear relationship. These quantities are small, in relation to the total variability in Y, when the regression function accounts for a large proportion of the total variation in Y. It seems reasonable then to measure the strength of a regression relationship by the following ratio:

$$(12.28) \qquad r^2 = \frac{SS \text{ due to regression}}{SS \text{ of } Y \text{ ignoring regression}}$$

Called the *coefficient of determination*, this ratio "determines" the proportion of the Y variability which the regression function explains. Clearly, $0 \leqslant r^2 \leqslant 1$, since SS cannot be negative nor can the numerator in (12.28) exceed the denominator. For other than simple linear regression, the customary symbol is R^2.

For simple linear regression, the coefficient of determination is:

$$r^2 = \hat{\beta}_1 CP_{XY}/SS_Y = (CP_{XY})^2/SS_X SS_Y \tag{12.29}$$

► ***Example 12-7***

The coefficient of determination for the regression in Example 12-1 is

$$r^2 = (7)^2/10(5.2) = 49/52 = 0.94$$

and in Example 12-2, the coefficient is

$$r^2 = (64.08)^2/103.36(71.01) = 0.56$$ ◄

(Problems 1(d), 3(d), 5(d))

*12.9.2 Adjustment of *Y* Values

For many regression studies, values of the independent variable(s) can be controlled by the researcher. Physical science experiments where X represents temperature, pressure, and so on, are examples of this type. Biological and social experiments using X as amounts of vitamin, time, stimulus, and so on, also fall into the controlled variable category.

Control of the independent variable is not possible or realistic in many other cases. If X represents the number of plants which reach maturity on experimental plots, it is usually difficult to have a uniform stand on all plots because some plants fail to germinate, others wash out, some weak plants do not mature, and so on. Obviously, differences in the yield, Y, per plot will be reflected by differences in X. With a regression analysis, the yields can be adjusted to a common X value.

In general, any Y values of a regression study can be placed on a common basis by an adjustment. All points to be adjusted are moved parallel to the regression equation until they coincide with the vertical line $X = x_a$, where x_a is the common basis point, the point of adjustment. Quite often, x_a is taken to be $\bar{x}$. Figure 12-6 illustrates the technique of adjusting.

By adjusting, the effect due to X is eliminated, whereby adjusted Y values may be readily compared. The adjusting equation can easily be derived from Figure 12-6; it is:

$$\text{Adj } y_i = y_i - \hat{\beta}_1(x_i - x_a) \tag{12.30}$$

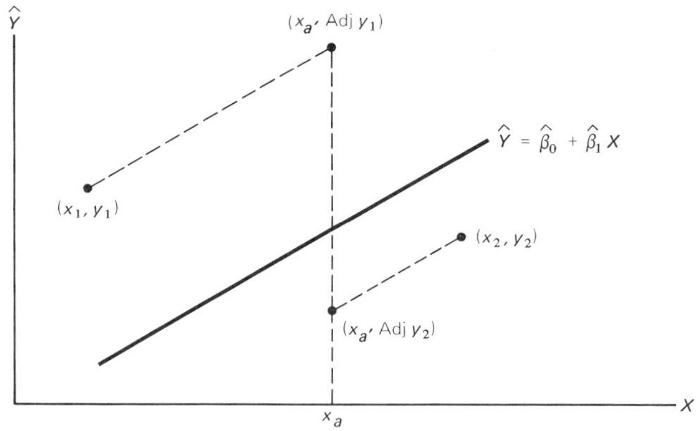

Figure 12-6 Regression Adjustment

► ***Example 12-8***

A young man about to graduate from college is fortunate in obtaining eight job offers. The positions are nearly identical with respect to fringe benefits and type of work, but differences in location and cost of living result in quite diverse salaries. Suppose location is of minor consequence; the deciding factor will be the amount of money that can be put into savings.

To arrive at a savings figure for each job, a cost-of-living index is obtained for each of the eight cities. The salaries and indices are:

City	Salary (Y)	Index (X)	Adjusted Salary
1	\$9,100	125.5	\$9,150
2	8,200	118.3	8,993
3	9,000	129.6	8,628
4	9,800	130.4	9,345
5	8,500	124.7	8,633
6	9,450	128.6	9,181
7	8,750	119.4	9,429
8	9,875	131.4	9,317

The LS equation for the regression of salaries on the index is

$$\hat{Y} = -3904.9 + 103.1X,$$

and the adjusted salaries, found in the last column above, are given by

$$\text{Adj } y_i = y_i - 103.1(x_i - 126)$$

Since the highest adjusted salary is \$9,429, a salary of \$8,750 at a city with index of 119.4 would be best, being comparable to a salary of $9,429 where the cost-of-living index is 126, and so on. ◀

(Problcm 13)

*12.9.3 Allocation of X Values

If one is free to choose which X values to use in a linear regression study, how should the choice be made in order to get the best regression line ("best" in the sense of minimum variance)? When estimation of the slope is most important, it is seen from (12.18) that we need to maximize SS_X. This can be accomplished by taking all X values at the extremes (exactly half or $\frac{n+1}{2}$ at one extreme, the remaining at the other extreme).

To estimate β_0 with minimum variance, the x's should be chosen so that $\bar{x}$ is 0 or as small as possible; see also (12.18). They cannot all be taken at the minimum value, since at least two different x's are needed to fit a line. This, then, is nearly the opposite of the allocation required for the best estimation of β_1, unless negative x's are possible. Symmetrically placing the x's about zero will result in the best estimation of both β_0 and β_1.

Another point needs to be taken into account. If the experiment covers a large range of X, one should use X values at the extremes and at several intermediate points. Thus, a prominent nonlinear relation would be more readily detected when many different x's are used than in studies using only two or a small number of x's.

*12.10 POWER OF TESTS

Since the only tests discussed in this chapter were Student T tests, all power calculations will follow the procedure outlined in Section 10.5. These probabilities can be easily obtained from Appendix B.1. The discussion here will be limited to power of tests about β_0, β_1, and $\mu_{Y|x}$. Tests about individual performances are not made.

Recall that power of a Student T test can be read from Appendix B.1 when the quantity $\delta = |\Delta_a|/\sigma_{\hat{\theta}}$ is specified. Here, Δ_a must be some multiple of σ_ε. Let β_0^*, β_1^*, and $\mu_{Y|x}^*$ denote null values of tests about β_0, β_1, and $\mu_{Y|x}$, respectively. Further, let β_0', β_1', and $\mu'_{Y|x}$ denote values specified by the alternative when making a test about β_0, β_1, and $\mu_{Y|x}$, respectively. Then, the required quantities are:

$$\delta = \frac{|\beta_0' - \beta_0^*|}{\sigma_\varepsilon \sqrt{1/n + \bar{x}^2/SS_X}} \tag{12.31}$$

$$\delta = \frac{|\beta_1' - \beta_1^*|}{\sigma_\varepsilon} \sqrt{SS_X} \tag{12.32}$$

$$\delta = \frac{|\mu'_{Y|x} - \mu^*_{Y|x}|}{\sigma_\varepsilon \sqrt{1/n + (x - \bar{x})^2/SS_X}} \tag{12.33}$$

► ***Example 12-9***

For the test of β_1 given in Example 12-5, we find the power at the alternatives $\beta_1' = 0.5 + 0.1\sigma_\varepsilon$; $\beta_1' = 0.5 + 0.2\sigma_\varepsilon$; $\beta_1' = 0.5 + 0.5\sigma_\varepsilon$. For these values,

$$\delta = 0.1\sqrt{103.36} = 1.02$$

$$\delta = 0.2(10.2) = 2.04$$

$$\delta = 0.5(10.2) = 5.10$$

The corresponding values of power can then be read from Appendix B.1:

$$P_{0.5+0.1\sigma_\varepsilon} = 0.06$$

$$P_{0.5+0.2\sigma_\varepsilon} = 0.28$$

$$P_{0.5+0.5\sigma_\varepsilon} = 0.99 \blacktriangleleft$$

(Problems 15, 17)

PROBLEMS

1. The following statistics are given:

$$n = 16 \qquad x. = 48 \qquad y. = 128$$

$$SS_X = 150 \qquad SS_Y = 1630 \qquad CP_{XY} = 450$$

(a) Find the LS predicting equation for the linear regression of Y on X.
(b) Give the regression variance.
(c) Estimate the variance of $\hat{\beta}_1$.
(d) Evaluate the coefficient of determination.

2. In a regression study, the following values were recorded:

$$n = 25 \qquad x. = 300 \qquad y. = 175$$

$$CP_{XY} = 200 \qquad SS_X = 100 \qquad SS_Y = 975$$

(a) Find the LS predicting equation for the linear regression of Y on X.
(b) Calculate the regression variance.
(c) Estimate the variance of $\hat{\beta}_1$.
(d) Determine the coefficient of determination.

3. Regression calculations were:

$$n = 18 \qquad x. = 180 \qquad y. = 270$$

$$\sum x_i y_i = 2780 \qquad \sum x_i^2 = 1840 \qquad \sum y_i^2 = 4370$$

(a) Find the LS linear regression equation.
(b) Determine the regression variance.
(c) Give the standard error of the slope.
(d) Evaluate the coefficient of determination.

4. Given the following calculations:

$$n = 20 \qquad \sum x_i = 120 \qquad \sum y_i = 330$$

$$\sum x_i y_i = 2340 \qquad \sum x_i^2 = 780 \qquad \sum y_i^2 = 9645$$

(a) Find the LS predicting equation for the linear regression of Y on X.
(b) Give the regression variance.
(c) Estimate the standard error of $\hat{\beta}_1$.
(d) Evaluate the coefficient of determination.

5. Take the following to be data from a regression experiment (Y being the dependent variable):

x	y	x	y
0	2.1	3	4.0
0	2.4	4	4.2
1	2.8	4	4.1
1	2.3	5	4.5
2	3.0	6	5.0
3	3.5	6	4.5

(a) Plot the data on graph paper and make a "guesstimate" of the LS line.
(b) Calculate the LS line and compare with your guesstimate.
(c) Give the regression variance.
(d) What proportion of the variability in Y is explained by the linear regression function found in (b)?

6. Data recorded in a regression study were (Y being the dependent variable):

x	y	x	y
1	2	3	5
4	8	5	7
2	6	6	7
5	6	4	8
8	10	4	6

(a) Plot the data on graph paper and make a "guesstimate" of the LS line.
(b) Calculate the equation of the LS line and compare with your guesstimate.
(c) Give the regression variance.
(d) What proportion of the variability in Y is explained by the linear regression function found in (b)?

7. To study the effect of biotin on the weight of turkeys, Dobson [1] placed sixty newly hatched turkeys in groups of six each. All groups received the same diet and care except for the level of biotin added per kilogram of feed. After three weeks, the average weight, Y, in grams, per group was as follows:

Biotin level	Average weight
0	206
25	222
50	225
75	296
100	324
125	375
150	363
175	408
200	407
225	434

(a) Plot the data to investigate the type of relationship between the average three-week weight and level of biotin added.
(b) Approximate the true relationship by a linear regression function.
(c) Determine the regression variance.
(d) Evaluate the coefficient of determination.
(e) Is there evidence of a positive slope? Let $\alpha = 0.05$.

8. In the production of a certain chemical, the yield, Y, can be increased by increasing the amount of a particular catalytic agent used. Twenty trials were made with different amounts of the catalyst. Analysis of the yields, measured in grams, and the amounts, X, measured in milligrams, revealed:

$$n = 20 \qquad x. = 194 \qquad y. = 2767$$

$$\sum x_i y_i = 28{,}210 \qquad \sum x_i^2 = 2385 \qquad \sum y_i^2 = 388{,}522$$

(a) Find the LS line from which the yield, Y, can be predicted knowing the amount of catalyst used.
(b) Calculate the regression variance.
(c) Does the slope differ significantly from zero? Let $\alpha = 0.05$.
(d) Predict the yield of chemical if no catalyst is used and estimate its standard error.
(e) Give a 95% interval estimate of the mean yield of chemical produced when $X = 10$ milligrams of catalyst is used.
(f) Discuss assumptions required in (c), (d), and (e).

9. A physical quantity related to internal stresses in ferromagnetic material is the so-called magnetomechanical damping peak height, denoted here by Y. Roberts and Barrand [6] studied the behavior of Y for nickel-copper alloys at different Centigrade temperatures, X, and obtained:

x	y	x	y	x	y
−92	0.375	23	0.224	125	0.081
−50	0.331	40	0.216	162	0.056
−40	0.300	52	0.204	176	0.035
−3	0.251	78	0.146		

(a) Find the LS line for the linear regression of Y on X.
(b) Evaluate the coefficient of determination.
(c) Is the slope significantly different from zero? Let $\alpha = 0.01$.
(d) Give a 95% interval estimate of the mean of Y when $X = 0°$ Centigrade. Do likewise for $X = 100°$.
(e) Predict the value of Y when $X = 20°$ and estimate its standard error.

10. Let the following calculations be from a study of the regression of Y on X:

$$n = 10 \qquad x. = 20 \qquad y. = 40$$

$$\sum x_i y_i = -82 \qquad \sum x_i^2 = 121 \qquad \sum y_i^2 = 516$$

(a) Find the LS predicting equation.
(b) Subdivide SS_Y into two components—see (12.14)—and explain what each of these means in this problem.
(c) Determine the regression variance.
(d) Calculate the coefficient of determination. What does this value mean?
(e) Estimate the variance of $\hat{\beta}_0$.
(f) Estimate the variance of $\hat{\beta}_1$.
(g) Test that the slope equals -1.5. Use a two-sided alternative and let $\alpha = 0.05$.
(h) Set and interpret a 95% *CI* for the mean of Y when $X = 3$.
(i) What assumptions are required for (g) and (h) to be statistically valid? Are any of these assumptions required in (a) through (f)?

11. Total sales, Y, of life insurance, in billions of dollars, for years 1951 through 1962 were reported in [3] as follows:

x	y	x	y
1951	$27.5	1957	$66.7
1952	31.5	1958	66.8
1953	36.2	1959	70.8
1954	45.4	1960	74.4
1955	48.4	1961	79.0
1956	55.3	1962	79.6

(a) Determine the linear function explaining the regression of Y on X.
(b) Calculate s_e^2.
(c) Evaluate the coefficient of determination.
(d) Suppose a person performed the analysis in (a) during January 1963 for the sole purpose of predicting total sales in 1963. Comment on this procedure.
(e) Make the prediction in (d) and compare with the actual value, $Y = 89.6$.

12. Analyses performed by Kerr [4] on fragments of a 14-inch gas pipeline that ruptured gave the data below. Energy values at initial failure, measured in foot-pounds and denoted by Y, and testing temperatures, in degrees Centigrade and denoted by X, were:

x	y	x	y
20	13	60	23
20	15	60	25
20	13	80	34
20	10	80	39
40	16	80	31
40	17	80	35
40	18	100	39
40	22	100	42
60	32	100	42
60	27	100	59

(a) Find the LS equation for the linear regression of Y on X.
(b) Calculate the regression variance.
(c) Evaluate the coefficient of determination.
(d) Is there any evidence, at the 5% level, that the slope β_1 is greater than 0.25?
(e) Set a 95% CI for the mean initial energy value at 50° Centigrade. Do likewise for 75°.
(f) Discuss assumptions required in the statistical analysis.

***13.** (a) Adjust the life insurance sales in Problem 11, to the year 1957.
(b) Are the adjusted values ranked in the same manner as the unadjusted values?

***14.** (a) Adjust the average weight gains of turkeys in Problem 7 to a biotin level of 100.
(b) Are the adjusted values ranked in the same manner as the unadjusted values?

***15.** For the test in Problem 10(g) determine the power if β_1 is actually
(a) $-1.5 + 0.1\sigma_\varepsilon$ (b) $-1.5 - 0.2\sigma_\varepsilon$ (c) $-1.5 + 0.3\sigma_\varepsilon$

***16.** Draw the power curve for the test in Problem 12(d).

***17.** For the test in Problem 7(e), determine the power if β_1 is actually
(a) $0.01\sigma_\varepsilon$ (b) $0.015\sigma_\varepsilon$ (c) $0.02\sigma_\varepsilon$

REFERENCES

[1] Dobson, D. C., "Biotin Requirement of Turkey Poults," *Poultry Science*, **49**, 2 (1970).
[2] Draper, N. R., and H. Smith, *Applied Regression Analysis*, Wiley, New York, 1966.
[3] Institute of Life Insurance, *Life Insurance Fact Book*, New York, 1967.
[4] Kerr, J. G., "Charpy Test Correlations from Unusual Pipeline Failure," *Welding Journal, Research Supplement*, **41**, 6 (1962).
[5] Ostle, B., *Statistics in Research*, 2nd ed., Iowa State University Press, Ames, Iowa, 1963.
[6] Roberts, J. T. A., and P. Barrand, "Anisotropy Effects in the Magnetochemical Damping of Nickel-Copper Alloys," *Journal of Physics D: Applied Physics*, **3**, 9 (1970).
[7] Williams, E. J., *Regression Analysis*, Wiley, New York, 1959.

Chapter 13

CORRELATION ANALYSIS

Regression techniques are used in the study of functional relations when the primary interest lies in predicting from a model. If there is little or no interest in prediction but a measure of the strength of the relationship is desired, *correlation* procedures are more appropriate.

When dealing with a set of variables having a joint distribution (that is, when all variables are subject to random errors), the strength of the relationship is usually of more interest than a predicting model. Thus, in studying measurements obtained from two different techniques, it might be more reasonable to ask how closely these measurements are related than to make predictions of one from the other. A biologist might wish to determine how the length and weight of fish are correlated, though he would not plan to predict weight from length.

A second, and perhaps more important, difference between regression and correlation analyses is the type of relationship being measured. While we directed our attention to only the linear case of regression, we pointed out that nonlinear models are available and are being used more and more. Correlation, on the other hand, is concerned only with the strength of the *linear* relationship. This fact cannot be emphasized too strongly, since a common error is to infer a lack of relationship among variables if a low correlation measure is obtained when, in fact, a strong curvilinear relationship could exist. Problem 13.7 illustrates this.

Cause-and-effect is not necessarily implied by a significant correlation. The remarks on the cause-and-effect situation in the treatment of regression (see Section 12.1) are equally applicable here.

It has been indicated that the coefficient of determination (Section 12.9.1) can be used as an indicator of the strength of nonlinear relationships. No other measures of the strength of curvilinear relationships will be studied in this book.

In this chapter the discussion will be restricted to the bivariate case. Correlation in multivariate distributions will not be considered. For simplicity of notation, we will use X and Y for the two variables. Some authors use X_1 and X_2, and others use Y_1 and Y_2.

13.1 THE BIVARIATE NORMAL DISTRIBUTION

Suppose that random variables X and Y have a joint distribution as illustrated in Figure 13-1. Such a distribution is called the *bivariate normal.* While it may not be completely obvious, five parameters are needed to specify a bivariate normal distribution: two means, μ_X and μ_Y; two variances, σ_X^2 and σ_Y^2; and, finally, a parameter to give the distribution its orientation. We take the fifth parameter as a measure of the degree of linear association between X and Y. Some authors prefer to take the fifth parameter to be σ_{XY}, the covariance between X and Y. For most purposes, it is better to measure the linear association by taking the covariance relative to the product of the two standard deviations. This new measure is the *correlation coefficient*,

$$\rho_{XY} = \sigma_{XY}/\sigma_X\sigma_Y, \tag{13.1}$$

which is a dimensionless quantity such that $-1 \leqslant \rho_{XY} \leqslant +1$. For brevity, we use ρ when the meaning is clear from the context.

Planes parallel to the X, Y plane (that is, planes that are horizontal) and intersecting a bivariate normal surface such as the one in Figure 13-1 give contours which are ellipses. If these contours are circles, $\rho = 0$. On the other hand, if $\rho = 0$, the ellipses have axes parallel to the X and Y axes.

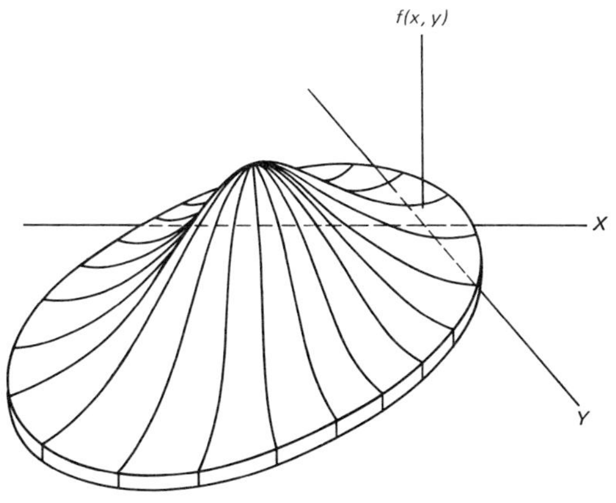

Figure 13-1 A Bivariate Normal Distribution

The surface in Figure 13-1 is defined mathematically by

$$f(x, y) = \frac{1}{2\pi\sigma_X\sigma_Y\sqrt{1-\rho^2}} exp\left\{\frac{-1}{2(1-\rho^2)}\left(\frac{(x-\mu_X)^2}{\sigma_X^2} - \frac{2\rho(x-\mu_X)(y-\mu_Y)}{\sigma_X\sigma_Y} + \frac{(y-\mu_Y)^2}{\sigma_Y^2}\right)\right\} \tag{13.2}$$

where $-\infty < x, y < +\infty$.

Equation (13.2) will not be used directly in any of our analyses and hence need not be memorized. Some properties of the bivariate normal family are worth mentioning, however, since they have a direct bearing on regression and correlation analyses. If the effect of either variable is eliminated in the distribution (13.2), the remaining variable is univariate normal; for example, $Y \sim N(\mu_Y, \sigma_Y^2)$. Secondly, in the bivariate normal distribution (13.2), the regression of Y on X and the regression of X on Y are each linear. Finally, the variance $\sigma_{Y|X}^2$ is constant for all x, and the variance $\sigma_{X|Y}^2$ is constant for all y.

Note that the regression assumptions are embodied in the above properties. This partially explains an earlier remark that X's need not be fixed in a regression study.

13.2 *CORRELATION IN BIVARIATE DISTRIBUTIONS*

The correlation coefficient, ρ, defined in (13.1) is not restricted to the bivariate normal; it exists and has a similar interpretation for any bivariate distribution. There are no units attached to ρ, which is a pure number. If the value of ρ is either $+1$ or -1, the two variables are perfectly related in a linear manner. Any value of ρ between these extreme values indicates the variables are linearly related to a lesser degree, with $\rho = 0$ implying no *linear* relationship. Variables having $\rho = 0$ are said to be uncorrelated.

It can be shown that variables independent in the statistical sense are always uncorrelated. The converse is generally not true. In the case of the bivariate normal, however, $\rho = 0$ does imply independence.

Positive values of ρ mean that both variables tend to increase or decrease together. Negative values of ρ mean that an increase in either variable is generally accompanied by a decrease in the other variable. Figure 13-2 gives contours corresponding to some correlation coefficients.

Now suppose a random sample of n paired observations is obtained from a bivariate distribution having correlation coefficient ρ. The usual estimator of ρ_{XY} is the sample correlation coefficient denoted by R_{XY}, and whose sample values are

$$r_{XY} = CP_{XY}/\sqrt{SS_X SS_Y} \tag{13.3}$$

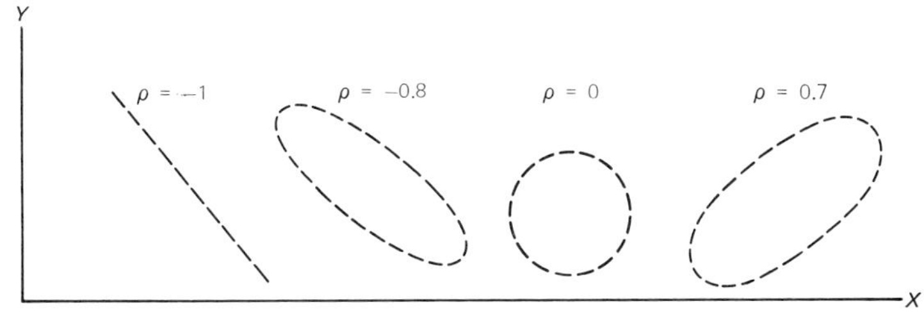

Figure 13-2 Some Correlation Contours

Here the terms on the right-hand side are the usual sums of squares and cross products. We omit the subscripts on r_{XY} when the meaning is clear from the context. A computational form is

(13.4) $$r = \frac{\sum x_i y_i - x.y./n}{\sqrt{(\sum x_i^2 - x.^2/n)(\sum y_i^2 - y.^2/n)}}$$

For most distributions, R_{XY} is a biased estimator of ρ. The distribution of R_{XY} is extremely complex even under the assumption that X and Y have a bivariate normal distribution.

The square of r given in (13.3) is, of course, the coefficient of determination defined in the last chapter.

► ***Example 13-1***

Consider the following data a random sample from a bivariate distribution:

x	y	x^2	xy	y^2
4	7	16	28	49
6	10	36	60	100
3	8	9	24	64
5	9	25	45	81
6	8	36	48	64
7	9	49	63	81
6	10	36	60	100
5	8	25	40	64
8	10	64	80	100
3	6	9	18	36

$x. = 53$ $y. = 85$ $\sum x_i^2 = 305$ $\sum x_i y_i = 466$ $\sum y_i^2 = 739$

$$SS_X = 305 - 280.9 = 24.1$$

$$SS_Y = 739 - 722.5 = 16.5$$

$$CP_{XY} = 466 - 450.5 = 15.5$$

$$r = \frac{15.5}{\sqrt{(16.5)(24.1)}} = \frac{15.5}{19.94} = 0.777 ◀$$

▶ ***Example 13-2***

Two criteria used at some colleges and universities in evaluating admission applications are the student's high school grade-point average and his scores on a battery of ACE (American Council of Education) tests. For 103 high school seniors who applied to a certain university, the following is obtained with respect to

$$W = \text{Quantitative ACE score}$$

$$X = \text{Linguistic ACE score}$$

$$Y = \text{High school average}$$

$$SS_W = 5562.48 \qquad CP_{WX} = 1059.42$$

$$SS_X = 24910.55 \qquad CP_{WY} = 209.32$$

$$SS_Y = 68.14 \qquad CP_{XY} = 612.33$$

Thus, for the applicants to this university, their two ACE scores have correlation estimated by

$$r_{WX} = 1059.42/\sqrt{5562.48(24910.55)} = 0.09,$$

their high school grade point averages and Quantitative ACE scores have correlation estimated by

$$r_{WY} = 209.32/\sqrt{5562.48(68.14)} = 0.34,$$

and their high school grade point averages and Linguistic ACE scores have correlation estimated by

$$r_{XY} = 612.33/\sqrt{24910.55(68.14)} = 0.47 ◀$$

(Problems 1(a), 3(a), (b), 5, 7)

13.3 INFERENCE ABOUT ρ IN THE BIVARIATE NORMAL

This section considers tests and *CI* about the parameter ρ of a bivariate normal distribution. No treatment of this inference will be undertaken for other distributions.

Recall that zero correlation is equivalent to independence in the bivariate normal. Thus, for this particular distribution, testing the independence of X and Y is equivalent to testing $H_0: \rho = 0$. (In the usual spirit of statistics, assuming an *approximate* bivariate normal distribution for X and Y results in *approximate* inference.)

Testing the special hypothesis $H_0 : \rho = 0$ has a particularly simple structure under bivariate normality. When $\rho = 0$ in a bivariate normal, it can be shown that the sample correlation coefficient R_{XY} has a simple behavior which may be described by

$$T_{n-2} = R_{XY}\sqrt{(n-2)/(1-R_{XY}^2)} \tag{13.5}$$

Because of (13.5), the Student T distribution can be used to test $H_0 : \rho = 0$ in a bivariate normal distribution. Or, since the right-hand side of (13.5) depends only on R_{XY} for fixed n, it is possible to give critical points for this test in terms of the magnitude of r. These points will be denoted by $r_{\nu;\gamma}$; that is, $P[R_{XY} \leqslant r_{\nu;\gamma} | \rho = 0] = \gamma$. Table A.12 gives points $r_{\nu;\gamma}$ for $\gamma = 0.95$, 0.975, 0.99, and 0.995, and for $\nu = n - 2$. The first three steps of a $100\alpha\%$ test are

(13.6)

① ⓐ $H_0 : \rho \leqslant 0$ $H_1 : \rho > 0$ ① ⓑ $H_0 : \rho \geqslant 0$ $H_1 : \rho < 0$ ① ⓒ $H_0 : \rho = 0$ $H_1 : \rho \neq 0$

②, ③ H_0 is rejected if

ⓐ $r \geqslant r_{\nu;1-\alpha}$ ⓑ $r \leqslant -r_{\nu;1-\alpha}$ ⓒ $|r| \geqslant r_{\nu;1-\alpha/2}$

► ***Example 13-3***

Using the data in Example 13-2, we shall assume that each of the 3 pairs of variables follows an approximate bivariate normal distribution. We wish to test that the two ACE scores represent independent measures of students' ability. Thus, at the 5% level:

① $H_0 : \rho_{WX} = 0$ versus $H_1 : \rho_{WX} \neq 0$
②, ③ Reject H_0 if $|r_{WX}| \geqslant r_{101;0.975} \doteq 0.1966$
④ $r_{WX} = 0.09$
Therefore, we fail to reject H_0.
⑤ There is no evidence at the 5% level that the two ACE scores are related linearly. ◄

To test hypotheses concerning nonzero correlation, $H_0 : \rho = \rho_0$, where $\rho_0 \neq 0$, an approximate procedure is generally employed. This is necessary, as we have indicated earlier, because the sample correlation coefficient R_{XY} does not have a workable distribution even in the bivariate normal case, unless $\rho = 0$.

Fisher [1] proposed the following transformation

$$Z(r) = \tfrac{1}{2}\log_e[(1+r)/(1-r)] \tag{13.7}$$

and has shown for large sample sizes that $Z(r)$ is approximately normal with mean and variance:

(13.8)
$$Z(\rho) = \tfrac{1}{2}\log_e[(1+\rho)/(1-\rho)]$$
$$\sigma^2_{Z(r)} = 1/(n-3)$$

Thus, an approximate standard normal variable is

(13.9)
$$Z \doteq [Z(r) - Z(\rho)]\sqrt{n-3}$$

whereby the usual Z test is appropriate.

Values of $Z(r)$ and $Z(\rho)$ may be read directly from Table A.11. Negative signs may be affixed to each pair of entries in the table; that is,

(13.10)
$$Z(-r) = -Z(r)$$

► ***Example 13-4***

Suppose it is of interest to determine whether the students entering the college mentioned in Example 13-2 gave high school grade-point averages and quantitative ACE scores with $\rho = 0.5$. Assuming these scores have an approximate bivariate normal distribution, a 1% test would be

① $H_0: \rho_{WY} = 0.5$ versus $H_1: \rho_{WY} \neq 0.5$

② Under H_0, $Z \doteq \dfrac{Z(r) - Z(0.5)}{\sqrt{1/100}}$

③ Reject H_0 if $|Z| \geqslant 2.576$

④ $r_{WY} = 0.34$, with the aid of Table A.11

$$Z(0.34) = 0.3541 \qquad \text{and} \qquad Z(0.5) = 0.5493$$
$$Z \doteq 10[0.3541 - 0.5493] = -1.952$$

Therefore, we fail to reject H_0.

⑤ There is no evidence at the 1% level that the correlation between high school average and quantitative ACE scores is other than 0.5 for this group of applicants. ◄

Confidence intervals for the correlation, ρ, of a bivariate normal may be obtained in two ways: (i) using Appendix B.5, or (ii) using Fisher's log transformation in large sample cases.

Appendix B.5 contains pairs of curves for various sample sizes to give two-sided *CI* with $\alpha = 0.10, 0.05, 0.02$, or 0.01. One-sided *CI* can be obtained by using one curve of a given pair and halving α.

▶ *Example 13-5*

Using the data of Example 13-2 and Appendix B.5, a 95% *CI* for the correlation between linguistic ACE scores and high school averages for the applicants involved is found to be approximately (from the curves for $n = 100$): $0.28 < \rho_{XY} < 0.60$. ◀

For sample sizes not covered by Appendix B.5, an approximate $100(1 - \alpha)\%$ *CI* for the correlation in a bivariate normal distribution may be obtained as follows:

(i) Obtain $Z(r)$ from Table A.11.
(ii) Compute

$$Z(r) \pm \frac{Z_{1-\alpha/2}}{\sqrt{n-3}}, \tag{13.11}$$

which establishes a $100(1 - \alpha)\%$ *CI* for the transformed parameter $Z(\rho)$.
(iii) Convert the limits obtained in (13.11) with the aid of Table A.11.

▶ *Example 13-6*

For comparative purposes, we calculate a 95% *CI* for the correlation between the linguistic ACE scores and high school averages in Examples 13-2 and 13-5 by Fisher's log transformation. From Table A.11, $Z(0.47) = 0.5101$, whereby

$$0.5101 \pm \frac{1.96}{10} = 0.3141; 0.7061$$

or

$$0.3141 < Z(\rho) < 0.7061$$

Reconverting these limits by Table A.11, we obtain approximately: $0.305 < \rho < 0.609$, which is fairly close to the graphic results given in Example 13.5. ◀

(Problems 1(b), (c), 3, 11)

*13.4 *TESTING THE EQUALITY OF CORRELATIONS FROM TWO BIVARIATE NORMALS*

When independent samples are obtained from two bivariate normal populations, there may be interest in judging whether the two populations have significantly different correlations. Usually the same pair of variables defines both populations which are distinguished because of time, location, or another factor. Thus, in the study of ACE scores and high school averages, there may be some reason for checking whether these variables are correlated the same for students from high schools in two different states, or for students in-state and out-of-state.

The procedure for testing the equality of two such correlations, say ρ_1 and ρ_2, is an extension of Fisher's Z test for one correlation. Specifically,

① ⓐ $H_0: \rho_1 \leqslant \rho_2$ ① ⓑ $H_0: \rho_1 \geqslant \rho_2$ ① ⓒ $H_0: \rho_1 = \rho_2$
$H_1: \rho_1 > \rho_2$ $H_1: \rho_1 < \rho_2$ $H_1: \rho_1 \neq \rho_2$

(13.12) ② Under H_0, $$Z \doteq \frac{Z(r_1) - Z(r_2)}{\sqrt{\dfrac{1}{n_1 - 3} + \dfrac{1}{n_2 - 3}}}$$

③ H_0 is rejected if
ⓐ $Z \geqslant z_{1-\alpha}$ ⓑ $Z \leqslant z_\alpha$ ⓒ $|Z| \geqslant z_{1-\alpha/2}$

Note that this is an approximate test and that the sample sizes need not be equal.

*13.5 POWER OF TESTS

Power calculations for the test (13.6) of zero correlation in a bivariate normal are quite difficult. This is because the test statistic given in (13.5) has a very complex distribution when $\rho \neq 0$ ([4], p. 208). We omit further discussion of this topic.

Approximate power for the two approximate Z tests can be obtained with the aid of the cumulative standard normal distribution or, more easily, from the curve for $v = \infty$ in Appendix B.1.

For the test of $H_0: \rho = \rho_0$, where $\rho_0 \neq 0$, using the test statistic in (13.9), let ρ_0' denote a value of ρ specified by H_1. Then the approximate power can be read from Appendix B.1 using the graph designated by α ($\alpha/2$ for two-sided tests), the curve designated by $v = \infty$, and

(13.13) $$\delta = \sqrt{n - 3}|Z(\rho_0) - Z(\rho_0')|$$

For the test of $H_0: \rho_1 = \rho_2$ outlined in (13.12), the approximate power is read from Appendix B.1 using the graph designated by α ($\alpha/2$ for two-sided tests), the curve designated by $v = \infty$, and

(13.14) $$\delta = \frac{Z(\rho_1') - Z(\rho_2')}{\sqrt{1/(n_1 - 3) + 1/(n_2 - 3)}}$$

where ρ_1' and ρ_2' are alternative values of ρ_1 and ρ_2.

▶ ***Example 13-7***

Referring to Example 13-4, if ρ_{WY} is actually 0.4, what are the chances of rejecting H_0? If ρ_{WY} is 0.35? By (13.13) and Table A.11, we have when $\rho'_{WY} = 0.4$,

$$Z(\rho_0) - Z(\rho'_0) = 0.5493 - 0.4236 = 0.1257,$$

giving $\delta = 10(0.1257) = 1.257$, and from Appendix B.1, using $\alpha/2 = 0.005$, the power is $P_{0.4} = 0.10$.

When $\rho'_{WY} = 0.35$, $Z(\rho_0) - Z(\rho'_0) = 0.5493 - 0.3654$, giving $\delta = 10(0.1839) = 1.839$ and power $P_{0.35} = 0.26$. ◀

(Problem 17)

PROBLEMS

1. Values obtained in a correlation study were:

$$n = 30 \qquad x. = 900 \qquad y. = 2{,}100$$

$$CP_{XY} = 160 \qquad SS_X = 144 \qquad SS_Y = 400$$

(a) Calculate the correlation coefficient.
Assuming X and Y have a bivariate normal distribution,
(b) set a 95% CI for ρ.
(c) test $H_0: \rho = 0$ versus $H_1: \rho > 0$ at the 5% level.

2. Given the following statistics, which were calculated in a correlation study:

$$n = 25 \qquad x. = 740 \qquad y. = 890$$

$$CP_{XY} = 250 \qquad SS_X = 450 \qquad SS_Y = 200$$

(a) Estimate the correlation between X and Y.
(b) What meaning can be attached to the value found in (a)?

3. Calculations in a correlation analysis were:

$$n = 18 \qquad x. = 42 \qquad y. = 27$$

$$\sum x_i y_i = 140 \qquad \sum x_i^2 = 209 \qquad \sum y_i^2 = 121.5$$

(a) Estimate the correlation between X and Y.
(b) Interpret the value found in (a).

4. The following calculations were made in a correlation problem:

$$n = 25 \qquad x_i = 150 \qquad y. = 225$$

$$\sum x_i y_i = 1{,}423 \qquad \sum x_i^2 = 1{,}080 \qquad \sum y_i^2 = 2{,}065$$

(a) Calculate the correlation coefficient.
(b) What interpretation can be attached to this measure?
Assuming that X and Y have a bivariate normal distribution,
(c) set a 95% CI for the population correlation.
(d) test that the population correlation is zero. Use a two-sided alternative and let $\alpha = 0.01$.

5. Calculate the correlation coefficient for the following data:

x	4	5	7	7	5	8
y	8	6	1	3	6	0

6. (a) Calculate the correlation coefficient for the following data:

x	6	2	4	3	2	6	1	4
y	9	1	6	5	4	7	3	5

(b) What interpretation can be attached to this measure?
Assuming that X and Y have a bivariate normal distribution,
(c) set a 95% *CI* for the population correlation.
(d) test that the population correlation is zero. Use a two-sided alternative and let $\alpha = 0.01$.

7. Consider the following data:

X	−3	−2	0	+2	+3
Y	9	4	0	4	9

(a) Calculate the correlation coefficient.
(b) The variables X and Y are not independent, since they satisfy the quadratic relationship $Y = X^2$ perfectly. Interpret and discuss the meaning of the value obtained in (a).

8. For a class of 26 students, analysis of their final exam scores, denoted by Y (200 points possible), and their overall total scores denoted by X (700 points possible), revealed:

$$x_. = 12{,}561 \qquad y_. = 3{,}234$$

$$\sum x_i y_i = 1{,}634{,}216 \qquad \sum x_i^2 = 6{,}399{,}037 \qquad \sum y_i^2 = 461{,}004$$

(a) Calculate the correlation coefficient.
(b) Is a student's final exam score as good an indicator of performance in this class as the overall total score? Explain.
(c) Is the correlation significantly different from 0? Make the appropriate test at the 5% level.
(d) What assumptions are required for (c) to be statistically valid?

9. Monthly production of creamed cottage cheese, in millions of pounds, denoted by X, and ice cream, in millions of gallons, denoted by Y, for 1969 were reported in [2] as follows:

x	y	x	y
70.9	54.9	80.7	80.1
72.4	54.1	75.5	77.0
81.6	61.6	73.9	69.4
79.8	64.3	77.6	63.1
85.2	68.5	69.8	49.9
77.6	75.7	69.3	54.1

(a) Without plotting the data or making any calculations, would a high correlation between X and Y be expected? Explain.
(b) Calculate the coefficient of correlation between X and Y.

10. A biologist is investigating the relationship between the calcium content, X, and the potassium content, Y, of turnip greens. Samples are obtained from 35 randomly selected plants and analyzed, giving, in milligrams per gram:

$$n = 35 \qquad x. = 154 \qquad y. = 492$$

$$\sum x_i y_i = 2{,}235 \qquad \sum x_i^2 = 821 \qquad \sum y_i^2 = 7{,}214$$

(a) Estimate the correlation between X and Y.
(b) Test the significance of this correlation at the 1% level.
(c) What assumptions are required for (b) to be statistically valid?

11. Samples taken at random from fifteen batches of a polymer made by a given process exhibited the following values for solubility, X, and tensile strength, Y:

x	y	x	y
8.76	8,684	8.77	8,674
8.81	8,667	8.79	8,671
8.75	8,688	8.76	8,679
8.77	8,675	8.80	8,665
8.75	8,678	8.79	8,670
8.77	8,679	8.78	8,672
8.76	8,680	8.77	8,677
8.80	8,672		

(a) Estimate the correlation between solubility and tensile strength for this polymer.
(b) Set a 95% *CI* for this correlation.
(c) At the 1% level, test that the correlation differs significantly from 0.

12. Twenty-five pellets of uranium dioxide were selected at random from a prepared batch with the following measurements given (see [1]):

X = density, in grams per cubic centimeter

Y = per cent porosity

x	y	x	y	x	y
10.673	2.55	10.600	3.21	10.612	3.10
10.675	2.53	10.616	3.07	10.592	3.29
10.661	2.66	10.600	3.21	10.560	3.58
10.651	2.75	10.614	3.09	10.649	2.77
10.661	2.66	10.614	3.09	10.639	2.86
10.660	2.67	10.612	3.10	10.651	2.75
10.648	2.78	10.624	2.99	10.638	2.87
10.612	3.01	10.630	2.94	10.642	2.83
				10.653	2.73

(a) Estimate the correlation between the density and porosity of uranium dioxide pellets.
(b) Set a 99% CI for this correlation.
(c) Test $H_0: \rho \geqslant -0.95$ versus $H_1: \rho < -0.95$ at the 5% level.
(d) Discuss the assumptions required in (b) and (c).

13. Refer to the data given in Problem 17 of Chapter 7.
(a) Estimate the coefficient of correlation between monthly gas and electric values.
(b) Does the value obtained in (a) appear to be reasonable for these two variables, given that the values were for consecutive months? Explain.

14. Refer to the data given in Problem 12 of Chapter 11. Before accepting the X-ray method as a substitute for the direct method (for a particular solid), it must be demonstrated that the densities obtained under the two methods are highly correlated.
(a) For the idium–cadmium solid, estimate the correlation between densities by the direct method and densities from the X-ray method. (You may code the data linearly without affecting the value of the correlation coefficient.)
(b) Test that the correlation between these variables is no more than 0.9. Let $\alpha = 0.05$.
(c) What assumptions are required for (b) to be valid statistically?

15. Refer to the data given in Problem 13 of Chapter 11.
(a) Calculate the coefficient of correlation between amylase measurements from the two methods.
(b) Is there evidence that the true correlation is above 0.9? Let $\alpha = 0.01$.
(c) Discuss the assumptions required in your analysis.

***16.** Find the power of the test in Problem 14 if the unknown correlation is actually
(a) 0.85 (b) 0.80 (c) 0.75 (d) 0.50

***17.** Find the power of the test in Problem 15 if the unknown correlation is actually
(a) 0.92 (b) 0.95 (c) 0.98 (d) 0.99

REFERENCES

[1] Ainscough, J. B., and M. J. Wheeler, "The High-Temperature Thermal Conductivity of Sintered Uranium Dioxide," *British Journal of Applied Physics*, Series 2, **1** (1968).
[2] *Dairy Record*, **70**, 23 (1970).
[3] Fisher, R. A., "Frequency Distribution of the Values of the Correlation Coefficient in Samples From an Indefinitely Large Population," *Biometrika*, **10** (1915).
[4] Graybill, F. A., *Introduction to Linear Statistical Models*, Vol. I, McGraw-Hill, New York, 1961.

Chapter 14

INFERENCE FOR ENUMERATIVE DATA

A large number of investigations are concerned with enumeration (counting) variables. Such variables have a range of values which is a subset of the positive integers and zero, thus forming a special subset of the discrete variables. Families such as the binomial and Poisson belong to this category.

Typical questions related to enumeration variables are the following: What proportion of the days do more than 50 people trade a particular stock on the New York Stock Exchange? Is there any evidence that second-generation offspring appear in a ratio other than 9:3:3:1? Between what limits might a businessman find the average number of bad checks cashed in his establishment per week?

As a general rule, statistical inference about discrete distributions is characterized by summations. To illustrate a summation that might be encountered, the following occurs in a certain hypothesis test for a binomial distribution:

$$\alpha = \sum_{x=20}^{50} \binom{50}{x} (1/4)^x (3/4)^{50-x} \tag{14.1}$$

Setting a *CI* for the parameter p of a binomial distribution is even more complicated: it requires finding a root between 0 and 1 of two different polynomials in p, one having degree equal to the sample size n (see [2] for a discussion of these two and related inference problems).

Fortunately, the scope of existing tables is now great enough so that probabilities like the one in (14.1) can be read directly or obtained after a few simple calculations. Graphs and tables are also available for obtaining confidence limits, such as the roots of polynomials mentioned above. Appendixes A and B contain selected portions of tables and graphs which will be extremely useful for inferences in the discrete case.

Many problems in discrete inference which are not covered by the existing tables and graphs have been solved under *large-sample theory*. When based on a large sample, inference about the *mean* of any distribution is always possible by virtue of the Central Limit Theorem. Other large-sample distributions give approximate results for other phases of inference. As usual, one must be willing to accept the approximate results in exchange for the time and effort saved by not developing a more exact procedure.

Both exact and approximate inference will be discussed for the Poisson and binomial families. Power of the Poisson and binomial tests is presented in a separate section. Inference about multinomial data is limited to the classical approximate tests. Details of derivations, for both exact and approximate inference, are beyond the scope of this book and will be omitted.

14.1 INFERENCE FOR THE POISSON FAMILY

14.1.1 Exact Inference

If X denotes a Poisson variable, X measures the number of successes per observational unit. Its distribution depends upon a single parameter, λ, the average number of successes per unit. The usual estimator of λ is the sample mean $\hat{\lambda} = \bar{X}$, which is an unbiased, consistent, and minimum variance estimator of λ. From (8.13) and (7.38), we see that $\hat{\lambda}$ has a variance of

$$\sigma_{\bar{X}}^2 = \lambda/n \tag{14.2}$$

which would be estimated unbiasedly by

$$\hat{\sigma}_{\bar{X}}^2 = \bar{X}/n \tag{14.3}$$

Good quality inference about λ is therefore expected when using $\bar{X}$, or some function thereof. The following, borrowed from distribution theory, holds the key to this inference.

- *Theorem 14-1: Let X have a Poisson distribution with parameter λ. In random samples of n units from this distribution, the observed number of successes, $X. = \sum X_i$, has a Poisson distribution with parameter $\lambda^* = n\lambda$.*

Theorem 14-1 is important because inference about λ can be based on $X.$ (a special transformation of $\bar{X}$) and existing Poisson tables. It might be pointed out that using $X.$ simply forces us to make inference about its mean value, λ^*, which is related to λ in the same way as $X.$ is to $\bar{X}$, the sample mean.

Using the number of successes $x.$ observed from a random sample of n Poisson units, a $100(1 - \alpha)\%$ CI for λ is specified by the limits

(14.4) $$L = \lambda_1/n \qquad \text{and} \qquad U = \lambda_2/n,$$

where λ_1 and λ_2 are the values for which the following two Poisson probabilities hold:

(14.5) $$F(x. - 1; \lambda_1) = 1 - \alpha/2$$
$$F(x.; \lambda_2) = \alpha/2$$

If the coverage is available, the equations in (14.5) can be solved with the aid of cumulative Poisson tables. Appendix B.3 gives curves from which λ_1 and λ_2 can be read directly. The curves can be used for two-sided CI for $\alpha = 1, 2, 5$, and 10% and one-sided CI for $\alpha = 0.5, 1, 2.5$ and 5%. Graphical results are generally as good as those obtained from (14.4) and (14.5), since interpolation is usually required in the Poisson tables. Coverage beyond that of the graphs is provided by the approximation discussed in the next subsection.

An example illustrates each of the two above procedures.

► ***Example 14-1***

Because of the limited size of the Poisson tables in Appendix A, the following fictitious data are used merely for showing how the tables are used to set a 99% CI for λ. For $n = 6$ units drawn from a Poisson distribution, $x. = 9$ successes were observed. Using the first equation in (14.5) and Table A.4,

$$F(8; 3) = 0.9962 \qquad \text{and} \qquad F(8; 4) = 0.9786,$$

which means that $3 < \lambda_1 < 4$. Interpolating linearly gives $\lambda_1 = 3.08$. Using the second equation in (14.5) and Table A.4,

$$F(9; 20) = 0.005$$

Thus, 99% limits on λ are

$$L = \frac{3.08}{6} = 0.51 \qquad \text{and} \qquad U = \frac{20}{6} = 3.33$$ ◄

► ***Example 14-2***

Suppose that the number of daily withdrawals from a certain savings and loan company is a Poisson variable. A random sample of 12 days showed 86 withdrawals. Referring to Appendix B.3 we see that $\lambda_1 = 68$ and $\lambda_2 = 106$.

Therefore, a 95% *CI* for the expected number of daily withdrawals is specified by the limits

$$L = \frac{68}{12} = 5.67 \qquad \text{and} \qquad U = \frac{106}{12} = 8.83$$

(Interpolating in the GE tables, these parameters were found to be $\lambda_1 = 68.7$ and $\lambda_2 = 106.5$.) ◀

Tests about the Poisson parameter can also be made quite easily with the aid of cumulative Poisson tables. Using Theorem 14-1, a $100\alpha\%$ test about λ has the following structure:

(14.6)

① ⓐ $H_0: \lambda \leqslant \lambda_0$ $H_1: \lambda > \lambda_0$

① ⓑ $H_0: \lambda \geqslant \lambda_0$ $H_1: \lambda < \lambda_0$

① ⓒ $H_0: \lambda = \lambda_0$ $H_1: \lambda \neq \lambda_0$

② Under H_0, the number of successes $X.$ has a Poisson distribution with parameter $\lambda_0^* = n\lambda_0$.

③ Reject H_0 if

ⓐ $F(x. - 1; \lambda_0^*) > 1 - \alpha$

ⓑ $F(x.; \lambda_0^*) < \alpha$

ⓒ either $F(x.; \lambda_0^*) < \alpha/2$ or $F(x. - 1; \lambda_0^*) > 1 - \alpha/2$.

▶ ***Example 14-3***

A publisher claims an average of no more than 3 typographical errors on the front page of his daily newspaper. A random sample of 5 daily newspapers revealed 28 typographical errors. Can the claim be rejected at the 10% level?

Assuming a Poisson distribution of typographical errors on the front pages of this newspaper, a 10% test of this claim is

① $H_0: \lambda \leqslant 3$ versus $H_1: \lambda > 3$.
② Under H_0, $X.$ is a Poisson variable with parameter $\lambda^* = 15$.
③ H_0 is to be rejected if $F(x. - 1; 15) > 0.90$.
④ $x. = 28$ and from Table A.4

$$F(27; 15) = 0.9983$$

Therefore, H_0 is rejected.

⑤ The average number of typographical errors on the front page of the daily newspaper exceeds 3, unless a 10% error has occurred. ◀

▶ ***Example 14-4***

Most airlines allow extra reservations for flights, since they expect a certain number of cancellations because of missed connections, failures to show, and the like. A certain airline has a policy of allowing 5 extra fares to be sold on its No. 12A flight. Does this appear to be an unreasonable number if eight of its No. 12A flights that were sold to capacity had cancellations of

4, 5, 7, 4, 6, 7, 3, 6? Assuming the cancellations on its No. 12A flight are approximately Poisson, a 5% test is

① $H_0: \lambda = 5$ versus $H_1: \lambda \neq 5$.
② Under H_0, $X.$ is Poisson with parameter $\lambda^* = 40$.
③ H_0 will be rejected if

$$F(x. - 1; 40) > 0.975 \qquad \text{or} \qquad \text{if } F(x.; 40) < 0.025$$

④ $x. = 42$ and from Table A.4

$$F(41; 40) = 0.6033$$

$$F(42; 40) = 0.6618$$

Therefore, we fail to reject H_0.
⑤ There is no evidence at the 5% level to indicate that the extra fare policy should be changed for this flight. ◀

(Problems 1, 5)

14.1.2 Approximate Inference

In observing a large number of units from a Poisson distribution, the Central Limit Theorem allows us to make inference about the parameter λ, the mean value. More exactly,

(14.7)
$$Z \doteq \frac{\bar{X} - \lambda}{\sqrt{\lambda/n}} = \frac{X. - n\lambda}{\sqrt{n\lambda}}$$

This approximation actually depends on the size of $n\lambda$ rather than on n only; it is quite good whenever $n\lambda > 10$ (which certainly is the case for coverage not afforded by existing graphs and tables).

Under the hypothesis $\lambda = \lambda_0$, the test statistic is

(14.8)
$$Z \doteq \frac{X. - n\lambda_0}{\sqrt{n\lambda_0}}$$

The usual Z test would be made.

An approximate $100(1 - \alpha)\%$ *CI* for λ can be obtained with the aid of (9.6) and (14.7). The *CI* is the set of values for λ which satisfy the inequality

(14.9)
$$-z_{1-\alpha/2} < \frac{\bar{X} - \lambda}{\sqrt{\lambda/n}} < z_{1-\alpha/2}$$

Because the middle term of (14.9) contains λ in both numerator and denominator, to different powers, it is necessary to solve a quadratic equation. This solution gives approximate limits

(14.10)
$$L, U = \bar{x} \pm z_{1-\alpha/2}\sqrt{\bar{x}/n}$$

where terms of the order $(1/n)z_{1-\alpha/2}^2$ have been ignored in the simplification. For most cases, the sample size n will be large enough to make such terms negligible. The error in the limits (14.10) is roughly $(1/2n)z_{1-\alpha/2}^2$.

▶ ***Example 14-5***

The normal approximation to the Poisson will be applied to the data of Example 14-2 to set a 95% *CI* for λ. In practice, the normal approximation would probably not be used, since Appendix B.3 covers the situation. We are applying the approximation for illustrative and comparative purposes. The limits for a 95% approximate *CI* are, by (14.9),

$$L, U = \frac{86}{12} \pm 1.96\sqrt{\frac{86}{144}} = \frac{86 \pm 1.96(9.27)}{12} = 5.65; 8.68$$

(Since n was small in this experiment, $z_{0.975}^2/2n = 1/6$; hence we can expect some discrepancy between this answer and the one obtained in Example 14-2.) ◀

(Problems 5(b), 17, 19)

14.2 INFERENCE FOR THE BINOMIAL FAMILY

14.2.1 Exact Inference

If the random variable X denotes the number of successes observed in n Bernoulli trials, the usual estimator of p is

(14.11) $$\hat{p} = X/n$$

This estimator possesses the three properties of Section 9.1. Since $\hat{p}$ is simply the variable X coded, it follows from (8.5) that

(14.12) $$\sigma_{\hat{p}}^2 = pq/n$$

which is estimated unbiasedly by

(14.13) $$s_{\hat{p}}^2 = \hat{p}\hat{q}/(n-1)$$

Exact inference for the binomial family follows the same general lines as that for the Poisson. A distinct difference is that the factor n does not appear in many formulas for the binomial case. Because X counts the number of successes in an experiment, and in view of (14.11), the distribution of X is appropriate for inference about p.

A $100(1-\alpha)\%$ *CI* for the proportion p of successes occurring in n Bernoulli trials is specified by the limits

(14.14) $$L = p_1 \quad \text{and} \quad U = p_2$$

where p_1 and p_2 are the proportions for which the following two binomial probabilities hold:

(14.15) $\quad F(x-1; n, p_1) = 1 - \alpha/2 \quad$ and $\quad F(x; n, p_2) = \alpha/2$

Appendix B.4 gives curves from which p_1 and p_2 can be read directly. Specific curves for two-sided *CI* are given for $n = 10, 15, 20, 30, 50, 100, 250, 1{,}000$ and $\alpha = 0.01, 0.05$. The curves can be used for one-sided *CI*, with the same values of n and half the stated value of α. Interpolation can be performed in these graphs, or the equations in (14.15) may be used with the Harvard Tables. For $n > 1{,}000$, the normal approximation can be used; this is discussed in Section 14.2.3.

► ***Example 14-6***

Investigating the activity of a certain low-volume stock listed on the American Stock Exchange, it was found over a 4-week period (20 days) that on 7 days there were more than 10 trades (more than 10 trades on a given day is taken as a success). We shall assume that the daily tradings of this stock are independent with the probability of more than 10 trades being the same for every day. These assumptions would be unreasonable if there were trends due to a possible merger, failure to merge, anticipated news of a new product announcement, and any other factor. A 99% *CI* for the proportion of days of more than ten trades is found, by entering Appendix B.4 with $\alpha = 0.01$, $\hat{p} = 0.35$, and $n = 20$, to be

$$L = 0.11 \qquad \text{and} \qquad U = 0.66$$ ◄

A test of hypothesis about a binomial proportion can be made with cumulative probabilities (as in the Poisson case). Under the hypothesis that $p = p_0$, the number of successes has a binomial distribution with parameters n and p_0. Thus, $100\alpha\%$ tests about p have as the first three steps:

(14.16)

① ⓐ $H_0: p \leqslant p_0$, $H_1: p > p_0$ ① ⓑ $H_0: p \geqslant p_0$, $H_1: p < p_0$ ① ⓒ $H_0: p = p_0$, $H_1: p \neq p_0$

② Under H_0, the number of successes, X, has the $B(n, p_0)$ distribution.

③ Reject H_0 if

ⓐ $F(x-1; n, p_0) > 1 - \alpha$

ⓑ $F(x; n, p_0) < \alpha$

ⓒ $F(x-1; n, p_0) > 1 - \alpha/2$ or $F(x; n, p_0) < \alpha/2$

Approximate tests about p are discussed in Section 14.2.3.

▶ ***Example 14-7***

A tire manufacturer claims that no more than 10% of its tires sold as "blemishes" have noticeable defects. From a shipment of "blemishes," a tire dealer takes a random sample of 20 and finds 3 with noticeable defects. A 5% test is:

① $H_0: p \leqslant 0.10$
$H_1: p > 0.10$
② If H_0 is true, X is binomial with $p = 0.10$.
③ Reject H_0 if $F(x - 1; 0.10) > 0.95$.
④ $x = 3$ and from Table A.3, $F(2; 0.10) = 0.6769$. Therefore, H_0 is not rejected.
⑤ There is no sample evidence that more than 10% of the "blemish" tires have noticeable defects. ◀

▶ ***Example 14-8***

To check whether a particular coin is biased, has $p \neq 1/2$, a person tosses it 50 times. On 33 tosses, heads are observed. A 10% test is:

① $H_0: p = 1/2$ versus $H_1: p \neq 1/2$.
② Under H_0, X, the number of heads in 50 tosses of the coin, has the B(50, 1/2) distribution.
③ H_0 is to be rejected if

$$F(x - 1; 50, 1/2) \geqslant 0.95 \qquad \text{or} \qquad \text{if } F(x; 50, 1/2) < 0.05$$

④ $x = 33$ and from Table A.3

$$F(32; 50, 1/2) = 0.9923 \qquad \text{and} \qquad F(33; 50, 1/2) = 0.9836$$

Therefore, H_0 is rejected.
⑤ The coin is not well balanced, the chances of heads and tails are not equal, unless a 10% error has been made. ◀

(Problems 4(a), 10(a), (b), 25(a), (b))

14.2.2 Normal Approximation of Binomial Probabilities

According to genetic theory, the second-generation plants of a certain parental cross are to have red and white flowers in the 3:1 ratio. If a florist plants 2,000 seeds of such a cross and they all germinate, what is the probability the florist has exactly 500 white-flowered plants for sale? at least 530 white-flowered plants for sale? Assuming the conditions of the binomial are met, these exact probabilities are given by

(14.17)
$$P(X = 500) = \binom{2{,}000}{500}(1/4)^{500}(3/4)^{1{,}500}$$
$$P(X \geqslant 530) = \binom{2{,}000}{530}(1/4)^{530}(3/4)^{1{,}470} \text{ plus 1,470 similar terms}$$

Neither of the probabilities in (14.17) can be read from existing tables, and to compute them, even with logarithms, would be extremely time-consuming.

Fortunately, binomial probabilities such as those in (14.17) can be approximated by a normal distribution. Viewing X/n as a mean number (proportion) of successes observed, it follows that the approximation is a direct consequence of the Central Limit Theorem. We now formally state this without proof.

- *Theorem 14-2: Let the random variable X have the binomial distribution with parameters n and p. Then, as n approaches infinity, the distribution of* $\frac{X - np}{\sqrt{npq}}$ *approaches the* N(0, 1) *distribution. Thus, for large n*

(14.18)
$$Z \doteq \frac{X - np}{\sqrt{npq}}$$

As a consequence of Theorem 14-2, we can say for large n that X has an approximate N(np, npq) distribution. The size of n needed for a good approximation depends upon the value of p. Good results are obtained when the following two conditions hold simultaneously:

(14.19)
$$n \geqslant 25 \qquad \text{and} \qquad \min(np, nq) \geqslant 5$$

Since the binomials are symmetric when $p = 1/2$, the approximation is very good whenever p is close to 1/2.

Representing a B(n, p) distribution by a rectangular bar graph, as in Figure 14-1, the area of the bar centered over $X = c$ represents $P(X = c)$.

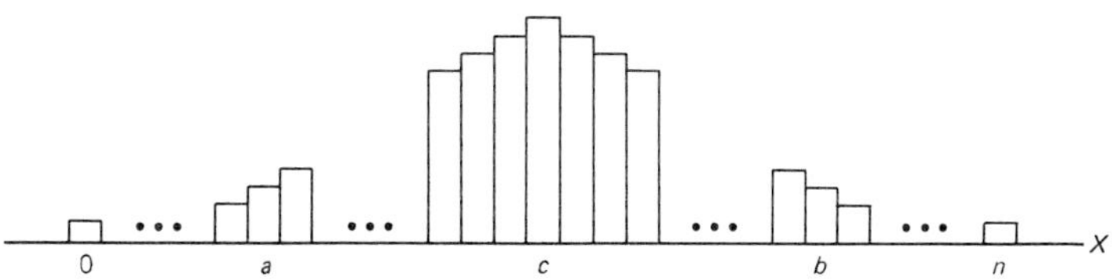

Figure 14-1 The B(n, p) Distribution

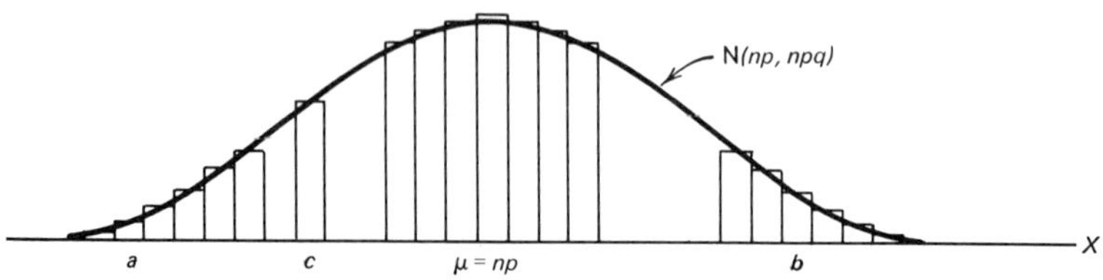

Figure 14-2 Normal Approximation to the Binomial

The B(n, p) curve and the approximating normal curve, the N(np, npq) distribution are shown in Figure 14-2. With respect to the normal distribution, $P(X = c)$ is given approximately by the area from $c - 1/2$ to $c + 1/2$; that is,

(14.20) $$P(X = c) \doteq P(c - 1/2 \leqslant Y \leqslant c + 1/2)$$

where $X \sim$ B(n, p) and $Y \sim$ N(np, npq). Extending this argument to two or more adjacent values of X, it is easily seen with the aid of Figure 14-2 that

(14.21) $$P(a \leqslant X \leqslant b) \doteq P(a - 1/2 \leqslant Y \leqslant b + 1/2)$$

The right-hand sides of (14.20) and (14.21) would be evaluated by standardizing Y and using Table A.5. The value of $\pm 1/2$ are called continuity corrections. Failure to use these factors results in the omission of one-half the probability at $X = a$ and $X = b$; when calculating $P(X = c)$, the omission of $\pm 1/2$ would always give a zero probability.

The two probabilities posed at the beginning of this section are evaluated in the next example, which is followed by a comparative example.

▶ ***Example 14-9***

Assuming the number of white-flowered plants, X, among sets of 2,000 second-generation plants is binomial with parameters $n = 2{,}000$ and $p = 1/4$, the approximating distribution is the normal with $\mu = 2{,}000(1/4) = 500$ and $\sigma^2 = 2{,}000(1/4)(3/4) = 375$. Thus,

$$P(X = 500) \doteq P\left(\frac{499.5 - 500}{\sqrt{375}} < Z < \frac{500.5 - 500}{\sqrt{375}}\right)$$

$$= P(-0.026 < Z < 0.026)$$

$$= 2(0.5104) - 1 = 0.0208$$

and

$$P(X \geqslant 530) \doteq P\left(\frac{529.5 - 500}{\sqrt{375}} < Z < \frac{2000.5 - 500}{\sqrt{375}}\right)$$

$$= P(1.52 < Z < 77.5)$$

$$= 1 - 0.9357 = 0.0643$$

The florist has about a 2% chance of getting exactly 500 white-flowered plants among 2,000 and slightly over a 6% chance of having more than 530 white-flowered plants for sale. ◀

▶ ***Example 14-10***

The quality of the normal approximation to the binomial will be illustrated for two different distributions. Specifically, $P(X = 20)$ and $P(10 \leqslant X \leqslant 28)$ will be calculated for the B(50, 1/4) and B(50, 1/2) distributions by the normal approximation and compared with the exact values found in Table A.3. When $X \sim$ B(50, 1/4), $\mu = 12.5$ and $\sigma = 3.06$, and

$$P(X = 20) \doteq P\left(\frac{19.5 - 12.5}{3.06} < Z < \frac{20.5 - 12.5}{3.06}\right)$$

$$= P(2.29 < Z < 2.62)$$

$$= 0.9956 - 0.9890 = 0.0066$$

$$P(10 \leqslant X \leqslant 28) \doteq P\left(\frac{9.5 - 12.5}{3.06} < Z < \frac{28.5 - 12.5}{3.06}\right)$$

$$= P(-0.98 < Z < 5.23)$$

$$= 0.8365$$

Exact values are 0.0077 and 0.8363.

When $X \sim$ B(50, 1/2), $\mu = 25$ and $\sigma = 3.54$, whereby

$$P(X = 20) \doteq P\left(\frac{19.5 - 25}{3.54} < Z < \frac{20.5 - 25}{3.54}\right)$$

$$= P(-1.55 < Z < -1.27)$$

$$= 0.9394 - 0.8980 = 0.0414$$

$$P(10 \leqslant X \leqslant 28) \doteq P\left(\frac{9.5 - 25}{3.54} < Z < \frac{28.5 - 25}{3.54}\right)$$

$$= P(-4.38 < Z < 0.99)$$

$$= 0.8389$$

Exact values are 0.0418 and 0.8389. The approximation is expected to be better in the latter case, since the distribution is symmetric when $p = 1/2$. ◀

(Problems 7(a), 21, 22)

14.2.3 Approximate Inference

Two large-sample procedures for making inference about the binomial will be discussed. One of these, based on normal theory, provides approximate tests and *CI* about p. A second technique, developed by Pearson [4], is the chi-square test for proportions.

From Theorem 14-2, the quantity $\dfrac{X - np}{\sqrt{npq}}$ is an approximate Z variable for large n. Adhering to the large sample conditions (14.19), the $100\alpha\%$ test about p would have the following structure:

(14.22)

① ⓐ $H_0: p \leqslant p_0$, $H_1: p > p_0$ ① ⓑ $H_0: p \geqslant p_0$, $H_1: p < p_0$ ① ⓒ $H_0: p = p_0$, $H_1: p \neq p_0$

② Under H_0, $Z \doteq \dfrac{X - np_0}{\sqrt{np_0(1 - p_0)}}$

③ Reject H_0 if
ⓐ $Z \geqslant z_{1-\alpha}$ ⓑ $Z \leqslant z_\alpha$ ⓒ $|Z| \geqslant z_{1-\alpha/2}$

Since the quality of approximate inference depends upon the ability to obtain good approximation of probabilities, it follows, in view of remarks made in the preceding section, that the Z test described above gives good results under conditions (14.19). The power of this Z test is discussed in Section 14.5.2.

► ***Example 14-11***

A candidate believes that more than 3/4 of the voters favor a particular issue. To check this contention, campaign workers get the opinions of 2,500 voters chosen at random. They found 1,953 who favored the issue. An approximate 5% test is

① $H_0: p \leqslant 3/4$ versus $H_1: p > 3/4$

② Under H_0, $Z \doteq \dfrac{X - n(3/4)}{\sqrt{n(3/4)(1/4)}}$

③ Reject H_0 if $Z \geqslant 1.645$

④ $Z \doteq \dfrac{1953 - 1875}{\sqrt{2500(3/16)}} = 3.60$

Therefore, H_0 is rejected.

⑤ More than 3/4 of the voters favor the issue, unless a 5% error has been made. ◀

An approximate $100(1 - \alpha)\%$ *CI* for p can be obtained from (9.6) and (14.18). The *CI* is the set of values for p which satisfy the inequality

(14.23) $$-z_{1-\alpha/2} < \frac{X - np}{\sqrt{npq}} < z_{1-\alpha/2}$$

The same problem is encountered in solving (14.23) as with the large sample *CI* for the Poisson; see the remarks following (14.9). Dropping terms $(1/n)z_{1-\alpha/2}^2$ the solution of (14.23) gives approximate limits

(14.24) $$L, U = \hat{p} \pm z_{1-\alpha/2}\sqrt{\hat{p}(1 - \hat{p})/n}$$

Again, the error in limits (14.24) is roughly $(1/2n)z_{1-\alpha/2}^2$.

For testing $H_0: p = p_0$ versus $H_1: p \neq p_0$ in a binomial experiment, a substitute for the large sample Z test is an equivalent chi-square test. (With certain modifications, this chi-square test could be used when the alternative is one-sided, but the Z test is more suitable and is recommended.) The chi-square test readily extends to the multinomial case, which is discussed in the next section; the Z test is not suitable for extension to multinomial analyses.

Common notations for the chi-square test of binomial proportions are:

p_1 = proportion in class 1 (successes)
p_2 = proportion in class 2 (failures)
O_1 = number observed in class 1
O_2 = number observed in class 2
E_1 = number expected in class 1
E_2 = number expected in class 2.

Note that $O_1 + O_2 = n$ and $p_1 + p_2 = 1$. The hypotheses $H_0: p = p_0$ versus $H_1: p \neq p_0$ are equivalent to

$$H_0: p_1 = p_{1,0} \text{ versus } H_1: p_1 \neq p_{1,0}$$
$$\quad p_2 = p_{2,0} \qquad\qquad\quad p_2 \neq p_{2,0}$$

where $p_{1,0}$ and $p_{2,0}$ are postulated values of the two class proportions. If H_0 is true, the number of observations expected in the first class is $E_1 = np_{1,0}$

and, in the second class, $E_2 = np_{2,0}$. Thus, $E_1 + E_2 = n$ and the approximate chi-square statistic is

(14.25)
$$\chi_1^2 = \sum_{i=1}^{2} \frac{(O_i - E_i)^2}{E_i}$$

When there is close agreement between the sample and hypothesized results, each $(O_i - E_i)$ will be small; hence, the value of χ_1^2 will be small. In contrast, samples that deviate greatly from the hypothesized give large values of $(O_i - E_i)$ and, thereby, large values of χ_1^2.

In summary, the approximate χ^2 test of a binomial proportion would be:

(14.26)

① $H_0: p_i = p_{i,0}$ for $i = 1, 2$
$H_1: p_i \neq p_{i,0}$ for $i = 1, 2$

② Under H_0, $\sum_{i=1}^{2} \frac{(O_i - E_i)^2}{E_i} \doteq \chi_1^2$

③ Reject H_0 if $\chi_1^2 \geqslant \chi_{1;1-\alpha}^2$

It should be noted that this chi-square test has a one-sided critical region even though the alternative is two-sided. Establishing the equivalence of these approximate Z and χ^2 tests makes use of the fact that

(14.27)
$$\chi_1^2 = Z^2,$$

a well-known relationship that has been established in mathematical statistics.

Because of the equivalence of the approximate Z and chi-square tests, it follows that results of these tests will be quite good when the binomial conditions (14.19) hold. This means that each expected value, E_i, should be at least 5.

The use of continuity corrections when making approximate tests and *CI* has been a subject of some debate recently. We will not use continuity corrections except to calculate probabilities as in Section 14.2.2. Continuity corrections should never be used in calculating chi-squares when the degrees of freedom are greater than one or when the chi-squares are to be added.

► ***Example 14-12***

To check whether 1% of the jewelers' screws of a certain type have defective heads (unpolished, faulty slots, or other flaws), a random sample of 1,500 is selected from those available for sale. There are 21 with defective heads. Because of the large stockpile from which the sample was selected, the independence and constant probability conditions of the binomial are approximately satisfied. A 5% approximate normal test is:

① $H_0: p_1 = 0.01$ versus $H_1: p_1 \neq 0.01$
$\quad p_2 = 0.99 \qquad\qquad\qquad p_2 \neq 0.99$

② Under H_0, $Z \doteq \dfrac{X - 15}{\sqrt{n(0.01)(0.99)}}$

③ H_0 is rejected if $|Z| \geqslant 1.96$

④ $Z \doteq \dfrac{21 - 15}{\sqrt{14.85}} = \dfrac{6}{3.85} = 1.56$

Therefore, we fail to reject H_0.

⑤ There is no sample evidence at the 5% level to say that other than 1% of the jewelers' screws of the given stockpile have defective heads.

Equivalently, a 5% approximate chi-square test would be

① Same

② Under H_0, $\chi_1^2 \doteq \sum_{i=1}^{2} \dfrac{(O_i - E_i)^2}{E_i}$

③ H_0 is to be rejected if $\chi_1^2 \geqslant 3.84 = (1.96)^2$

④ $O_1 = 21$, $O_2 = 1{,}479$, $E_1 = 15$, $E_2 = 1{,}485$

$$\chi_1^2 \doteq \frac{36}{15} + \frac{36}{1{,}485} = \frac{36}{14.85} = 2.42 = (1.56)^2$$

Therefore, H_0 is not rejected.

⑤ Same ◀

(Problems 3, 7(b), 13)

14.3 INFERENCE IN MULTINOMIAL EXPERIMENTS

14.3.1 Inference about Proportions in a Single Multinomial Experiment

Analyses of multinomial data are easily accomplished through straightforward modifications of the large-sample techniques of the binomial case. Other than point estimation, we will limit our discussion to approximate inference under these procedures.

Confidence intervals will not be discussed. Large-sample *CI* for single-class proportions can be made by reducing the data to a binomial set; that is, two classes: the class of interest and all other classes combined. Then (14.24) can be used to obtain an approximate *CI*.

When a single sample of multinomial data is available, a problem of particular interest is that of testing hypotheses which specify the class proportions, that is, attempting to assess the reasonableness of a particular multinomial model.

For $i = 1, 2, \ldots, k$, let

$$\begin{aligned} p_i &= \text{proportion for class } i \\ p_{i,0} &= \text{postulated value of the proportion for class } i \\ O_i &= \text{number observed in class } i \\ E_i &= \text{number expected in class } i \end{aligned} \tag{14.28}$$

Restrictions on these quantities are

$$\begin{aligned} \sum_{i=1}^{k} p_i &= \sum_{i=1}^{k} p_{i,0} = 1 \\ \sum_{i=1}^{k} O_i &= \sum_{i=1}^{k} E_i = n \end{aligned} \tag{14.29}$$

An unbiased estimator of the i-th class proportion p_i is

$$\hat{p}_i = O_i/n, \tag{14.30}$$

which has a variance of

$$\sigma^2_{\hat{p}_i} = p_i(1 - p_i)/n \tag{14.31}$$

Since the O_i are related, as seen in (14.29), the $\hat{p}_i$ are also related. The covariance of $\hat{p}_i$ and $\hat{p}_j$ can be shown to be

$$\sigma_{\hat{p}_i\hat{p}_j} = -p_i p_j/n; \quad \text{for } i \neq j \tag{14.32}$$

Unbiased estimates of these variability measures are

$$\begin{aligned} s^2_{\hat{p}_i} &= \hat{p}_i(1 - \hat{p}_i)/(n - 1) \\ s_{\hat{p}_i\hat{p}_j} &= -\hat{p}_i\hat{p}_j/(n - 1); \qquad \text{for } i \neq j \end{aligned} \tag{14.33}$$

In practice, when n is large, the denominators in (14.33) are frequently replaced by n.

▶ ***Example 14-13***

In a consumer survey, 200 women are asked to indicate their preferred grocery-shopping time: mornings, afternoons, or evenings, denoted by classes 1, 2, and 3, respectively. The results are $O_1 = 52$, $O_2 = 86$, and $O_3 = 62$. Estimates of the three class proportions are $\hat{p}_1 = 0.26$, $\hat{p}_2 = 0.43$, and $\hat{p}_3 = 0.31$. Using n rather than $n - 1$ in (14.33), the variances, standard errors, and covariances are estimated by the following values:

$$s_{\hat{p}_1}^2 = 0.26(0.74)/200 = 0.000962; \qquad s_{\hat{p}_1} = 0.0981$$
$$s_{\hat{p}_2}^2 = 0.43(0.57)/200 = 0.0001226; \qquad s_{\hat{p}_2} = 0.0111$$
$$s_{\hat{p}_3}^2 = 0.31(0.69)/200 = 0.0001070; \qquad s_{\hat{p}_3} = 0.0104$$
$$s_{\hat{p}_1\hat{p}_2} = -0.26(0.43)/200 = -0.000559$$
$$s_{\hat{p}_1\hat{p}_3} = -0.26(0.31)/200 = -0.000403$$
$$s_{\hat{p}_2\hat{p}_3} = -0.43(0.31)/200 = -0.000667$$

A 95% approximate *CI* for the proportion of women who prefer afternoon grocery shopping is, by (14.24) and (14.33):

$$\hat{p}_2 \pm z_{0.975}s_{\hat{p}_2} = 0.43 \pm 1.96(0.0111)$$
$$= 0.43 \pm 0.02$$

Or, with 95% confidence, we can say between 41 and 45 per cent of the group sampled prefer afternoon grocery shopping. ◀

An obvious generalization of the large-sample chi-square test of binomial proportions will handle an important class of statistical tests in multinomial experiments. Specifically:

(14.34)

① $H_0: p_i = p_{i,0}$; for $i = 1, 2, \ldots, k$
$H_1: p_i \neq p_{i,0}$; for at least two i

② If H_0 is true, $E_i = np_{i,0}$ and

$$\chi_{k-1}^2 \doteq \sum_{i=1}^{k} (O_i - E_i)^2/E_i$$

③ H_0 is rejected if $\chi_{k-1}^2 \geqslant \chi_{k-1;1-\alpha}^2$

Again, this test has a one-sided critical region, even though H_1 is multidirectional. Because of the restrictions (14.29), when one proportion differs from the postulated value, at least one other proportion must differ from its postulated value. No distinction is made in H_1 whether the proportions are above or below the postulated values, hence, many directionals are actually tested at the same time.

As in the binomial case, the approximate chi-square test is quite good whenever $E_i \geqslant 5$ for all i. Failure of this condition to hold tends to inflate the calculated value of χ^2. One remedy which is frequently proposed is to combine any class having $E_i < 5$ with an adjacent class. If $1 < E_i < 5$ for only one class and $E_i \geqslant 5$ for all other classes, many statisticians feel it is unnecessary to combine the troublesome class.

▶ ***Example 14-14***

In the consumer study of Example 14-13, it was believed that the proportions should be 1/4, 1/2, and 1/4 for classes 1, 2, and 3, respectively. Is there any evidence at the 10% level that the proportions differ from the 1:2:1 ratio?

① $H_0: p_1 = p_3 = 1/4, p_2 = 1/2$
H_1: At least two p_i differ from the stated values.

② Under H_0, $\chi_2^2 \doteq \sum_{i=1}^{3} (O_i - E_i)^2/E_i$

③ Reject H_0 if $\chi_2^2 \geqslant 4.61$

④ $O_1 = 52, O_2 = 86, O_3 = 62$
$E_1 = 50, E_2 = 100, E_3 = 50$
$\chi_2^2 \doteq (52 - 50)^2/50 + (86 - 100)^2/100 + (62 - 50)^2/50$
$= 4.92$
Therefore, H_0 is rejected.

⑤ Unless a 10% error has occurred, we conclude that the preference times are not 1/4, 1/2, and 1/4 for morning, afternoon, and evening, respectively. ◀

(Problems 9, 15)

*14.3.2 Tests about Proportions in Several Related Multinomial Experiments

Suppose that m similar multinomial distributions (each having the same k defining classes) are being investigated, the resulting individual multinomial experiments being independent of each other. Such would be the case if the same experiment was conducted at different locations, at different times, and so on. For example, the consumer survey considered in the previous section might be conducted in different cities. A manufacturing process might be conducted at different plant sites, the number of different items produced forming a multinomial. An agricultural investigation of the proportion of a certain crop infected with rust might be carried out at each of several experiment stations.

Four questions of interest are discussed.

(1) Does each individual multinomial support a fixed hypothesis about the class proportions?

The data of each multinomial experiment would be subjected to the multinomial chi-square test of Section 14.3.1. That is, we test the null hypothesis

$$H_0^{(1)}: p_{ij} = p_{j,0} \qquad \text{for fixed } i \text{, and for } j = 1, 2, \ldots, k,$$

where p_{ij} is the j-th class proportion for the i-th multinomial and $p_{j,0}$ is a *specified* null value. The significance level applies to each individual test, not to the collective set of tests of $H_0^{(1)}$.

If we make a single test that the multinomials simultaneously support a fixed hypothesis, we obtain a significance level that applies to the overall set of multinomials. Thus we ask a second question:

(2) Do the multinomials simultaneously support a fixed hypothesis about the class proportions?

Here we test

$$H_0^{(2)}: p_{ij} = p_{j,0} \qquad \text{for all } i, \text{ and for } j = 1, 2, \ldots, k,$$

where p_{ij} and $p_{j,0}$ have the same meaning as in the earlier case. Note the difference: in (2), the test is for all i, while in (1), the test is for a single i at a time. The statistic used to test $H_0^{(2)}$ is the sum of the individual chi-square statistics used to test the hypotheses $H_0^{(1)}$. This is called a *total chi-square* and has degrees of freedom equal to the sum of degrees of freedom, namely, $m(k - 1)$.

Now consider a third question:

(3) If each individual multinomial supports $H_0^{(1)}$, do the multinomials when pooled together support the same hypothesis about the class proportions?

Here a *pooled chi-square test* is performed in the same manner as the individual chi-squares but using the "pooled total" for each class. The pooled chi-square has $(k - 1)$ degrees of freedom, the same as the individual chi-squares, since the number of classes is still k. The hypothesis tested by the pooled chi-square is

$$H_0^{(3)}: p_{\cdot j} = p_{j,0} \qquad \text{for } j = 1, 2, \ldots, k,$$

where $p_{\cdot j}$ denotes the proportion belonging to the j-th class when the individual multinomials are combined ($p_{j,0}$ is the same specified null value as before).

Now consider a fourth and final question:

(4) Do the individual multinomial experiments which *do not* follow the specified hypothesis $H_0^{(1)}$ all follow the same multinomial distribution or are they heterogeneous?

If only one individual multinomial experiment does not follow the specified hypothesis, we know immediately that it differs significantly from the rest. We therefore suppose that $w \geqslant 2$ individual multinomials differ significantly from the specified distribution. We know that these w multinomials differ significantly from the set of multinomials that support the specified hypothesis, but do these w differ among themselves or are they members of the same (but unspecified) multinomial?

Using the data for the w individual experiments which differ significantly from the specified multinomial (renumbering them $1, 2, \ldots, w$, if necessary), we calculate a "homogeneity chi-square" as follows:

$$\chi_v^2 = \sum_{i=1}^{w} \sum_{j=1}^{k} (O_{ij} - E_{ij})^2 / E_{ij} \tag{14.35}$$

where

$$\begin{aligned} E_{ij} &= n_i C_j / n^* \\ n_i &= \text{sample size for the } i\text{-th multinomial} \\ n^* &= \text{sample size for the } w \text{ multinomials} \\ C_j &= \text{"pooled total" for the } j\text{-th class of the } w \text{ multinomials} \\ O_{ij} &= \text{number observed in the } j\text{-th class of the } i\text{-th multinomial} \\ v &= (w - 1)(k - 1) \end{aligned} \tag{14.36}$$

The hypothesis tested by the homogeneity chi-square is

$H_0^{(4)}$: All multinomials not supporting $H_0^{(1)}$ have identical distributions

Equivalently,

$$H_0^{(4)}: p_{ij} = p_{\cdot j}^* \qquad \text{for } i = 1, 2, \ldots, w \text{ and } j = 1, 2, \ldots, k$$

where $p_{\cdot j}^*$ is an *unspecified* proportion for the j-th class.

► ***Example 14-15***

In a large factory, where a number of assembly lines are producing steel shafts the production is presumed to be 80% usable, 15% in need of reworking (too long), and 5% scrap (too short). From each assembly line, 200 steel shafts were inspected. Do the following data from six assembly lines support this hypothesis at the 5% level of significance?

Line	Usable	Need reworking	Scrap	Individual χ^2	*df*
1	160	30	10	0	2
2	150	25	25	23.96	2
3	180	20	0	15.83	2
4	175 (160)	15 (30)	10 (10)	8.91	2
5	150	45	5	10.62	2
6	160	35	5	3.33	2
Totals	975 (960)	170 (180)	55 (60)	$\chi_{12}^2 = 62.65$	12

Expected numbers are in parentheses. With equal numbers, $n = 200$, the expected numbers are the same for each assembly line. Thus there is statistical evidence, at the 5% level of significance, that assembly lines 2, 3, 4, and 5 are not producing according to the postulated proportions. Pooled chi-square is not appropriate in view of significantly different productions.

Using the data for lines 2, 3, 4, and 5, their homogeneity is now investigated:

Line	Usable	Need reworking	Scrap
2	150	25	25
3	180 (163.75)	20 (26.25)	0 (10)
4	175	15	10
5	150	45	5
Totals	655	105	40

From (14.35) and (14.36), the homogeneity chi-square is

$$\begin{aligned}\chi_6^2 &= (150 - 163.75)^2/163.75 + \cdots + (5 - 10)^2/5 \\ &= 59.45\end{aligned}$$

Since the 5% critical point of χ_6^2 is 12.6, there is sufficient evidence to say that lines 2, 3, 4, and 5 are producing in different manners. ◀

(Problems 12, 27)

14.4 CONTINGENCY TABLES

When a binomial or multinomial experiment is conducted, the data are recorded as belonging to two or more classes defined by a single factor (criterion) such as "voting preference." If every population and sample element were categorized according to two factors, the resulting array would be a *two-way contingency table*. If M factors were used in the classification of elements, the result would be an *M-way contingency table.*

In a sense, then, an M-way contingency table can be looked upon as a multinomial in M dimensions. The two-way tables are by far the most common ones and will be the only ones discussed. (For the extension to $M > 2$ dimensions, refer to [3], p. 131.)

For the two-way tables, customary notation is:

i is the subscript for rows (factor A)
j is the subscript for columns (factor B)
O_{ij} = number of elements observed in the (i, j)-th cell
r = number of rows
c = number of columns

$R_i = O_{i\cdot} =$ observed total for the i-th row
$C_j = O_{\cdot j} =$ observed total for the j-th column
$E_{ij} =$ number of elements expected in the (i, j)-th cell
$n = O_{\cdot\cdot} =$ total number of observations.

The first three steps of the usual contingency test are:

① H_0: Factors A and B represent independent criteria of classification
H_1: Factors A and B represent dependent criteria of classification

(14.37) ② If H_0 is true,

$$\sum_{i=1}^{r} \sum_{j=1}^{c} (O_{ij} - E_{ij})^2/E_{ij} \doteq \chi^2_\nu, \qquad \text{where } \nu = (r-1)(c-1) \text{ and } E_{ij} = R_iC_j/n$$

③ Reject H_0 if $\chi^2_\nu \geqslant \chi^2_{\nu;1-\alpha}$

If H_0 is true, the two factors are independent, whereby the observations occur according to the following probabilities:

$$P(\text{obs. is in cell } i, j) = P(\text{obs. is in } i\text{-th row and } j\text{-th column}) = P(\text{obs. is in row } i)P(\text{obs. is in column } j) \tag{14.38}$$

The probabilities on the right-hand side of (14.38) are *estimated*, respectively, by R_i/n and C_j/n. Thus the probability or proportion of sample observations falling in the i, j-th cell is estimated by R_iC_j/n^2, so the expected number in the i, j-th cell is *estimated* by

$$n(R_iC_j/n^2) = R_iC_j/n = E_{ij} \tag{14.39}$$

The fact that the expected numbers must be estimated does not destroy the approximation by chi-square. This estimation is necessary because cell proportions or probabilities are not specified in a contingency analysis, whereas in a multiple multinomial analysis cell proportions are specified. This serves as a means of distinguishing multiple multinomial from contingency analyses.

► ***Example 14-16***

It is of interest to determine whether preference of a bond issue is related to annual salary, for people of a certain locality. Interviews of 120 randomly selected persons gave:

Annual Salary	Preference: Favor	Oppose	No Opinion	R_i
Less than \$5,000	15 (10.5)	9 (8.0)	6 (11.5)	30
\$5,000–\$10,000	16 (21.0)	12 (16.0)	32 (23.0)	60
Over \$10,000	11 (10.5)	11 (8.0)	8 (11.5)	30
C_j	42	32	46	120

The entries in parentheses are the E_{ij}. The 5% contingency test is:

① H_0: Preference of bond issue is independent of annual salary.
H_1: Preference of bond issue is not independent of annual salary.

② If H_0 is true, $\chi_4^2 \doteq \sum_{i=1}^{3} \sum_{j=1}^{3} (O_{ij} - E_{ij})^2/E_{ij}$

③ Reject H_0 if $\chi_4^2 \geqslant 9.49$

④ $\chi_4^2 \doteq (15 - 10.5)^2/10.5 + \cdots + (8 - 11.5)^2/11.5$
$= 20.25/10.5 + \cdots + 12.25/11.5$
$= 12.59$

Therefore, H_0 is rejected.

⑤ Preference of the bond issue depends upon a person's salary, unless a 5% error has been made. ◀

(Problems 11, 18)

*14.5 *POWER OF TESTS*

Power calculations for binomial and Poisson tests are discussed in this section, which considers both exact and approximate tests. The power of approximate chi-square tests will not be treated.

*14.5.1 Power of Poisson Tests

For the exact test about the mean value of a Poisson distribution outlined in (14.6), the power can be calculated with the aid of cumulative Poisson tables. As a first step, we determine critical points with respect to $X.$, the total number of successes. (These points could be used instead of the cumulative function to make the test.) Thus, for a test of size α, we need the following:

case ⓐ: the smallest x_1 such that

$$F(x_1 - 1; \lambda_0^*) \geqslant 1 - \alpha$$

case ⓑ: the largest x_2 such that

(14.40) $$F(x_2; \lambda_0^*) \leqslant \alpha$$

case ⓒ: the smallest x_3 and largest x_4 such that

$$F(x_3 - 1; \lambda_0^*) \geqslant 1 - \alpha/2 \quad \text{and} \quad F(x_4; \lambda_0^*) \leqslant \alpha/2$$

Then the power of the test (14.6) at the point λ_a (a value of λ specified by the alternative hypothesis) is given by the following cumulative Poisson probabilities:

case ⓐ: $P_{\lambda_a} = 1 - F(x_1 - 1; \lambda_a^*)$

(14.41) case ⓑ: $P_{\lambda_a} = F(x_2; \lambda_a^*)$

case ⓒ: $P_{\lambda_a} = 1 - F(x_3 - 1; \lambda_a^*) + F(x_4; \lambda_a^*)$

where $\lambda_a^* = n\lambda_a$.

For the large-sample Z test using (14.8) the power may be calculated as in Section 10.5. Now the power is given by one of the following cumulative standard normal probabilities:

case ⓐ: $P_{\lambda_a} = F[gz_\alpha + n\Delta_a/\sqrt{\lambda_a^*}]$

(14.42) case ⓑ: $P_{\lambda_a} = F[gz_\alpha - n\Delta_a/\sqrt{\lambda_a^*}]$

case ⓒ: $P_{\lambda_a} = F[gz_{\alpha/2} + n\Delta_a/\sqrt{\lambda_a^*}] + F[gz_{\alpha/2} - n\Delta_a/\sqrt{\lambda_a^*}]$

where $g = \sqrt{\lambda_0/\lambda_a}$; $\Delta_a = \lambda_a - \lambda_0$; and $\lambda_a^* = n\lambda_a$.

▶ ***Example 14-17***

We will calculate power for some of the Poisson tests made earlier.

Using the data in Example 14-3, suppose we would like to know the chances of rejecting H_0 if, in fact, $\lambda = 4\,(=\lambda_a)$. Since $F(20; 15) > 0.90$, we have $x_1 = 21$ and $\lambda_a^* = 5(4) = 20$. Hence the power of this test at $\lambda_a = 4$ is

$$P_4 = 1 - F(20; 20) = 1 - 0.5591 = 0.4409.$$

That is, there is a 44% chance of rejecting $H_0: \lambda \leqslant 3$ when λ is really 4.

For the test in Example 14-4, consider $\lambda_a = 2.5$. Then $\lambda_a^* = 20$, and from Table A.4 we find $F(27; 40) \leqslant 0.025$ and $F(53; 40) \geqslant 0.975$. Thus the power of this test (probability of rejecting H_0) when λ is really 2.5 is

$$P_{2.5} = 1 - F(53; 20) + F(27; 20)$$
$$= 1 - 1.0000 + 0.9475 = 0.9475 \blacktriangleleft$$

(Problem 31)

*14.5.2 Power of Binomial Tests

For the exact test about a binomial proportion outlined in (14.16), power can be calculated with the aid of existing tables. As a first step, critical points are determined with respect to X, the number of successes. (Again, these points could be used in making the test.) Thus, for a test of size α, we need:

case ⓐ: the smallest x_1 such that

$$F(x_1 - 1; n, p_0) \geqslant 1 - \alpha$$

case ⓑ: the largest x_2 such that

$$F(x_2; n, p_0) \leqslant \alpha \tag{14.43}$$

case ⓒ: the smallest x_3 and largest x_4 such that

$$F(x_3 - 1; n, p_0) \geqslant 1 - \alpha/2 \quad \text{and} \quad F(x_4; n, p_0) \leqslant \alpha/2$$

Then, for an alternative proportion p_a, the power of test (14.16) is:

(14.44)

case ⓐ: $P_{p_a} = 1 - F(x_1 - 1; n, p_a)$

case ⓑ: $P_{p_a} = F(x_2; n, p_a)$

case ⓒ: $P_{p_a} = 1 - F(x_3 - 1; n, p_a) + F(x_4; n, p_a)$

For the large-sample approximate Z test (14.22), the power can be calculated as in Section 10.5. Here the power is given by one of the following cumulative standard normal probabilities:

(14.45)

case ⓐ: $P_{p_a} = F[hz_\alpha + n\Delta_a/\sqrt{np_aq_a}]$

case ⓑ: $P_{p_a} = F[hz_\alpha - n\Delta_a/\sqrt{np_aq_a}]$

case ⓒ: $P_{p_a} = F[hz_{\alpha/2} + n\Delta_a/\sqrt{np_aq_a}] + F[hz_{\alpha/2} - n\Delta_a/\sqrt{np_aq_a}]$

where $h = \sqrt{p_0q_0/p_aq_a}$ and $\Delta_a = p_a - p_0$.

▶ ***Example 14-18***

For the test made in Example 14-7, the chance of rejecting H_0 when the proportion is actually 0.20 is of interest. We find from Table A.3 that $F(4; 20, 0.10) > 0.95$; therefore, with $x_1 = 5$ and $p_a = 0.20$, the required probability is

$$P_{0.20} = 1 - F(4; 20, 0.20)$$

$$= 1 - 0.6296 = 0.3704$$

In other words, with the proposed test procedure, there is not a very good chance of rejecting H_0 even when the proportion is actually 0.20.

For the test in Example 14-8, we will find the power if $P(\text{head}) = 0.4(=p_a)$. From Table A.3, $F(31; 20, 1/2) > 0.95$ and $F(18; 20, 1/2) < 0.05$. Therefore, with $x_3 = 32$, $x_4 = 18$, and $p_a = 0.4$, the power is given by

$$P_{0.4} = 1 - 0.995 + 0.3356 = 0.3361 \blacktriangleleft$$

(Problem 29)

PROBLEMS

1. Examination of 15 square yards of carpet revealed 7 flaws. Set a 95% *CI* for the average number of flaws per square yard.
2. From past experience, it is known that 8% of all tires recapped by a given process fail to outlast the guarantee. If a retailer sells 500 of these recapped tires and replaces those which fail to outlast the guarantee, find the approximate probability that he must replace
   (a) between 30 and 60 tires.
   (b) less than 30 tires.
3. A new insecticide is supposed to kill at least 80% of the grasshoppers upon contact. If it is applied to 400 grasshoppers and 302 die upon contact, can the claim be rejected at the 1% level?
4. A book publisher needs some information on small orders that are damaged in shipment. By contacting customers receiving 400 randomly selected small orders, it was found that 60 small orders were damaged in shipment.
   (a) Set a 90% *CI* for the true proportion of small orders damaged in shipment.
   (b) At the 5% level, is there evidence that more than 10% of all small orders are damaged in shipment?
5. Suppose the number of passengers departing daily from a small airport is an approximate Poisson variable. For 20 days selected at random, the number of departures was 156.
   (a) Set a 99% *CI* for the average number of daily departures at this airport.
   (b) Is there evidence, at the 5% level, that the average number of daily departures at this airport exceeds 6?
6. Because of small flaws developing during the manufacture of a metallic crystal, one-tenth of all crystals produced by a given process are defective. A company manufactures 2,500 such crystals each week. Let X denote the number of defective crystals produced in a week. Approximate

   (a) $P(230 < X < 275)$. (b) $P(X = 250)$. (c) $P(X \geqslant 240)$.

7. A person claims to be gifted with ESP (extra-sensory perception). To investigate the claim, a psychologist performs the following trial: four cards, one of each suit, are placed face down on a table, and the subject is asked to select the diamond.
   (a) If 48 independent trials are performed and the person does not have ESP, what is the probability he gets exactly 16 correct? at least 12 correct?

(b) Suppose one is willing to admit that the person has ESP if the diamond is correctly chosen at least 60% of the time. If 150 independent trials are performed and the subject selects the correct card only 80 times, can the claim be refuted at the 5% level?

8. In making an appeal for subsidy, an ambulance company of a certain city claims at least 10 dry runs per week. "Dry runs" are false alarms or calls made at the first sight of an accident; in either case, the ambulance makes a trip for which no fee is collected. City commissioners, investigating the appeal, have found the following number of dry runs for ten randomly chosen weeks: 5, 12, 9, 7, 8, 13, 7, 9, 6, 8.

(a) Estimate the mean number of dry runs per week.
(b) Give the standard error of the estimate found in (a).
(c) Test the claim at the 5% level.

9. A professor claims to give grades in an introductory course as follows:

A's—10%	C's—48%	F's—12%
B's—15%	D's—15%	

In past terms, this professor gave the following grades in this introductory course:

A's—32	C's—133	F's—40
B's—40	D's— 55	

(a) Do these data support his claim? Let $\alpha = 5\%$.
(b) Estimate the proportion of C's given by this professor in this course and give its standard error.

10. In a sample of 400 elm trees, 234 had Dutch elm disease.

(a) Estimate the proportion of trees having Dutch elm disease.
(b) Give the standard error of this estimate.
(c) Is there any reason to doubt the contention that no more than one-half of the trees have Dutch elm disease? Let $\alpha = 0.10$.

11. In a study of mammalogy literature, Anderson and Van Gelder [1] gave the following data on the number of papers published in mammalogy:

	Number of Authors	
Length	Single	Multiple
Short papers (1–5 pages)	1,021	482
Medium papers (6–49 pages)	839	608
Long papers (over 49 pages)	71	21

Is there any relationship between the length and number of authors for mammalogy papers? Do the appropriate analysis with $\alpha = 0.05$.

*__12.__ A germination experiment was carried out with three light sources. Twelve hours of light were supplied each day, 1,600 seeds were used for each type of light. The results were:

Light	Number Germinated
Blue	1,326
Green	1,257
Yellow	1,384

Is the germination percentage influenced by the type of light? Do the appropriate analysis with $\alpha = 0.025$.

13. A random sample of 900 voters were surveyed, and 422 favored a given proposal. In order for the proposal to be adopted, at least 50% of the voters must vote in favor of it. At the 5% level, is there any reason to believe that the proposal will not be adopted?

14. A bookstore will special-order books not carried in stock. It is not unreasonable to assume that the weekly requests for special orders for books is approximately Poisson. During 10 randomly selected weeks, the number of special book orders was 42.
Set a 95% *CI* for the average number of special book orders
(a) per week.
(b) per month (four weeks).

15. In a study of human influence on system malfunctions, Robinson et al. [5] categorized 213 vehicle malfunctions as follows:

Primarily human	25
Primarily equipment	158
Combination human-equipment	30

(a) Do these data support the hypothesis that vehicle malfunctions are Human: Equipment:Combination in the 1:4:1 ratio? Let $\alpha = 0.05$.
(b) Estimate the three proportions and their standard errors.

16. When selling their crop, chili growers are paid according to the proportion of mature red and mature green pods. Before the harvest of a certain grower's crop, chili plants were selected until 2,000 pods were obtained. Inspection revealed 1,018 mature red pods, 721 mature green pods, and 261 pods that were not mature.
(a) Estimate the proportion of his crop which is mature red pods and give its standard error.
(b) Estimate the proportion of his crop which is mature green pods and give its standard error.
(c) Give a 95% interval estimate for the proportion of mature red pods in his crop. (Hint: Let mature red pods define one group and combine the other two types of pods into one group.)

17. Refer to Problem 24 of Chapter 7, and assume that the number of absences due to illness has an approximate Poisson distribution. Do the data given there provide evidence against the claim that the number of weekly absences due to illness is no greater than 4 on the average? Let $\alpha = 0.05$.

18. A barber keeps track of the type of haircuts he gives over a one-month period. He also finds out, while conversing with the customer, whether or not each person is a college student. He records

	Type of Haircut	
Customer	Crew-Cut	Non-Crew-Cut
College student	16	106
Non-college student	38	152

Is there any relationship between the type of haircut and the student status of his customers? Let $\alpha = 5\%$.

19. During a criminal trial, a psychologist testified that he sees an average of at least 8 paranoid patients per week. An examination of his records revealed that he saw 107 paranoid patients during 15 randomly selected weeks. At the 1% level, is there evidence against his claim?

20. A potato grower knows that 1/5 of his potato plants were infected with a certain type of rust during each of the past two seasons. He believes that this year's crop will have a significantly lower infection rate. Examination of 625 plants of this season showed 104 infected with rust.
(a) Estimate the proportion of this season's crop infected with rust.
(b) Estimate the standard error for the proportion found in (a).
(c) Estimate the proportion of this season's crop infected with rust, by a 99% *CI*.
(d) Is this year's infection significantly lower than that of the past two seasons? Let $\alpha = 5\%$.

21. In a particular taste testing experiment, the taster is presented with five identical appearing samples: 3 are brand A, 2 are brand B. After tasting each of the five samples, presented in a random order, the taster must identify the brand B samples. Suppose that a certain taster cannot tell the difference between brands A and B and guesses. In 100 such trials, what is the probability he correctly identifies the B samples
(a) at least 30 times?
(b) no more than 45 times?
(c) exactly 40 times?

22. The probability of dealing a bridge hand containing at least two aces is about 1/4. If a person plays 192 hands of bridge, what is the probability that the number of hands containing at least two aces is
(a) exactly 50?
(b) no more than 40?
(c) between 40 and 60?

23. Suppose the number of customers arriving for service (at a cafe, root beer stand, beauty shop, etc.) during 15-minute intervals has an approximate Poisson distribution. For 20 independent 15-minute intervals, the arrivals at a certain establishment were: 7, 16, 13, 9, 11, 17, 10, 12, 11, 16, 14, 8, 15, 10, 8, 13, 11, 16, 14, 12.

(a) Estimate the mean number of arrivals for a 15-minute interval.

(b) Give the standard error of this estimate.

(c) Set a 99% *CI* for the mean number of arrivals for 15-minute intervals.

(d) Test $H_0: \lambda = 10$ versus $H_1: \lambda > 10$ at the 5% level.

24. A large company is considering three health insurance plans. Each employee is asked to indicate his or her preference. The following is recorded:

Type of Worker	Plan A	Plan B	Plan C
Blue collar	103	82	242
Supervisory	12	11	30
Clerical	22	20	26
Administrative	20	8	12

Are the proportions of employees favoring the various health insurance plans independent of job classification? Do the appropriate analysis with $\alpha = 0.05$.

25. The motor vehicle division of a certain state wished to estimate the proportion of vehicles (licensed by the state) which have defective equipment. During safety inspections at several localities, 4,900 state-licensed vehicles were checked. Of these, 1,040 had defective equipment.

(a) Estimate the proportion of state-licensed vehicles having defective equipment.

(b) Estimate the variance of the estimate found in (a).

(c) Set a 95% *CI* for the true proportion of state-licensed vehicles having defective equipment.

26. A geneticist working with the coat color of rodents expects the second generation offspring to have coat colors Agouti, Black, and Albino in the 9:3:4 ratio when first generation parents of a certain genetic makeup are crossed. In the second generation of such a cross, there were 75 Agouti, 42 Black, and 43 Albino. At the 1% level, is this in accordance with genetic expectations?

***27.** The number of illegal parking tickets issued at each of the three main parking lots of a college campus over a period of one year was as follows:

Quarter	Lot A	Lot B	Lot C
Fall	145	118	133
Winter	196	145	181
Spring	127	104	108
Summer	83	98	77

At the 5% level, examine the hypothesis that 1/3 of the illegal parking tickets (given in the three main lots) are issued in each lot.

*28. Vehicle malfunctions given in Problem 15 were actually categorized by Robinson et al. [5] in the following two-way layout:

Type	Primarily Human	Primarily Equipment	Combination Human-Equipment
Operations vans	19	70	11
Remote vans	6	35	10
Equipment vans	0	53	9

Can vehicle malfunctions be regarded as Human : Equipment : Combination in the 1:4:1 ratio for each vehicle type? Do the appropriate analysis, letting $\alpha = 10\%$.

*29. Find the power of the test in Problem 3 if the unknown proportion is actually
(a) 0.75 (b) 0.70 (c) 0.72

*30. Find the power of the test in Problem 4(b) if the unknown proportion is actually
(a) 0.18 (b) 0.20 (c) 0.22

*31. Find the power of the test in Problem 5(b) if the unknown average daily departures are actually
(a) 7 (b) 7.5 (c) 8

*32. Find the power of the test in Problem 8(c) if the unknown average dry runs per week is actually
(a) 9 (b) 8 (c) 7.5

*33. Find the power of the test in Problem 13 if the unknown proportion of voters favoring the proposal is actually
(a) 0.44 (b) 0.45 (c) 0.46

*34. Find the power of the test in Problem 20(d) if the unknown proportion infected is actually
(a) 0.16 (b) 0.12 (c) 0.10

REFERENCES

[1] Anderson, S., and R. G. Van Gelder, "The History and Status of the Literature of Mammalogy," *BioScience*, **20**, 17 (1970).

[2] Mood, A. M., and F. A. Graybill, *Introduction to the Theory of Statistics*, 2nd ed., McGraw-Hill, New York, 1963.

[3] Ostle, B., *Statistics in Research*, 2nd ed., Iowa State University Press, Ames, Iowa, 1963.

[4] Pearson, Karl, "On a Criterion that a Given System of Deviations from the Probable in the Case of a Correlated System of Variables Is Such that It Can Reasonably Be Supposed to Have Arisen in Random Sampling," *Philosophical Magazine*, **50** (1900).

[5] Robinson, J. E., Jr., W. E. Deutsch, and J. G. Rogers, "The Field Maintenance Interface between Human Engineering and Maintainability Engineering," *Human Factors*, **12**, 3 (1970).

Chapter 15

ANALYSIS OF VARIANCE

A common problem facing researchers—and one to which we have already alluded several times— is making a distinction between various populations on the basis of their mean values. For example, a biologist may be interested in the average growth rate of several strains of bacteria in a particular environment. An engineer may wish to determine the strength of steel having different percentages of carbon. A horticulturist may be asked to choose the best variety (as measured by average yield) from a set of six new varieties of peaches.

15.1 INFERENCE ABOUT SEVERAL POPULATION MEANS

Chapter 11 discussed inference for two normal populations. We now extend some of those analyses to more than two populations.

Multiple-population procedures were studied under the regression topic. Recall that a basic assumption of regression is that the different population means are structured and related in a particular way (we dealt in detail only with linear relations). No such structure is assumed among the population means in this chapter; it is in this respect that the regression techniques differ from the techniques to be discussed.

We would ultimately like to determine which population means differ and by how much. We cannot make either of these statements from the procedures discussed herein; more advanced books on analysis of variance contain material of this nature. We will settle for much weaker conclusions: either that the means differ or they do not. These conclusions result from a test of

$$H_0: \mu_1 = \mu_2 = \cdots = \mu_k \qquad (15.1)$$
$$H_1: \text{at least two } \mu_i \text{ differ}$$

where $\mu_1, \mu_2, \ldots, \mu_k$ are the means of the k populations under investigation. When $k \geqslant 3$, the test of (15.1) cannot be made by any of the previously discussed procedures. Testing the equality of three or more population means can be accomplished by the analysis of variance, one of the most powerful and most widely used statistical techniques.

Definition 15-1

Analysis of variance (ANOVA) is the process of subdividing the total variability of experimental observations into portions attributable to recognized sources of variation.

It is actually the total sum of squares rather than the total or overall variance that is subdivided by the ANOVA. Without actually saying so, we performed an ANOVA in regression when we subdivided SS_Y into two parts: the *SS* due to regression and the *SS* of deviations about regression.

ANOVA results are frequently presented in tabular form. A partial ANOVA table for simple linear regression is given as Table 15-1. The column headed *MS* gives the *mean squares* obtained upon division of *SS* by *df*. It should be noted that the two mean squares do *not* add to s_Y^2, but the two entries in the *df* and *SS* columns do add to the total *df* and *SS*, respectively. This may be stated as a general rule in ANOVA.

•• *Rule A-1:* When *SS* are additive, so are their *df*.

Table 15-1 ANOVA for Simple Linear Regression

Source (of variation)	*df*	*SS*	*MS*
Due to regression	1	$\sum(\hat{Y}_i - \bar{Y})^2$	$\sum(\hat{Y}_i - \bar{Y})^2$
Deviation about regression	$n-2$	$\sum(Y_i - \hat{Y}_i)^2$	s_e^2
Total	$n-1$	$\sum(Y_i - \bar{Y})^2$	s_Y^2

15.2 SOME CONCEPTS OF EXPERIMENTAL ANALYSIS

Since an experiment is a planned scientific investigation designed to examine a set of populations, it is important that the populations be well defined. The various conditions (processes, techniques, operations, and so on) which distinguish the populations are called *treatments*; each treatment uniquely defines a population. The term *treatments* is therefore generic and might refer to different (i) amounts of vitamin, (ii) lengths of time, (iii) teaching methods, (iv) percentages of carbon in steel, (v) breeds of cattle, or (vi) psychological tests. There are really two types of populations: those defined by the treatment designations (populations of measurements for different breeds of cattle, for example) and those generated when each treatment is applied or allocated to a certain portion of the experimental material (populations of yields of wheat from plots treated with different amounts of fertilizer).

An *experimental unit* (EU) is the smallest entity providing a population measurement. When we investigate the effects of different amounts of a vitamin, an EU would probably be a single laboratory animal. A batch of steel might be an EU in case (iv). For teaching methods the EU might be individual persons or, more often, a whole class of students.

With the above terminology, we can say that each experimental population is the set of conceptual measurements from EU of the same treatment. An experiment may be described as a planned investigation designed to estimate treatment effects.

Residual variation is a measure of dispersion based on deviations exhibited by observations from experimental units of the same treatment. Two main sources of residual variation are (i) differences naturally inherent in the experimental material and (ii) the lack of uniformity in carrying out the experiment. In a sense, residual variability is a conglomerate of all extraneous (unexplained) variation. For some experiments, it is reasonable to assume that all treatments have roughly the same variability. In such cases, this within treatment variability is pooled into a single term, a principle encountered as early as Chapter 11, where we calculated the pooled estimator of variance.

In order to estimate residual variation, treatments must occur more than once. If the i-th treatment occurs r_i times, we sometimes say the i-th treatment has been *replicated* r_i times. While replication is mandatory for any estimation of residual variation, replication must increase in order to detect smaller and smaller deviations among means.

The ability of an experiment to detect existing differences is referred to as its *sensitivity*. The more sensitive of two experiments has the better chance of detecting a prescribed difference.

The function of *randomization* has been discussed earlier; see Chapter 2.

For a fixed set of treatments and a set of EU, there may be several ways of conducting the experiment. A particular experimental layout, including the arrangement of the EU and the allocation of treatments to the EU, is called an *experimental design.* When a choice of experimental designs is available, the experimenter selects the one having the highest sensitivity. This problem will not concern us, for we will be discussing only one experimental design, the so-called *one-way layout* or *completely randomized* (CR) design.

15.3 COMPLETELY RANDOMIZED DESIGNS

Suppose that k treatments are to be investigated. One may be the dummy treatment where nothing is applied to the EU; the inclusion of the dummy treatment may be done to establish a reference. A total of n EU are available for use in the experiment. The EU are homogeneous in all respects (or nearly so); that is, no source of variation can be recognized among the EU under any grouping or arrangement.

The i-th treatment is randomly assigned to r_i EU in an unrestricted fashion, except that no treatment may be assigned to an EU which has already been assigned a treatment. And, of course, $\sum_{i=1}^{k} r_i = n$.

Some of the more important advantages of the CR designs are:

(i) flexibility—Subject to available resources, there is no restriction on either the number of treatments or on the number of replications.
(ii) Ease of analysis—Of all designs, the CR designs are easiest to analyze, even with unequal replication of treatments.
(iii) Maximum *df* for estimation of residual variability—This is important only for small experiments or for experiments where the variation among EU is great.

The chief disadvantage of using a CR design is that it is relatively low in sensitivity when another experimental design should really be used. This stems from the fact that all *unrecognized* variation (which if recognized would dictate the use of another design) and all extraneous variation are included in the residual term.

15.4 THE LINEAR MODEL AND ASSUMPTIONS: CR DESIGN

Linear models have already been encountered in regression analyses, where the population means were assumed to be linearly related; see (12.1). Taking into account the deviation from the mean, an individual value was also given a linear specification; see (12.3).

Observations drawn from a single population could also be explained by a linear model,

$$X_j = \mu + \varepsilon_j \qquad (15.2)$$

where ε_j is the deviation of X_j from the population mean, μ. Thus, observations drawn from a set of k populations would have a set of k representations like (15.2). By introducing an index to distinguish the populations, the k representations can be combined into a single model,

$$X_{ij} = \mu_i + \varepsilon_{ij} \tag{15.3}$$

where

(15.4)
X_{ij} is the j-th observation obtained from the i-th population,
μ_i = mean value of the i-th population,
$\varepsilon_{ij} = X_{ij} - \mu_i$ = deviation of the (i, j)-th observation.

No linear relationship is postulated among the μ_i as in regression.

Since we would like to investigate the effect or contribution of particular treatments (populations), it will be necessary to modify model (15.3). This is easily accomplished by measuring the deviation of μ_i from some reference point, conventionally taken as an *overall mean value*, defined as

$$\mu = \frac{1}{n} \sum_{i=1}^{k} r_i \mu_i \qquad \text{with } n = \sum_{i=1}^{k} r_i \tag{15.5}$$

Then (15.3) becomes

$$\begin{aligned} X_{ij} &= \mu + (\mu_i - \mu) + \varepsilon_{ij} \\ &= \mu + \tau_i + \varepsilon_{ij} \end{aligned} \tag{15.6}$$

where $\tau_i = \mu_i - \mu$ is customarily referred to as the i-th treatment effect. Equation (15.6) is the *linear model for the CR design*. The term "linear" refers to the fact that all parameters (μ and the τ's in this case) enter the equation linearly, not to a linear relationship among the means.

All treatments of interest may or may not be included in the experiment. If all treatments occur every time the experiment is conducted, inference would be directed toward this "fixed" set of treatments; this is called the *fixed-effects* case. Otherwise, different experiments will include different sets of treatments; and inference will be directed toward the larger set from which the experimental treatments were selected; this is called the *random-effects* case. If a research worker is interested in six varieties of cotton and uses all six varieties in an experiment, he is conducting a fixed-effects experiment. On the other hand, an engineer might conduct an experiment using ten temperatures but would like to make inferences about all temperatures in the range covered; he is conducting a random-effects experiment.

For inference to be valid under the ANOVA procedures which we will discuss, the following assumptions must hold:

(i) The linear model (15.6) is appropriate.
(ii) The EU are as homogeneous as possible.
(iii) μ is a fixed constant, common to all observations.
(iv) The ε_{ij} are normally and independently distributed with zero mean and variance σ_ε^2 for all treatment populations.
(v) In the *fixed case* the τ_i are fixed constants such that

$$\sum_{i=1}^{k} r_i \tau_i = 0$$

In the *random case* the τ_i are normally and independently distributed with zero mean and variance σ_τ^2, and the τ_i are distributed independently of the ε's.

We emphasize that assumption (iv) specifies normality and homogeneous within population variances. Failure of any of these assumptions to hold results in approximate inference. A test of equality of variances can be made in the CR designs, but we will not discuss that here. For a discussion of this and related topics, see [2], [3], [5], and [8].

(Problems 9(a), 11(a))

15.5 *THE ANOVA TABLE FOR A CR DESIGN*

For convenience, the experimental data can be recorded in the following manner:

Observation	1	2		j		r_i	Treatment Total	Treatment Mean
Treatment 1	x_{11}	x_{12}	$\cdots$	x_{1j}	$\cdots$	x_{1r_1}	$x_{1\cdot}$	$\bar{x}_{1\cdot}$
2	x_{21}	x_{22}	$\cdots$	x_{2j}	$\cdots$	x_{2r_2}	$x_{2\cdot}$	$\bar{x}_{2\cdot}$
$\vdots$	$\vdots$						$\vdots$	$\vdots$
i	x_{i1}	x_{i2}	$\cdots$	x_{ij}	$\cdots$	x_{ir_i}	$x_{i\cdot}$	$\bar{x}_{i\cdot}$
$\vdots$	$\vdots$						$\vdots$	$\vdots$
k	x_{k1}	x_{k2}	$\cdots$	x_{kj}	$\cdots$	x_{kr_k}	$x_{k\cdot}$	$\bar{x}_{k\cdot}$
	Overall total and mean						$x_{\cdot\cdot}$	$\bar{x}_{\cdot\cdot}$

The total SS for X is to be partitioned into two parts: one explaining variation among treatment means, the other explaining the residual or within-treatment variability. Thus

$$\sum_{i,j} (x_{ij} - \bar{x}_{..})^2 = \sum_i r_i(\bar{x}_{i\cdot} - \bar{x}_{..})^2 + \sum_{i,j} (x_{ij} - \bar{x}_{i\cdot})^2 \tag{15.7}$$

The partitioning (15.7) as well as expanded forms of the SS therein are derived in Section 15.9. The results are:

$$\begin{aligned} C &= \text{general correction term} \\ &= x_{..}^2/n \end{aligned} \tag{15.8}$$

$$\begin{aligned} T_{XX} &= \text{treatment } SS \\ &= \sum_i r_i(\bar{x}_{i\cdot} - \bar{x}_{..})^2 \\ &= \sum_i x_{i\cdot}^2/r_i - C \end{aligned} \tag{15.9}$$

$$\begin{aligned} SS_X &= \text{total } SS \\ &= \sum_{i,j} (x_{ij} - \bar{x}_{..})^2 \\ &= \sum_{i,j} x_{ij}^2 - C \end{aligned} \tag{15.10}$$

$$\begin{aligned} R_{XX} &= \text{residual (within-treatment) } SS \\ &= \sum_{i,j} (x_{ij} - \bar{x}_{i\cdot})^2 \\ &= SS_X - T_{XX} \end{aligned} \tag{15.11}$$

The expanded forms should always be used in calculating the value of SS_X and T_{XX} with R_{XX} found by subtraction. When all $r_i = r$, some additional simplifications can be made in some of these formulas.

Examination of (15.8) through (15.11) reveals the following general rule which is valid for any experimental design.

> •• *Rule A-2:* The square of any total used in calculating any SS is divided by the number of observations forming that total.

The ANOVA table for a CR design has one of the forms shown in Table 15-2; the last part of the table is chosen to conform with the model used. "Source" is an abbreviation for the phrase *Source of Variation.* The $E(MS)$ column gives the *Expected Value of the Mean Squares* (expected values were discussed in Section 6.4). Derivations of the $E(MS)$ will not be given. The $E(MS)$ are essential for the derivation of exact tests, which are made by comparing two mean squares having the same expectation under the null hypothesis. The next section will consider the problem of hypothesis testing.

Table 15-2 ANOVA for CR Designs

Source	*df*	*SS*	*MS*	*E(MS)*	
				fixed	random*
Treatments (among treatment means)	$k-1$	T_{XX}	MS_T	$\sigma_\varepsilon^2 + \sum r_i\tau_i^2/(k-1)$	$\sigma_\varepsilon^2 + r_0\sigma_\tau^2$
Residual (within treatments)	$n-k$	R_{XX}	MS_R	σ_ε^2	σ_ε^2
Total	$n-1$	SS_X	XXX	XXX	XXX

* $r_0 = (n - \sum r_i^2/n)/(k-1)$

▶ *Example 15-1*

The annual growth was measured, in centimeters, for Douglas Fir trees treated with four different types of commercial plant food. Since there was no interest in any other plant foods, the fixed-effects model was appropriate. Twenty-four uniform trees were purchased, and six were randomly selected to receive each plant food. Each tree is an experimental unit. Two trees died before the end of the year, giving unequal replication of the treatments. The data were:

Tree		1	2	3	4	5	6	$x_{i\cdot}$	$\bar{x}_{i\cdot}$
	1	8	9	11	10	7		45	9
Plant	2	9	8	9	9	7	9	51	8.5
food	3	4	5	7	4	4	5	29	4.83
	4	7	6	7	8	9		37	7.4

$$x_{\cdot\cdot} = 162 \qquad 7.36 = \bar{x}_{\cdot\cdot}$$

$$C = (162)^2/22 = 1{,}192.9$$

$$T_{XX} = (45)^2/5 + (51)^2/6 + (29)^2/6 + (37)^2/5 - C$$

$$1{,}252.5 - 1{,}192.9 = 59.6$$

$$SS_X = 8^2 + 9^2 + \cdots + 9^2 - C$$

$$1{,}278.0 - 1{,}192.9 = 85.1$$

$$R_{XX} = 85.1 - 59.6 = 25.5$$

For this experiment, the ANOVA table has the following form:

Table 15-3 ANOVA for Annual Growth Under Four Food Plants

Source	*df*	*SS*	*MS*	*E(MS)*
Brand (treatments)	3	59.6	19.87	$\sigma_\varepsilon^2 + \Sigma r_i \tau_i^2/3$
Residual	18	25.5	1.42	σ_ε^2
Total	21	85.1	xxx	xxx

(Problems 1(a), 3(a), 7(a), 9(b), 11(b))

15.6 *TESTING THE EQUALITY OF TREATMENT MEANS*

The population defined by the *i*-th treatment is assumed to have the $N(\mu_i, \sigma_\varepsilon^2)$ distribution. Consequently, the test of

$$H_0\text{: All treatment means are equal}$$

is equivalent to

$$H_0\colon \mu_1 = \mu_2 = \cdots = \mu_k$$

But, since $\tau_i = \mu_i - \mu$, where μ is defined in (15.5), it is easy to see that an equivalent hypothesis is

$$H_0\colon \text{all } \tau_i = 0 \tag{15.12}$$

If the hypothesis (15.12) is true, it follows that $\sigma_\tau^2 = 0$ under the random model and $\sum r_i \tau_i^2 = 0$ under the fixed model. For either case, it is seen in Table 15-1 that MS_T and MS_R have the same expectation. From mathematical statistics then, we know that hypothesis (15.12) may be tested by the F distribution. Specifically, the test statistic is

$$F_{k-1,n-k} = \frac{MS_T}{MS_R} \tag{15.13}$$

To establish the critical region, it is noted that when hypothesis (15.12) does not hold, either $\sigma_\tau^2 > 0$ or $\sum r_i \tau_i^2 > 0$. Thus, one *expects* MS_T to be no smaller than MS_R; that is, the value of $F_{k-1,n-k}$ defined in (15.13) is expected to be at least unity. Therefore, the hypothesis of equal treatment means should be rejected only for large values of the test statistic in (15.13).

In summary, then, the first three steps of a $100\alpha\%$ ANOVA test of equal treatment means in a CR design would be:

(15.14)

① H_0: All treatment means are equal
H_1: At least two treatment means differ

② Under H_0, $F_{k-1,n-k} = \dfrac{MS_T}{MS_R}$

③ Reject H_0 if $F_{k-1,n-k} \geqslant F_{k-1,n-k;1-\alpha}$

Occasionally in practice, the value of an ANOVA F statistic is less than unity. Two reasons have been proposed for this: (i) random sampling fluctuations or (ii) an incorrect linear model. When the denominator degrees of freedom exceed the numerator degrees of freedom on F (the usual situation in analysis of variance), the probability is greater than 1/2 that the F statistic (15.13) has a value less than unity; that is, $P(F_{k-1,n-k} < 1 | H_0 \text{ true}) > 0.5$. Thus values of F slightly below 1 can legitimately be attributed to random sampling fluctuations. How might we decide that a value of F below 1 indicates an incorrect model? An acceptable criterion for concluding an incorrect model is that of obtaining a value of F below the lower α percentile; that is, $F_{k-1,n-k} \leqslant F_{k-1,n-k;\alpha}$.

► ***Example 15-2***

Under the ANOVA assumptions, a 5% test that the average annual growth of Douglas Fir trees is equal under the four plant foods would be:

① H_0: The mean annual growth is the same for the four plant foods.
H_1: The mean annual growth differs for at least two plant foods.

② If H_0 is true, $F_{3,18} = \dfrac{MS_T}{MS_R}$

③ Reject H_0 if $F_{3,18} \geqslant F_{3,18;0.95} = 3.16$

④ $F_{3,18} = \dfrac{19.87}{1.42} \doteq 14$

Therefore, H_0 is rejected

⑤ The average annual growth of Douglas Fir trees is not the same when treated with the four plant foods, unless a 5% error has occurred. ◄

(Problems 1(c), 3(b), 7(b), 9(c))

15.7 INFERENCE ABOUT THE DIFFERENCE OF TREATMENT MEANS

When the ANOVA reveals significant differences among the treatment means, it does not indicate where the differences lie or the magnitude of the differences. Various procedures beyond the ANOVA F test are available for investigating these differences. Most of these additional procedures are beyond the scope of this book and will not be discussed; refer to [1], [3], [4], [6], and [8].

15.7.1 Additional Inference: Fixed Model

When the hypothesis of equal treatment means is rejected, we shall complete the analysis by giving a table of the difference of any two estimated treatment means and their standard errors. It can be shown that the best estimator of $\mu_i - \mu_{i*}$, where $i \neq i^*$, is given by $\bar{X}_{i\cdot} - \bar{X}_{i*}$ which has an estimated variance of

(15.15) $$s^2_{\bar{X}_{i\cdot} - \bar{X}_{i*}} = MS_R(1/r_i + 1/r_{i*})$$

It is also easy to show that

(15.16) $$\bar{X}_{i\cdot} - \bar{X}_{i*} \sim \mathrm{N}(\mu_i - \mu_{i*}, \sigma^2_\varepsilon[1/r_i + 1/r_{i*}]),$$

whereby a $100(1 - \alpha)\%$ CI for $\mu_i - \mu_{i*}$ can be constructed by using the Student T variable:

(15.17) $$T_{n-k} = \frac{\bar{X}_{i\cdot} - \bar{X}_{i*} - (\mu_i - \mu_{i*})}{\sqrt{MS_R(1/r_i + 1/r_{i*})}}$$

If (15.17) is applied to form CI for the difference of several pairs of means, we have the simplest version of multiple comparisons, one of the topics discussed in the references indicated at the beginning of this section. The quantity $1 - \alpha$ related to each individual CI does not apply, however, to the set of CI taken collectively.

► ***Example 15-3***

A table of mean differences and their standard errors for the plant food experiment of Example 15-1 is:

i	i^*	$\bar{X}_{i\cdot} - \bar{X}_{i*}$	$s^2_{\bar{X}_{i\cdot} - \bar{X}_{i*}}$
1	2	0.5	0.72
1	3	4.17	0.72
1	4	1.6	0.75
2	3	3.67	0.69
2	4	1.1	0.72
3	4	−2.57	0.72

The 90% confidence limits for the mean difference $\mu_1 - \mu_3$ are

$$L, U = \bar{x}_{1\cdot} - \bar{x}_{3\cdot} \pm s_{\bar{X}_{1\cdot} - \bar{X}_{3\cdot}} t_{18;0.95} = 4.17 \pm 0.72(1.734)$$
$$= 4.17 \pm 1.25$$
$$= 2.92; 5.42 ◀$$

(Problems 1(b), 7(d), 11(d))

15.7.2 Additional Inference: Random Model

Under the random model, we make inferences about a larger set of treatments than we used in the experiment. When $H_0: \sigma_\tau^2 = 0$ is rejected in favor of $H_1: \sigma_\tau^2 > 0$, we shall complete the analysis by giving an estimate of σ_τ^2. This estimate is

$$\hat{\sigma}_\tau^2 = \frac{MS_T - MS_R}{r_0} \qquad (15.18)$$

where r_0 is the constant defined in Table 15-2. Probabilistic inferences from $\hat{\sigma}_\tau^2$ to σ_τ^2 become very involved because $MS_T - MS_R$ has a very complex distribution (see [7], p. 410).

(Problem 3(c))

*15.8 *POWER OF THE ANOVA TEST*

Under the fixed model, the power of the ANOVA F test can be found from the graphs in Appendix B.7. Curves are given for $\alpha = 0.01$ and 0.05. Each curve is further indexed by ν_1, the numerator degrees of freedom, and by ν_2, the denominator degrees of freedom of the F test. A final quantity is required to obtain the power. Usually denoted by ϕ, it measures differences among the treatment means.

For a CR design, the relevant quantities are

$$\nu_1 = k - 1$$
$$\nu_2 = n - k \qquad (15.19)$$
$$\phi^2 = \sum_{i=1}^{k} r_i \tau_i^2 / k\sigma_\varepsilon^2$$

where $\tau_i = \mu_i - \mu$, these differences taken as multiples of σ_ε. There are many specifications of the μ_i which give the same value of ϕ and consequently the same power.

Curves given in Appendix B.7 are for $\nu_1 = 1(1)8$ and $\nu_2 = 6(1)10, 12, 15, 20, 30, 60, \infty$. The curves have been set up for equal replications but can be used for unequal replications to give approximate power.

▶ ***Example 15-4***

In the plant food experiment investigating annual growth of trees (Examples 15-1 and 15-2) $r_1 = r_4 = 5$ and $r_2 = r_3 = 6$. Suppose we would like to know the chances of detecting differences among the mean annual growths if

$$\mu_1 = \mu + \sigma_\varepsilon/2 \qquad \mu_3 = \mu + 3\sigma_\varepsilon/4$$

$$\mu_2 = \mu - \sigma_\varepsilon \qquad \mu_4 = \mu - \sigma_\varepsilon/2$$

For this specification of the μ_i

$$\begin{aligned}\phi^2 &= \tfrac{1}{4}[5(1/2)^2 + 6(-1)^2 + 6(3/4)^2 + 5(-1/2)^2] \\ &= 2.97, \text{ thus } \phi = 1.72\end{aligned}$$

From Appendix B.7, with $\alpha = 5\%$, $\nu_1 = 3$, and $\nu_2 = 18$, the approximate power is seen to be about 0.73. Any other specifications of the μ_i giving $\phi = 1.72$ has the same power.

Consider the case where μ_i are specified in a slightly different manner, as follows:

$$\mu_1 = \mu_2 ; \mu_3 = \mu_1 + 2\sigma_\varepsilon ; \mu_4 = \mu_1 - 3\sigma_\varepsilon .$$

Then, by (15.5),

$$\begin{aligned}\mu = \frac{1}{n}\sum r_i\mu_i &= \frac{1}{22}[5\mu_1 + 6\mu_1 + 6(\mu_1 + 2\sigma_\varepsilon) + 5(\mu_1 - 3\sigma_\varepsilon)] \\ &= \mu_1 - 3\sigma_\varepsilon/22\end{aligned}$$

and

$$\begin{aligned}\phi^2 &= \tfrac{1}{4}[5(3/22)^2 + 6(3/22)^2 + 6(47/22)^2 + 5(-63/22)^2] \\ &= 1.71\end{aligned}$$

Thus $\phi = 1.31$, and the approximate power is 0.45. ◀

Under the random model, the power of the ANOVA F test is simply a probability from an F distribution and may be found from an extensive F tabulation. Specifically, the power of the ANOVA F test under the random model is

(15.20) $$\text{Power} = P\left[F_{k-1,n-k} \geq \frac{\sigma_\varepsilon^2}{\sigma_\varepsilon^2 + r\sigma_\tau^2} F_{k-1,n-k;1-\alpha}\right]$$

where r is the number of observations per treatment used in the experiment. We must take σ_τ^2 to be a multiple of σ_ε^2 to evaluate (15.20). For unequal numbers of observations per treatment, the power is approximate. We shall not give an example of these power calculations, since the random model does not occur very often in practice.

(Problems 13, 15, 17)

*15.9 DERIVATIONS OF SUMS-OF-SQUARES FORMULAS

In this section SS_X is partitioned as shown in (15.7), after which expanded forms of the components of (15.7) are derived.

Employing a common trick when partitioning SS, the addition and subtraction of treatment means is incorporated in the total deviations as follows:

$$\begin{aligned} SS_X &= \sum_{i,j} (x_{ij} - \bar{x}_{i\cdot} + \bar{x}_{i\cdot} - \bar{x}_{\cdot\cdot})^2 \\ &= \sum_{i,j} [(x_{ij} - \bar{x}_{i\cdot})^2 + (\bar{x}_{i\cdot} - \bar{x}_{\cdot\cdot})^2 - 2(x_{ij} - \bar{x}_{i\cdot})(\bar{x}_{i\cdot} - \bar{x}_{\cdot\cdot})] \\ &= \sum_{i,j} (x_{ij} - \bar{x}_{i\cdot})^2 + \sum_{i,j} (\bar{x}_{i\cdot} - \bar{x}_{\cdot\cdot})^2 \end{aligned} \tag{15.21}$$

since

$$\begin{aligned} \sum_{i,j} (x_{ij} - \bar{x}_{i\cdot})(\bar{x}_{i\cdot} - \bar{x}_{\cdot\cdot}) &= \sum_i (\bar{x}_{i\cdot} - \bar{x}_{\cdot\cdot}) \sum_j (x_{ij} - \bar{x}_{i\cdot}) \\ &= \sum_i (\bar{x}_{i\cdot} - \bar{x}_{\cdot\cdot})[x_{i\cdot} - r_i\bar{x}_{i\cdot}] \\ &= 0 \end{aligned} \tag{15.22}$$

Thus (15.7) is established.

The expanded forms are easily derived as follows:

$$\begin{aligned} SS_X &= \sum_{i,j} (x_{ij} - \bar{x}_{\cdot\cdot})(x_{ij} - \bar{x}_{\cdot\cdot}) \\ &= \sum_{i,j} (x_{ij} - \bar{x}_{\cdot\cdot})x_{ij} - \bar{x}_{\cdot\cdot} \sum_{i,j} (x_{ij} - \bar{x}_{\cdot\cdot}) \\ &= \sum_{i,j} x_{ij}(x_{ij} - \bar{x}_{\cdot\cdot}) - 0 \\ &= \sum_{i,j} x_{ij}^2 - \bar{x}_{\cdot\cdot} \sum_{i,j} x_{ij} \\ &= \sum_{i,j} x_{ij}^2 - x_{\cdot\cdot}^2/n \end{aligned} \tag{15.23}$$

and

$$\begin{aligned} T_{XX} &= \sum_i r_i(\bar{x}_{i\cdot} - \bar{x}_{\cdot\cdot})^2 = \sum_i r_i(\bar{x}_{i\cdot} - \bar{x}_{\cdot\cdot})(\bar{x}_{i\cdot} - \bar{x}_{\cdot\cdot}) \\ &= \sum_i r_i\bar{x}_{i\cdot}(\bar{x}_{i\cdot} - \bar{x}_{\cdot\cdot}) - \bar{x}_{\cdot\cdot} \sum_i r_i(\bar{x}_{i\cdot} - \bar{x}_{\cdot\cdot}) \\ &= \sum_i r_i(x_{i\cdot}/r_i)(x_{i\cdot}/r_i - x_{\cdot\cdot}/n) \\ &= \sum_i (x_{i\cdot}^2/r_i - x_{i\cdot}x_{\cdot\cdot}/n) \\ &= \sum_i (x_{i\cdot}^2/r_i) - x_{\cdot\cdot}^2/n \end{aligned} \tag{15.24}$$

since

$$\sum_i r_i(\bar{x}_{i\cdot} - \bar{x}_{\cdot\cdot}) = \sum_i r_i(x_{i\cdot}/r_i - x_{\cdot\cdot}/n) = 0 \tag{15.25}$$

PROBLEMS

1. In a CR experiment, the following values were recorded:

$$k = 6 \qquad r_i = 10; \qquad i = 1, 2, \ldots, 6$$

$$T_{XX} = 510 \qquad E_{XX} = 540$$

(a) Set up the ANOVA table, assuming the fixed-effects model.
(b) Give the variance of the difference of any two estimated treatment means.
(c) Test the equality of treatment means at the 1% level.

2. Suppose the following information is given for a CR experiment:

$$k = 5 \qquad r_i = 8; \qquad i = 1, 2, \ldots, 5$$

$$x_{1\cdot} = 130 \qquad x_{3\cdot} = 94 \qquad x_{5\cdot} = 126$$

$$x_{2\cdot} = 108 \qquad x_{4\cdot} = 112 \qquad \sum_{i,j} x_{ij}^2 = 8735$$

(a) Set up the ANOVA table, assuming the fixed-effects model.
(b) Estimate the variance of the difference of any two estimated treatment means.
(c) Test the equality of treatment means at the 1% level.
(d) Discuss assumptions required in the above.

3. A CR experiment gave the following information:

$$k = 6 \qquad r_i = 5; \qquad i = 1, 2, \ldots, 6$$

$$x_{1\cdot} = 12 \qquad x_{3\cdot} = 9 \qquad x_{5\cdot} = 16$$

$$x_{2\cdot} = 18 \qquad x_{4\cdot} = 15 \qquad x_{6\cdot} = 20$$

$$\sum_{i,j} x_{ij}^2 = 310$$

(a) Set up the ANOVA table, assuming random treatment effects.
(b) Test the hypothesis $H_0: \sigma_\tau^2 = 0$ at the 5% level.
(c) Estimate the variance components σ_ε^2 and σ_τ^2.

4. The results of a CR experiment are summarized below:

i	r_i	$x_{i\cdot}$
1	8	85
2	6	75
3	5	50
4	8	90

$$\sum_{i,j} x_{ij}^2 = 3{,}402.3$$

(a) Give the ANOVA under the random model.
(b) Test the hypothesis $H_0: \sigma_\tau^2 = 0$ at the 1% level.
(c) Estimate the variance components, σ_ε^2 and σ_τ^2.

5. From a CR experiment the following data were recorded:

i	r_i	$x_{i\cdot}$
1	7	35
2	7	56
3	9	90
4	9	45

$$\sum_{i,j} x_{ij}^2 = 1{,}898$$

(a) Give the ANOVA under the fixed model.
(b) Test the equality of treatment means at the 1% level.
(c) Set a 99% CI for $\mu_2 - \mu_1$.
(d) Give a table of mean differences and standard errors.

6. For a CR experiment, the following are given:

$$k = 8 \qquad r_i = 5; \qquad i = 1, 2, \ldots, 8$$

$$\sum_i x_{i\cdot}^2 = 37{,}106 \qquad x_{\cdot\cdot} = 526 \qquad \sum_{i,j} x_{ij}^2 = 7{,}689$$

(a) Set up the ANOVA table, assuming fixed effects of treatments.
(b) Give the standard error of the difference of any two estimated treatment means.
(c) Test the equality of treatment means at the 5% level.
(d) If $x_{1\cdot} = 72$ and $x_{2\cdot} = 56$, set a 99% CI for $\mu_1 - \mu_2$.
(e) Discuss assumptions required for the above inferences to be valid statistically.

7. For a CR experiment, the number of observations were $r_1 = 7$, $r_2 = 5$, $r_3 = 8$, $r_4 = 5$ for treatments 1, 2, 3, and 4, respectively.
(a) Complete the following ANOVA table:

Source	*df*	*SS*	*MS*
Treatments			
Residual			20
Total		640	xxx

(b) Test the equality of treatment means at the 5% level.
(c) What assumptions are required for part (b) to be valid?
(d) Given that $x_{1\cdot} = 49$ and $x_{4\cdot} = 75$, set a 99% CI for $\mu_1 - \mu_4$.

8. Given below are ascorbic acid contents for three varieties of ripe peaches grown in a certain locality. All measurements were converted to milligrams per hundred grams.

Variety 1		Variety 2		Variety 3	
5.34	5.50	7.12	6.80	6.28	6.40
5.58	5.42	6.89	6.91	6.01	6.12
5.26	5.47	6.93	6.76	6.27	6.24
5.47	5.71	6.82	6.97	6.15	6.31
5.39	5.62	7.06	6.88	6.38	6.37

(a) Give the linear model for this experiment. Explain each term appearing in the model. Assume fixed effects.
(b) Give assumptions necessary for valid inferences under this model.
(c) Set up the ANOVA table under this model.
(d) Test, at the 5% level, that the average ascorbic acid content is the same for the three varieties.
(e) Give a table of mean differences and their standard errors.
(f) Set a 95% *CI* for $\mu_2 - \mu_3$.

9. The amounts of impurities remaining in commercial grade alcohol was measured for each of eight samples taken under each of four filtration methods. The results, in milligrams per hundred centimeters, were:

Method 1		Method 2		Method 3		Method 4	
15.4	15.3	15.8	15.8	15.6	16.5	16.4	15.9
14.8	14.8	15.7	15.6	15.9	15.2	16.5	16.1
14.9	15.0	16.4	16.3	16.3	16.8	16.0	16.1
14.2	14.6	16.0	16.3	15.7	14.9	15.6	15.8

(a) Give the linear model for this experiment. Define all terms used. Assume the fixed-effects model.
(b) Set up the ANOVA table.
(c) Are there significant differences among the average amount of impurities for the four filtration methods? Let $\alpha = 0.01$.
(d) Discuss assumptions required in your analysis.

10. The data given in Problem 9 of Chapter 11 represent weights of newly formed pupae under controlled environments having relative humidities (RH) of 70% and 20%. A third environment, having 1% RH, was also investigated. Fifteen newly formed pupae for this environment had weights of:

13.7	17.3	15.1	16.8	19.8
15.7	18.4	15.4	17.2	17.1
13.7	14.8	14.2	16.5	11.5

(a) Set up the ANOVA for the three environment experiments, assuming fixed effects.

(b) Is the average weight of newly formed pupae the same for the three humidities used? Let $\alpha = 0.05$.
(c) Give a table of mean differences and their standard errors.
(d) Is there a significant difference in average weight of newly formed pupae under 20% RH and 1% RH? Let $\alpha = 0.05$.

11. An experiment was conducted to examine the effect of potassium on the weight gain of baby chicks. Diet, environment, and so on, were all identical except for the percentage of potassium. Two-week weight gains, in grams, were:

0.5% Potassium		2% Potassium		3% Potassium	
44	54	56	53	62	70
51	45	54	62	67	64
50	51	58	55	68	72
48	47	67	58	66	74
46	52	60	57	65	63

(a) Give the linear model for this experiment. Explain all terms used. Assume the fixed effects model.
(b) Set up the ANOVA table.
(c) Does the average weight gain (for two weeks) differ for the three percentages of potassium? Let $\alpha = 0.01$.
(d) Is there a significant difference in average two week weight gain for 2% and 3% potassium diets? Let $\alpha = 0.05$.

12. In a greenhouse experiment, five types of light were used: natural, white, green, blue, and yellow. Eight tobacco plants each 10 mm tall were subjected to each type of light with the following measurements on stem growth (in mm):

Natural		White		Green		Blue		Yellow	
4.3	4.2	4.1	4.0	3.6	2.9	3.1	2.9	3.8	3.9
3.9	4.1	4.2	4.6	3.3	3.4	3.2	3.2	3.5	3.4
4.1	4.2	4.0	4.8	3.3	3.0	2.7	2.9	3.6	3.1
3.8	4.1	4.4	4.3	3.2	2.8	3.0	3.1	3.4	3.7

(a) Give the linear model for this experiment and the assumptions necessary for valid inferences, assuming fixed effects.
(b) Set up the ANOVA table.
(c) Test the equality of mean stem growth under the five types of light. Let $\alpha = 0.05$.
(d) Set and interpret a 95% *CI* for the difference between the mean stem growth under white and blue light.

***13.** Find the power of the test in Problem 8(d) if

(a) $\mu_1 = \mu_3$, $\mu_2 = \mu_1 + \sigma_\varepsilon$

(b) $\mu_1 = \mu_3$, $\mu_2 = \mu_1 + 3\sigma_\varepsilon/2$

(c) $\mu_1 = \mu - \sigma_\varepsilon/3$, $\mu_2 = \mu + \sigma_\varepsilon$, $\mu_3 = \mu - 2\sigma_\varepsilon/3$

*__14.__ Find the power of the test in Problem 9(c) if

(a) $\mu_1 = \mu_3$
$\mu_2 = \mu_4 = \mu_1 + \sigma_\varepsilon$

(b) $\mu_1 = \mu + \sigma_\varepsilon/2$
$\mu_2 = \mu - \sigma_\varepsilon/3$
$\mu_3 = \mu + \sigma_\varepsilon/3$
$\mu_4 = \mu - \sigma_\varepsilon/2$

*__15.__ Find the power of the test in Problem 11(c) if

(a) $\mu_1 = \mu + \sigma_\varepsilon/2$
$\mu_2 = \mu + \sigma_\varepsilon/2$
$\mu_3 = \mu - \sigma_\varepsilon$

(b) $\mu_1 = \mu + \sigma_\varepsilon/4$
$\mu_2 = \mu - \sigma_\varepsilon/2$
$\mu_3 = \mu + \sigma_\varepsilon/4$

*__16.__ Find the power of the test in Problem 4(b) if

(a) $\sigma_\tau^2 = 5.72\sigma_\varepsilon^2$ (b) $\sigma_\tau^2 = 0.04\sigma_\varepsilon^2$ (c) $\sigma_\tau^2 = 0.15\sigma_\varepsilon^2$

*__17.__ Find the power of the test in Problem 3(b) if

(a) $\sigma_\tau^2 = 2.7\sigma_\varepsilon^2$ (b) $\sigma_\tau^2 = 0.15\sigma_\varepsilon^2$ (c) $\sigma_\tau^2 = 1.2\sigma_\varepsilon^2$

REFERENCES

[1] Bancroft, T. A., *Topics in Intermediate Statistical Methods*, Vol. 1, Iowa State University Press, Ames, Iowa, 1968.
[2] Fryer, H. C., *Experimental Statistics*, Allyn and Bacon, Boston, Mass., 1966.
[3] Hicks, C. R., *Fundamental Concepts in the Design of Experiments*, Holt, Rinehart, and Winston, New York, 1964.
[4] Miller, R. G., Jr., *Simultaneous Statistical Inference*, McGraw-Hill, New York, 1966.
[5] Ostle, B., *Statistics in Research*, 2nd ed., Iowa State University Press, Ames, Iowa, 1963.
[6] Scheffe, H., *The Analysis of Variance*, Wiley, New York, 1959.
[7] Searle, S. R., *Linear Models*, Wiley, New York, 1971.
[8] Steel, R. G. D., and G. D. Torrie, *Principles and Procedures of Statistics*, McGraw-Hill, New York, 1960.

APPENDIX A

Table A.1 Ten Thousand Random Digits[†]

One hundred rows and one hundred columns, numbered 00 through 99.

[†] Reproduced with permission from the Rand Corporation compilation *A Million Random Digits*, published by the Free Press of Glencoe, Illinois, 1955.

Table A.1 *continued*

Row	Column 00		10		20		30		40	
00	54392	07845	33083	71868	13189	13599	99206	99774	54314	42075
	75821	10038	59767	46323	61884	92962	24307	06376	07152	53439
	43584	93879	12320	14385	70972	39555	14965	38891	02843	56290
	17989	45534	16466	68599	17620	27242	30242	05380	51907	40711
	42467	97706	13745	28098	42855	08575	23647	34579	58876	59689
	62548	22038	58252	91964	19904	66081	66756	13716	07534	02832
	29552	83811	87451	96494	72611	64579	34108	45525	55983	71837
	03525	44319	07138	45534	62873	43478	12644	89139	29972	73807
	85243	39194	15678	96715	79262	18675	26706	01637	63331	96427
	71131	73231	96253	78794	82796	71811	03463	51871	09935	18251
10	20087	33014	97489	80314	52742	57442	55479	09364	46593	51756
	38025	70529	96638	29976	74208	62463	04677	81470	94059	89869
	83340	33506	83958	15597	64615	82522	87314	04346	79388	41563
	79559	88404	40585	03826	92045	50770	95341	29739	93849	52856
	73453	25706	80775	20621	58031	35515	25783	75041	82242	09494
	59013	38774	57848	21137	22619	43826	31669	42968	86479	53527
	96113	24095	86064	68315	34708	09850	35592	24226	15068	47729
	90137	94259	47185	23732	59788	16428	74661	09130	04019	61634
	12473	05358	19746	01434	63508	63731	45555	34415	76621	81807
	83361	71572	83490	36205	04123	18404	41061	20485	74961	33900
20	67811	62410	91381	83100	58632	57566	03789	03799	37174	53086
	72599	84073	47057	67696	78921	16438	68950	64109	36124	43100
	58098	73435	27815	07421	51000	89888	42420	26531	86832	90415
	51173	41787	76256	10888	68127	64520	35318	33398	64395	59239
	21282	53655	79856	79147	53290	46262	21837	50597	08781	29643
	83547	49206	92574	52608	34816	58470	04700	72491	18996	98159
	11247	03952	63947	39155	17097	37464	12519	44124	66959	51768
	67899	00309	50065	50579	83913	51567	96682	55620	63999	38832
	84960	02657	59928	02880	88271	42053	65197	81917	56095	18219
	04975	14180	99510	92158	25813	56151	32551	84970	35156	51347
30	91537	40912	93878	38690	07719	54787	19297	99056	94028	98719
	56077	08915	03293	58922	43797	79107	46693	60477	64468	50847
	23078	22197	56577	30004	62619	40152	62580	89005	51684	75369
	00486	64250	00987	74869	73893	03800	50631	14210	47238	64701
	06246	83119	56622	31085	16671	09584	37725	32039	99250	34194
	09894	33817	63197	27035	05916	76191	05815	68160	19150	15417
	17963	94684	12375	15370	17987	19915	84817	37722	89939	96492
	60456	39502	34263	25981	77178	81197	18849	42322	02474	54810
	38260	42265	76214	96168	94430	59356	94601	91592	46238	81937
	73671	80503	98867	94741	98343	64314	71615	48734	56306	65164
40	55877	17159	66277	22442	34540	92876	84720	33689	68653	98464
	54730	73171	22269	36581	04782	36747	63094	77187	39634	59673
	23746	25334	55660	36079	64782	66588	95112	83567	68304	31726
	73708	72897	37640	86548	88998	87862	62121	88493	70447	14601
	30152	13088	54153	37829	18335	56976	18422	02506	43295	58238
	52631	62206	00688	82062	01704	52987	89502	16222	04471	51346
	27988	69512	05533	55913	61456	44502	47787	12919	65440	34596
	62226	36438	02043	03935	33488	73327	36507	51085	55191	43009
	97778	14106	63497	52067	55958	99178	26456	54272	87571	38940
	58010	34758	90641	99098	01066	36500	35247	15044	02369	51189

Table A.1 *continued*

Row	Column 00		10		20		30		40	
50	24810	61498	24935	86366	58262	44565	91426	86742	61747	79346
	75555	42967	02810	16754	08813	40079	62385	21488	38665	94197
	49718	05437	73341	17414	91868	24102	76123	67138	43728	43627
	04194	83554	98004	14744	63132	75018	75167	24090	02458	78215
	66155	24222	91229	63841	03271	56726	36817	51182	94336	20894
	10801	76783	05312	30807	40006	04465	70163	17305	46414	76468
	40422	75576	82884	27651	58495	87538	23570	66469	46900	95568
	95572	12125	12054	17028	03599	73764	48694	85960	89763	58305
	57276	49498	88937	08659	46840	83231	75611	94911	28467	67928
	80566	94963	83787	87636	89511	64735	86699	66988	91224	72484
60	44517	55108	17435	33109	60343	46193	66019	43713	24097	52921
	55424	87650	13896	90005	99458	20153	86688	13650	75201	79447
	80506	78301	97762	16434	62430	28438	13602	63236	81431	75641
	03646	54402	75413	39128	82975	73849	27269	73444	26120	06824
	14537	53791	43951	51326	33274	54833	80802	66976	04878	35832
	01644	33630	71247	59273	07811	33546	88628	06469	86257	39298
	39387	94217	77995	53285	13354	84980	83590	63494	06036	18502
	74962	49489	54662	93588	50466	55026	62458	06195	07995	71054
	21165	45577	46383	38855	21561	89332	94248	09703	78397	38770
	58519	95396	73607	72106	76597	85596	99075	39195	99605	66179
70	46982	79519	22294	15676	83484	98279	79200	02640	22501	43073
	58463	67619	18006	05028	32441	83599	28915	05362	21612	64681
	43055	00020	39254	68439	27399	24259	04641	50935	07112	55117
	84073	38387	14337	90766	60436	65757	57590	17880	13776	35810
	93542	37270	09361	62404	74056	52964	67372	81398	01482	97589
	54467	20234	52813	85296	14542	73241	74848	39001	97598	76641
	43608	42832	93917	67031	50220	94089	64858	27691	16719	99870
	64808	01692	46424	64722	87162	06582	01452	14980	17397	07403
	78703	93006	59651	48404	82284	66405	89818	00989	56112	78144
	14886	70359	32158	30401	20829	22534	88848	07669	25100	48602
80	69280	61856	78974	91485	01583	11620	53740	32705	80391	56749
	99680	99636	54107	79588	90845	21652	58875	13171	68531	18550
	01662	21554	63836	41530	21864	81711	68921	61749	36051	78024
	67852	69123	14280	17647	65125	82427	61594	32015	93473	05627
	13911	67691	97854	89950	40963	06697	82660	69097	65284	49808
	95822	09552	65950	34875	64250	41385	80133	70818	09286	30769
	44068	24928	27345	34235	44124	06435	06281	43723	97380	76080
	99222	66415	71069	62293	77467	35751	22548	23799	96272	58777
	08442	61287	72421	35777	61079	42462	17761	94518	98114	74031
	14967	60637	32097	28122	87708	19378	93372	23225	38453	80331
90	27864	15358	16499	91903	62987	98198	15036	23293	68241	44450
	99678	16125	52978	79815	85990	18659	00113	93253	49186	25165
	89143	79403	22324	54261	97830	42630	48494	09999	69961	39421
	34135	31532	42025	83214	83730	28249	25629	11494	70726	45051
	72117	97579	36071	29261	89937	78208	23747	56756	37453	51344
	19725	76199	08620	22682	52907	25194	84597	93419	95762	14991
	96997	66390	27609	41570	17749	23185	24475	56451	91471	33969
	44158	67618	15572	95162	95842	08301	11906	68081	40436	58735
	33839	40750	18898	61650	09970	47651	41205	65020	33537	01022
	53070	61630	84434	05732	18094	71669	41033	82402	16415	83958

Table A.1 *continued*

Column

Row	50		60		70		80		90	
00	28344	63846	63964	60895	12239	76694	12790	45894	54027	02972
	69127	84099	55078	52125	91803	70579	16626	21284	22855	93701
	59445	67135	84887	68826	03925	96875	26978	37266	24891	82182
	20763	77171	27443	91889	69429	32762	54925	38789	70005	73027
	35075	56594	85416	52016	55641	56810	84417	97336	31939	78465
	55470	27101	26764	49079	92349	76175	30719	49217	93468	93659
	16572	68913	05618	73677	95386	15694	41649	16016	46607	90959
	20523	79559	80778	90307	04276	18990	94101	04128	29130	41360
	68529	50841	11522	23568	35312	07039	63035	07494	88022	90791
	94262	24694	25772	70107	87059	44432	80189	22388	97284	92834
10	19831	99624	10258	75171	72811	17522	88025	90017	24567	81657
	41155	53777	00569	97580	51120	67568	91364	99467	30997	50844
	56080	19758	01017	01455	67651	90136	15988	31979	43592	97527
	58187	51211	28672	12648	80563	65905	78969	51970	18063	62322
	03254	17359	86883	88354	80416	28054	28288	85907	32791	40380
	26037	03644	18467	62861	87280	67718	60183	76294	71605	55833
	34186	42234	74217	68691	26209	35000	97018	36774	16745	82843
	52892	06172	05593	50933	28931	85321	16172	82618	28061	98199
	92535	70187	36851	10527	42984	29793	94684	80892	85985	66898
	69193	60470	31052	83452	82966	92021	82709	23202	43122	17808
20	12647	35381	76690	59805	86573	60979	30011	30730	39571	74864
	69596	21390	66413	18000	43994	76441	48752	94967	85206	67975
	15726	19693	94900	36143	37157	80508	85561	44142	30867	90038
	27168	31932	02011	23872	06614	06643	93896	53687	92365	33326
	53686	83524	89623	43222	02517	89809	02809	97484	56490	22073
	88813	42571	85747	73775	54741	72078	14563	94509	44745	35436
	04980	66033	92939	15100	46016	95380	21346	23919	16883	53421
	89213	31692	32545	40613	95669	42666	55063	70279	70187	15264
	26548	48158	30120	25200	68205	09995	20068	89549	80604	95164
	16013	60209	74308	72019	18893	91801	38397	07368	25020	63806
30	33715	97675	35548	03279	66728	41708	37992	21817	91041	78176
	26346	80668	13651	21915	46443	60865	61760	71761	48614	97139
	61626	39373	99569	59841	30477	83125	91477	46578	87937	12737
	36018	15490	06160	89695	60199	37806	45344	86880	21625	40029
	04573	61887	24100	08061	33078	69493	84334	08952	43234	75029
	86023	78670	78947	64605	31192	14907	71263	16920	06782	09441
	63133	71295	54976	99501	16347	14674	44431	39096	17033	97650
	07804	60898	10723	02066	52475	44102	40654	03176	70614	90818
	49258	99756	40168	73983	28561	14964	58785	02722	06584	72783
	15555	36779	18113	42406	15486	39654	32868	09180	07685	94741
40	95814	34777	79307	87541	16280	04587	07811	16501	52170	43579
	00088	93058	78023	44202	63622	73002	25544	41599	74846	08016
	28100	93027	17539	66586	19780	39782	58133	64434	67795	16583
	57095	82345	95509	10081	40356	92619	92043	08123	66157	58768
	07760	78619	73961	19414	21191	88205	91518	62443	92359	97477
	25749	53695	45208	64017	02908	04363	78442	24235	81163	07499
	43437	22611	24291	46022	46922	21663	03846	48422	32468	61649
	91356	72477	28465	77099	78934	33936	19888	94022	48204	18237
	16867	25864	50566	52419	99173	64341	38180	49665	25322	00641
	85827	46056	30676	49685	29418	90340	99098	09857	71346	40601

Row	Column 50		60		70		80		90	
50	38527	24693	85207	32770	23364	81313	05408	07885	30693	94400
	18787	47450	97853	03079	38231	90927	95246	56697	00496	37877
	43960	95514	33973	71102	23076	60827	55984	78408	48286	01941
	76201	36878	56283	60744	61743	28667	84532	37904	64504	98223
	56629	80708	53119	10708	24927	14562	50374	13142	18144	22596
	92577	71153	34303	98607	50692	49119	45005	86083	81982	42362
	23716	90573	43767	40812	65986	18229	51876	00505	74606	92539
	55364	70288	49371	35143	58877	35926	36474	76537	46490	20477
	82731	58687	80482	60110	99925	33485	71394	72308	47343	68773
	95093	29762	22429	53388	33704	71584	47657	04399	15731	48265
60	11940	64257	93046	39413	83891	60497	45824	97914	49776	20951
	68558	53000	09151	01803	98284	61152	00573	59228	05045	25375
	09951	53201	64373	96855	45123	84826	16654	12828	21490	78481
	99583	16189	25363	61740	41722	32100	04389	67756	39535	68386
	62818	85469	62578	27477	86558	30449	43353	42067	61056	06517
	01862	85044	27262	17069	29969	00666	92469	40706	47226	15006
	52047	53461	51206	19934	36813	11985	74686	12010	67377	54292
	66914	38216	67340	86343	94944	83406	75929	33230	77993	04539
	61197	83526	06483	40588	42988	13467	09083	18129	47303	01006
	12993	58421	89321	14629	82194	19430	92076	53807	05419	53774
70	59334	43660	57741	41267	50979	98734	13528	03912	15424	02141
	66465	30884	62864	06344	45794	26264	67098	31504	64705	18671
	31266	28603	16389	15052	27532	41081	50792	39126	55744	15126
	34505	83962	07615	97777	38760	59198	04861	18111	34908	91828
	85912	09384	80978	58917	27608	26147	07788	05637	81119	64210
	72357	78805	03809	78721	89761	27014	33821	51910	29857	58315
	26213	55746	06814	82095	09605	60296	00946	68198	44854	43960
	34613	09399	51224	89960	72017	63555	93683	85727	89832	65538
	46356	50373	41134	48749	66215	61906	07769	26493	51523	18063
	93141	62639	66786	87421	56814	15953	84187	27424	33382	77534
80	51171	94833	34011	61686	76573	84267	65801	58947	62893	11279
	81437	08972	82728	93896	22692	70711	93355	55120	05326	05135
	15053	46214	71335	49624	16579	75860	20554	21806	04877	35936
	57555	75056	62505	70605	39831	72600	85084	25888	67672	63603
	54571	77067	66142	51216	93796	16886	27490	15141	80861	17556
	18972	11011	92448	04353	03407	07990	52942	11668	02811	64412
	93623	89619	98261	70386	83651	08453	91471	91895	98037	85442
	02899	52218	05299	54454	54574	26019	44980	27133	73853	78183
	33489	56062	35157	75194	23957	15287	78660	68812	64174	76433
	19461	65172	74370	38697	80056	38879	69526	20989	47933	53747
90	74710	69781	02423	30372	12979	99184	63578	37345	18896	38419
	27367	58243	40114	19712	44569	03565	32097	80896	02127	87556
	45057	03468	26002	96998	74853	62667	49786	37192	94205	63455
	88398	46763	95748	74578	23926	83306	19522	28068	58691	81760
	29626	85201	49733	20686	85859	25864	27571	47269	34943	65612
	90785	71409	51627	22686	80822	00407	53412	35513	32296	15090
	78937	44387	67591	50550	47210	49785	57616	45017	85802	49940
	72255	96833	72380	70850	44318	59088	65852	36733	50138	88220
	83643	34294	64853	98588	83073	55199	61744	78187	90224	98356
	09295	75095	26430	67530	60518	53711	82022	67673	26896	89217

Table A.2 Binomial Coefficients†

Entries are $\binom{n}{r} = \frac{n!}{r!(n-r)!}$.

Thus, $\binom{11}{4} = 330$ and $\binom{20}{8} = 125{,}970$.

Missing entries may be obtained from the relationship $\binom{n}{r} = \binom{n}{n-r}$.

Thus, for $n = 15$ and $r = 12$: $\binom{15}{12} = \binom{15}{3} = 455$.

Entries were generated by the relationship $\binom{n+1}{r} = \binom{n}{r-1} + \binom{n}{r}$, which may be used to extend the table.

† Based on calculations performed by the computing staff of the Department of Experimental Statistics, New Mexico State University.

Table A.2 *continued*

							r						
n	0	1	2	3	4	5	6	7	8	9	10	11	12
0	1												
1	1												
2	1	2											
3	1	3											
4	1	4	6										
5	1	5	10										
6	1	6	15	20									
7	1	7	21	35									
8	1	8	28	56	70								
9	1	9	36	84	126								
10	1	10	45	120	210	252							
11	1	11	55	165	330	462							
12	1	12	66	220	495	792	924						
13	1	13	78	286	715	1287	1716						
14	1	14	91	364	1001	2002	3003	3432					
15	1	15	105	455	1365	3003	5005	6435					
16	1	16	120	560	1820	4368	8008	11440	12870				
17	1	17	136	680	2380	6188	12376	19448	24310				
18	1	18	153	816	3060	8568	18564	31824	43758	48620			
19	1	19	171	969	3876	11628	27132	50388	75582	92378			
20	1	20	190	1140	4845	15504	38760	77520	125970	167960	184756		
21	1	21	210	1330	5985	20349	54264	116280	203490	293930	352716		
22	1	22	231	1540	7315	26334	74613	170544	319770	497420	646646	705432	
23	1	23	253	1771	8855	33649	100947	245157	490314	817190	1144066	1352078	
24	1	24	276	2024	10626	42504	134596	346104	735471	1307504	1961256	2496144	2704156
25	1	25	300	2300	12650	53130	177100	480700	1081575	2042975	3268760	4457400	5200300

Table A.3 Individual and Cumulative Probabilities of Binomial Distributions†

Let X denote a binomial random variable with parameters n and p.

Entries in the $f(x)$ column represent the probability X has the value x:

$$f(x) = P(X = x) = \binom{n}{x} p^x (1 - p)^{n-x}$$

For $n = 6$ and $p = 0.3$,

$$f(4) = 0.0595 \qquad \text{and} \qquad f(6) = 0.0007$$

Entries in the $F(x)$ column represent the probability X has a value *no greater than* x:

$$F(x) = P(X \leqslant x)$$

For $n = 6$ and $p = 0.3$: $F(4) = 0.9891$ and $F(6) = 1.0$.

† Based on calculations performed by the computing staff of the Department of Experimental Statistics, New Mexico State University.

Table A.3 *continued*

n	x	p = 0.1 f(x)	p = 0.1 F(x)	p = 0.2 f(x)	p = 0.2 F(x)	p = 0.25 f(x)	p = 0.25 F(x)
3	0	0.7290	0.7290	0.5120	0.5120	0.4219	0.4219
	1	0.2430	0.9720	0.3840	0.8960	0.4219	0.8438
	2	0.0270	0.9990	0.0960	0.9920	0.1406	0.9844
	3	0.0010	1.0000	0.0080	1.0000	0.0156	1.0000
4	0	0.6561	0.6561	0.4096	0.4096	0.3164	0.3164
	1	0.2916	0.9477	0.4096	0.8192	0.4219	0.7383
	2	0.0486	0.9963	0.1536	0.9728	0.2109	0.9492
	3	0.0036	0.9999	0.0256	0.9984	0.0469	0.9961
	4	0.0001	1.0000	0.0016	1.0000	0.0039	1.0000
5	0	0.5905	0.5905	0.3277	0.3277	0.2373	0.2373
	1	0.3281	0.9185	0.4096	0.7373	0.3955	0.6328
	2	0.0729	0.9914	0.2048	0.9421	0.2637	0.8965
	3	0.0081	0.9995	0.0512	0.9933	0.0879	0.9844
	4	0.0005	1.0000	0.0064	0.9997	0.0146	0.9990
	5	0.0000	1.0000	0.0003	1.0000	0.0010	1.0000
6	0	0.5314	0.5314	0.2621	0.2621	0.1780	0.1780
	1	0.3543	0.8857	0.3932	0.6554	0.3560	0.5339
	2	0.0984	0.9842	0.2458	0.9011	0.2966	0.8306
	3	0.0146	0.9987	0.0819	0.9830	0.1318	0.9624
	4	0.0012	0.9999	0.0154	0.9984	0.0330	0.9954
	5	0.0001	1.0000	0.0015	0.9999	0.0044	0.9998
	6	0.0000	1.0000	0.0001	1.0000	0.0002	1.0000
7	0	0.4783	0.4783	0.2097	0.2097	0.1335	0.1335
	1	0.3720	0.8503	0.3670	0.5767	0.3115	0.4449
	2	0.1240	0.9743	0.2753	0.8520	0.3115	0.7564
	3	0.0230	0.9973	0.1147	0.9667	0.1730	0.9294
	4	0.0026	0.9998	0.0287	0.9953	0.0577	0.9871
	5	0.0002	1.0000	0.0043	0.9996	0.0115	0.9987
	6	0.0000	1.0000	0.0004	1.0000	0.0013	0.9999
	7	0.0000	1.0000	0.0000	1.0000	0.0001	1.0000
8	0	0.4305	0.4305	0.1678	0.1678	0.1001	0.1001
	1	0.3826	0.8131	0.3355	0.5033	0.2670	0.3671
	2	0.1488	0.9619	0.2936	0.7969	0.3115	0.6785
	3	0.0331	0.9950	0.1468	0.9437	0.2076	0.8862
	4	0.0046	0.9996	0.0459	0.9896	0.0865	0.9727
	5	0.0004	1.0000	0.0092	0.9988	0.0231	0.9958
	6	0.0000	1.0000	0.0011	0.9999	0.0038	0.9996
	7	0.0000	1.0000	0.0001	1.0000	0.0004	1.0000
	8	0.0000	1.0000	0.0000	1.0000	0.0000	1.0000
9	0	0.3874	0.3874	0.1342	0.1342	0.0751	0.0751
	1	0.3874	0.7748	0.3020	0.4362	0.2253	0.3003
	2	0.1722	0.9470	0.3020	0.7382	0.3003	0.6007
	3	0.0446	0.9917	0.1762	0.9144	0.2336	0.8343
	4	0.0074	0.9991	0.0661	0.9804	0.1168	0.9511
	5	0.0008	0.9999	0.0165	0.9969	0.0389	0.9900
	6	0.0001	1.0000	0.0028	0.9997	0.0087	0.9987
	7	0.0000	1.0000	0.0003	1.0000	0.0012	0.9999
	8	0.0000	1.0000	0.0000	1.0000	0.0001	1.0000
	9	0.0000	1.0000	0.0000	1.0000	0.0000	1.0000

Table A.3 *continued*

p = 0.3		p = 1/3		p = 0.4		p = 0.5	
f(x)	F(x)	f(x)	F(x)	f(x)	F(x)	f(x)	F(x)
0.3430	0.3430	0.2963	0.2963	0.2160	0.2160	0.1250	0.1250
0.4410	0.7840	0.4444	0.7407	0.4320	0.6480	0.3750	0.5000
0.1890	0.9730	0.2222	0.9630	0.2880	0.9360	0.3750	0.8750
0.0270	1.0000	0.0370	1.0000	0.0640	1.0000	0.1250	1.0000
0.2401	0.2401	0.1975	0.1975	0.1296	0.1296	0.0625	0.0625
0.4116	0.6517	0.3951	0.5926	0.3456	0.4752	0.2500	0.3125
0.2646	0.9163	0.2963	0.8889	0.3456	0.8208	0.3750	0.6875
0.0756	0.9919	0.0988	0.9877	0.1536	0.9744	0.2500	0.9375
0.0081	1.0000	0.0123	1.0000	0.0256	1.0000	0.0625	1.0000
0.1681	0.1681	0.1317	0.1317	0.0778	0.0778	0.0313	0.0313
0.3602	0.5282	0.3292	0.4609	0.2592	0.3370	0.1563	0.1875
0.3087	0.8369	0.3292	0.7901	0.3456	0.6826	0.3125	0.5000
0.1323	0.9692	0.1646	0.9547	0.2304	0.9130	0.3125	0.8125
0.0284	0.9976	0.0412	0.9959	0.0768	0.9898	0.1563	0.9688
0.0024	1.0000	0.0041	1.0000	0.0102	1.0000	0.0313	1.0000
0.1176	0.1176	0.0878	0.0878	0.0467	0.0467	0.0156	0.0156
0.3025	0.4202	0.2634	0.3512	0.1866	0.2333	0.0938	0.1094
0.3241	0.7443	0.3292	0.6804	0.3110	0.5443	0.2344	0.3438
0.1852	0.9295	0.2195	0.8999	0.2765	0.8208	0.3125	0.6563
0.0595	0.9891	0.0823	0.9822	0.1382	0.9590	0.2344	0.8906
0.0102	0.9993	0.0165	0.9986	0.0369	0.9959	0.0938	0.9844
0.0007	1.0000	0.0014	1.0000	0.0041	1.0000	0.0156	1.0000
0.0824	0.0824	0.0585	0.0585	0.0280	0.0280	0.0078	0.0078
0.2471	0.3294	0.2048	0.2634	0.1306	0.1586	0.0547	0.0625
0.3177	0.6471	0.3073	0.5706	0.2613	0.4199	0.1641	0.2266
0.2269	0.8740	0.2561	0.8267	0.2903	0.7102	0.2734	0.5000
0.0972	0.9712	0.1280	0.9547	0.1935	0.9037	0.2734	0.7734
0.0250	0.9962	0.0384	0.9931	0.0774	0.9812	0.1641	0.9375
0.0036	0.9998	0.0064	0.9995	0.0172	0.9984	0.0547	0.9922
0.0002	1.0000	0.0005	1.0000	0.0016	1.0000	0.0078	1.0000
0.0576	0.0576	0.0390	0.0390	0.0168	0.0168	0.0039	0.0039
0.1977	0.2553	0.1561	0.1951	0.0896	0.1064	0.0313	0.0352
0.2965	0.5518	0.2731	0.4682	0.2090	0.3154	0.1094	0.1445
0.2541	0.8059	0.2731	0.7414	0.2787	0.5941	0.2188	0.3633
0.1361	0.9420	0.1707	0.9121	0.2322	0.8263	0.2734	0.6367
0.0467	0.9887	0.0683	0.9803	0.1239	0.9502	0.2188	0.8555
0.0100	0.9987	0.0171	0.9974	0.0413	0.9915	0.1094	0.9648
0.0012	0.9999	0.0024	0.9998	0.0079	0.9993	0.0313	0.9961
0.0001	1.0000	0.0002	1.0000	0.0007	1.0000	0.0039	1.0000
0.0404	0.0404	0.0260	0.0260	0.0101	0.0101	0.0020	0.0020
0.1556	0.1960	0.1171	0.1431	0.0605	0.0705	0.0176	0.0195
0.2668	0.4628	0.2341	0.3772	0.1612	0.2318	0.0703	0.0898
0.2668	0.7297	0.2731	0.6503	0.2508	0.4826	0.1641	0.2539
0.1715	0.9012	0.2048	0.8552	0.2508	0.7334	0.2461	0.5000
0.0735	0.9747	0.1024	0.9576	0.1672	0.9006	0.2461	0.7461
0.0210	0.9957	0.0341	0.9917	0.0743	0.9750	0.1641	0.9102
0.0039	0.9996	0.0073	0.9990	0.0212	0.9962	0.0703	0.9805
0.0004	1.0000	0.0009	0.9999	0.0035	0.9997	0.0176	0.9980
0.0000	1.0000	0.0001	1.0000	0.0003	1.0000	0.0020	1.0000

Table A.3 *continued*

n	x	p = 0.1 f(x)	p = 0.1 F(x)	p = 0.2 f(x)	p = 0.2 F(x)	p = 0.25 f(x)	p = 0.25 F(x)
10	0	0.3487	0.3487	0.1074	0.1074	0.0563	0.0563
	1	0.3874	0.7361	0.2684	0.3758	0.1877	0.2440
	2	0.1937	0.9298	0.3020	0.6778	0.2816	0.5256
	3	0.0574	0.9872	0.2013	0.8791	0.2503	0.7759
	4	0.0112	0.9984	0.0881	0.9672	0.1460	0.9219
	5	0.0015	0.9999	0.0264	0.9936	0.0584	0.9803
	6	0.0001	1.0000	0.0055	0.9991	0.0162	0.9965
	7	0.0000	1.0000	0.0008	0.9999	0.0031	0.9996
	8	0.0000	1.0000	0.0001	1.0000	0.0004	1.0000
	9	0.0000	1.0000	0.0000	1.0000	0.0000	1.0000
	10	0.0000	1.0000	0.0000	1.0000	0.0000	1.0000
15	0	0.2059	0.2059	0.0352	0.0352	0.0134	0.0134
	1	0.3432	0.5490	0.1319	0.1671	0.0668	0.0802
	2	0.2669	0.8159	0.2309	0.3980	0.1559	0.2361
	3	0.1285	0.9444	0.2501	0.6482	0.2252	0.4613
	4	0.0428	0.9873	0.1876	0.8358	0.2252	0.6865
	5	0.0105	0.9978	0.1032	0.9389	0.1651	0.8516
	6	0.0019	0.9997	0.0430	0.9819	0.0917	0.9434
	7	0.0003	1.0000	0.0138	0.9958	0.0393	0.9827
	8	0.0000	1.0000	0.0035	0.9992	0.0131	0.9958
	9	0.0000	1.0000	0.0007	0.9999	0.0034	0.9992
	10	0.0000	1.0000	0.0001	1.0000	0.0007	0.9999
	11	0.0000	1.0000	0.0000	1.0000	0.0001	1.0000
	12	0.0000	1.0000	0.0000	1.0000	0.0000	1.0000
	13	0.0000	1.0000	0.0000	1.0000	0.0000	1.0000
	14	0.0000	1.0000	0.0000	1.0000	0.0000	1.0000
	15	0.0000	1.0000	0.0000	1.0000	0.0000	1.0000
20	0	0.1216	0.1216	0.0115	0.0115	0.0032	0.0032
	1	0.2702	0.3917	0.0576	0.0692	0.0211	0.0243
	2	0.2852	0.6769	0.1369	0.2061	0.0669	0.0913
	3	0.1901	0.8670	0.2054	0.4114	0.1339	0.2252
	4	0.0898	0.9568	0.2182	0.6296	0.1897	0.4148
	5	0.0319	0.9887	0.1746	0.8042	0.2023	0.6172
	6	0.0089	0.9976	0.1091	0.9133	0.1686	0.7858
	7	0.0020	0.9996	0.0545	0.9679	0.1124	0.8982
	8	0.0004	0.9999	0.0222	0.9900	0.0609	0.9591
	9	0.0001	1.0000	0.0074	0.9974	0.0271	0.9861
	10	0.0000	1.0000	0.0020	0.9994	0.0099	0.9961
	11	0.0000	1.0000	0.0005	0.9999	0.0030	0.9991
	12	0.0000	1.0000	0.0001	1.0000	0.0008	0.9998
	13	0.0000	1.0000	0.0000	1.0000	0.0002	1.0000
	14	0.0000	1.0000	0.0000	1.0000	0.0000	1.0000
	15	0.0000	1.0000	0.0000	1.0000	0.0000	1.0000
	16	0.0000	1.0000	0.0000	1.0000	0.0000	1.0000
	17	0.0000	1.0000	0.0000	1.0000	0.0000	1.0000
	18	0.0000	1.0000	0.0000	1.0000	0.0000	1.0000
	19	0.0000	1.0000	0.0000	1.0000	0.0000	1.0000
	20	0.0000	1.0000	0.0000	1.0000	0.0000	1.0000

p = 0.3		p = 1/3		p = 0.4		p = 0.5	
$f(x)$	$F(x)$	$f(x)$	$F(x)$	$f(x)$	$F(x)$	$f(x)$	$F(x)$
0.0282	0.0282	0.0173	0.0173	0.0060	0.0060	0.0010	0.0010
0.1211	0.1493	0.0867	0.1040	0.0403	0.0464	0.0098	0.0107
0.2335	0.3828	0.1951	0.2991	0.1209	0.1673	0.0439	0.0547
0.2668	0.6496	0.2601	0.5593	0.2150	0.3823	0.1172	0.1719
0.2001	0.8497	0.2276	0.7869	0.2508	0.6331	0.2051	0.3770
0.1029	0.9527	0.1366	0.9234	0.2007	0.8338	0.2461	0.6230
0.0368	0.9894	0.0569	0.9803	0.1115	0.9452	0.2051	0.8281
0.0090	0.9984	0.0163	0.9966	0.0425	0.9877	0.1172	0.9453
0.0014	0.9999	0.0030	0.9996	0.0106	0.9983	0.0439	0.9893
0.0001	1.0000	0.0003	1.0000	0.0016	0.9999	0.0098	0.9990
0.0000	1.0000	0.0000	1.0000	0.0001	1.0000	0.0010	1.0000
0.0047	0.0047	0.0023	0.0023	0.0005	0.0005	0.0000	0.0000
0.0305	0.0353	0.0171	0.0194	0.0047	0.0052	0.0005	0.0005
0.0916	0.1268	0.0599	0.0794	0.0219	0.0271	0.0032	0.0037
0.1700	0.2969	0.1299	0.2092	0.0634	0.0905	0.0139	0.0176
0.2186	0.5155	0.1948	0.4041	0.1268	0.2173	0.0417	0.0592
0.2061	0.7216	0.2143	0.6184	0.1859	0.4032	0.0916	0.1509
0.1472	0.8689	0.1786	0.7970	0.2066	0.6098	0.1527	0.3036
0.0811	0.9500	0.1148	0.9118	0.1771	0.7869	0.1964	0.5000
0.0348	0.9848	0.0574	0.9692	0.1181	0.9050	0.1964	0.6964
0.0116	0.9963	0.0223	0.9915	0.0612	0.9662	0.1527	0.8491
0.0030	0.9993	0.0067	0.9982	0.0245	0.9907	0.0916	0.9408
0.0006	0.9999	0.0015	0.9997	0.0074	0.9981	0.0417	0.9824
0.0001	1.0000	0.0003	1.0000	0.0016	0.9997	0.0139	0.9963
0.0000	1.0000	0.0000	1.0000	0.0003	1.0000	0.0032	0.9995
0.0000	1.0000	0.0000	1.0000	0.0000	1.0000	0.0005	1.0000
0.0000	1.0000	0.0000	1.0000	0.0000	1.0000	0.0000	1.0000
0.0008	0.0008	0.0003	0.0003	0.0000	0.0000	0.0000	0.0000
0.0068	0.0076	0.0030	0.0033	0.0005	0.0005	0.0000	0.0000
0.0278	0.0355	0.0143	0.0176	0.0031	0.0036	0.0002	0.0002
0.0716	0.1071	0.0429	0.0604	0.0123	0.0160	0.0011	0.0013
0.1304	0.2375	0.0911	0.1515	0.0350	0.0510	0.0046	0.0059
0.1789	0.4164	0.1457	0.2972	0.0746	0.1256	0.0148	0.0207
0.1916	0.6080	0.1821	0.4793	0.1244	0.2500	0.0370	0.0577
0.1643	0.7723	0.1821	0.6615	0.1659	0.4159	0.0739	0.1316
0.1144	0.8867	0.1480	0.8095	0.1797	0.5956	0.1201	0.2517
0.0654	0.9520	0.0987	0.9081	0.1597	0.7553	0.1602	0.4119
0.0308	0.9829	0.0543	0.9624	0.1171	0.8725	0.1762	0.5881
0.0120	0.9949	0.0247	0.9870	0.0710	0.9435	0.1602	0.7483
0.0039	0.9987	0.0092	0.9963	0.0355	0.9790	0.1201	0.8684
0.0010	0.9997	0.0028	0.9991	0.0146	0.9935	0.0739	0.9423
0.0002	1.0000	0.0007	0.9998	0.0049	0.9984	0.0370	0.9793
0.0000	1.0000	0.0001	1.0000	0.0013	0.9997	0.0148	0.9941
0.0000	1.0000	0.0000	1.0000	0.0003	1.0000	0.0046	0.9987
0.0000	1.0000	0.0000	1.0000	0.0000	1.0000	0.0011	0.9998
0.0000	1.0000	0.0000	1.0000	0.0000	1.0000	0.0002	1.0000
0.0000	1.0000	0.0000	1.0000	0.0000	1.0000	0.0000	1.0000
0.0000	1.0000	0.0000	1.0000	0.0000	1.0000	0.0000	1.0000

Table A.3 *continued*

n	x	p = 0.1 f(x)	p = 0.1 F(x)	p = 0.2 f(x)	p = 0.2 F(x)	p = 0.25 f(x)	p = 0.25 F(x)
25	0	0.0718	0.0718	0.0038	0.0038	0.0008	0.0008
	1	0.1994	0.2712	0.0236	0.0274	0.0063	0.0070
	2	0.2659	0.5371	0.0708	0.0982	0.0251	0.0321
	3	0.2265	0.7636	0.1358	0.2340	0.0641	0.0962
	4	0.1384	0.9020	0.1867	0.4207	0.1175	0.2137
	5	0.0646	0.9666	0.1960	0.6167	0.1645	0.3783
	6	0.0239	0.9905	0.1633	0.7800	0.1828	0.5611
	7	0.0072	0.9977	0.1108	0.8909	0.1654	0.7265
	8	0.0018	0.9995	0.0623	0.9532	0.1241	0.8506
	9	0.0004	0.9999	0.0294	0.9827	0.0781	0.9287
	10	0.0001	1.0000	0.0118	0.9944	0.0417	0.9703
	11	0.0000	1.0000	0.0040	0.9985	0.0189	0.9893
	12	0.0000	1.0000	0.0012	0.9996	0.0074	0.9966
	13	0.0000	1.0000	0.0003	0.9999	0.0025	0.9991
	14	0.0000	1.0000	0.0001	1.0000	0.0007	0.9998
	15	0.0000	1.0000	0.0000	1.0000	0.0002	1.0000
	16	0.0000	1.0000	0.0000	1.0000	0.0000	1.0000
	17	0.0000	1.0000	0.0000	1.0000	0.0000	1.0000
	18	0.0000	1.0000	0.0000	1.0000	0.0000	1.0000
	19	0.0000	1.0000	0.0000	1.0000	0.0000	1.0000
	20	0.0000	1.0000	0.0000	1.0000	0.0000	1.0000
	21	0.0000	1.0000	0.0000	1.0000	0.0000	1.0000
	22	0.0000	1.0000	0.0000	1.0000	0.0000	1.0000
	23	0.0000	1.0000	0.0000	1.0000	0.0000	1.0000
	24	0.0000	1.0000	0.0000	1.0000	0.0000	1.0000
	25	0.0000	1.0000	0.0000	1.0000	0.0000	1.0000
30	0	0.0424	0.0424	0.0012	0.0012	0.0002	0.0002
	1	0.1413	0.1837	0.0093	0.0105	0.0018	0.0020
	2	0.2277	0.4114	0.0337	0.0442	0.0086	0.0106
	3	0.2361	0.6474	0.0785	0.1227	0.0269	0.0374
	4	0.1771	0.8245	0.1325	0.2552	0.0604	0.0979
	5	0.1023	0.9268	0.1723	0.4275	0.1047	0.2026
	6	0.0474	0.9742	0.1795	0.6070	0.1455	0.3481
	7	0.0180	0.9922	0.1538	0.7608	0.1662	0.5143
	8	0.0058	0.9980	0.1106	0.8713	0.1593	0.6736
	9	0.0016	0.9995	0.0676	0.9389	0.1298	0.8034
	10	0.0004	0.9999	0.0355	0.9744	0.0909	0.8943
	11	0.0001	1.0000	0.0161	0.9905	0.0551	0.9493
	12	0.0000	1.0000	0.0064	0.9969	0.0291	0.9784
	13	0.0000	1.0000	0.0022	0.9991	0.0134	0.9918
	14	0.0000	1.0000	0.0007	0.9998	0.0054	0.9973
	15	0.0000	1.0000	0.0002	0.9999	0.0019	0.9992
	16	0.0000	1.0000	0.0000	1.0000	0.0006	0.9998
	17	0.0000	1.0000	0.0000	1.0000	0.0002	0.9999
	18	0.0000	1.0000	0.0000	1.0000	0.0000	1.0000
	19	0.0000	1.0000	0.0000	1.0000	0.0000	1.0000
	20	0.0000	1.0000	0.0000	1.0000	0.0000	1.0000
	21	0.0000	1.0000	0.0000	1.0000	0.0000	1.0000
	22	0.0000	1.0000	0.0000	1.0000	0.0000	1.0000
	23	0.0000	1.0000	0.0000	1.0000	0.0000	1.0000
	24	0.0000	1.0000	0.0000	1.0000	0.0000	1.0000
	25	0.0000	1.0000	0.0000	1.0000	0.0000	1.0000
	26	0.0000	1.0000	0.0000	1.0000	0.0000	1.0000
	27	0.0000	1.0000	0.0000	1.0000	0.0000	1.0000
	28	0.0000	1.0000	0.0000	1.0000	0.0000	1.0000
	29	0.0000	1.0000	0.0000	1.0000	0.0000	1.0000
	30	0.0000	1.0000	0.0000	1.0000	0.0000	1.0000

Table A.3 *continued*

$p = 0.3$ $f(x)$	$p = 0.3$ $F(x)$	$p = 1/3$ $f(x)$	$p = 1/3$ $F(x)$	$p = 0.4$ $f(x)$	$p = 0.4$ $F(x)$	$p = 0.5$ $f(x)$	$p = 0.5$ $F(x)$
0.0001	0.0001	0.0000	0.0000	0.0000	0.0000	0.0000	0.0000
0.0014	0.0016	0.0005	0.0005	0.0000	0.0001	0.0000	0.0000
0.0074	0.0090	0.0030	0.0035	0.0004	0.0004	0.0000	0.0000
0.0243	0.0332	0.0114	0.0149	0.0019	0.0024	0.0001	0.0001
0.0572	0.0905	0.0313	0.0462	0.0071	0.0095	0.0004	0.0005
0.1030	0.1935	0.0658	0.1120	0.0199	0.0294	0.0016	0.0020
0.1472	0.3407	0.1096	0.2215	0.0442	0.0736	0.0053	0.0073
0.1712	0.5118	0.1487	0.3703	0.0800	0.1536	0.0143	0.0216
0.1651	0.6769	0.1673	0.5376	0.1200	0.2735	0.0322	0.0539
0.1336	0.8106	0.1580	0.6956	0.1511	0.4246	0.0609	0.1148
0.0916	0.9022	0.1264	0.8220	0.1612	0.5858	0.0974	0.2122
0.0536	0.9558	0.0862	0.9082	0.1465	0.7323	0.1328	0.3450
0.0268	0.9825	0.0503	0.9585	0.1140	0.8462	0.1550	0.5000
0.0115	0.9940	0.0251	0.9836	0.0760	0.9222	0.1550	0.6550
0.0042	0.9982	0.0108	0.9944	0.0434	0.9656	0.1328	0.7878
0.0013	0.9995	0.0040	0.9984	0.0212	0.9868	0.0974	0.8852
0.0004	0.9999	0.0012	0.9996	0.0088	0.9957	0.0609	0.9461
0.0001	1.0000	0.0003	0.9999	0.0031	0.9988	0.0322	0.9784
0.0000	1.0000	0.0001	1.0000	0.0009	0.9997	0.0143	0.9927
0.0000	1.0000	0.0000	1.0000	0.0002	0.9999	0.0053	0.9980
0.0000	1.0000	0.0000	1.0000	0.0000	1.0000	0.0016	0.9995
0.0000	1.0000	0.0000	1.0000	0.0000	1.0000	0.0004	0.9999
0.0000	1.0000	0.0000	1.0000	0.0000	1.0000	0.0001	1.0000
0.0000	1.0000	0.0000	1.0000	0.0000	1.0000	0.0000	1.0000
0.0000	1.0000	0.0000	1.0000	0.0000	1.0000	0.0000	1.0000
0.0000	1.0000	0.0000	1.0000	0.0000	1.0000	0.0000	1.0000
0.0000	0.0000	0.0000	0.0000	0.0000	0.0000	0.0000	0.0000
0.0003	0.0003	0.0001	0.0001	0.0000	0.0000	0.0000	0.0000
0.0018	0.0021	0.0006	0.0007	0.0000	0.0000	0.0000	0.0000
0.0072	0.0093	0.0026	0.0033	0.0003	0.0003	0.0000	0.0000
0.0208	0.0302	0.0089	0.0122	0.0012	0.0015	0.0000	0.0000
0.0464	0.0766	0.0232	0.0355	0.0041	0.0057	0.0001	0.0002
0.0829	0.1595	0.0484	0.0838	0.0115	0.0172	0.0006	0.0007
0.1219	0.2814	0.0829	0.1668	0.0263	0.0435	0.0019	0.0026
0.1501	0.4315	0.1192	0.2860	0.0505	0.0940	0.0055	0.0081
0.1573	0.5888	0.1457	0.4317	0.0823	0.1763	0.0133	0.0214
0.1416	0.7304	0.1530	0.5848	0.1152	0.2915	0.0280	0.0494
0.1103	0.8407	0.1391	0.7239	0.1396	0.4311	0.0509	0.1002
0.0749	0.9155	0.1101	0.8340	0.1474	0.5785	0.0806	0.1808
0.0444	0.9599	0.0762	0.9102	0.1360	0.7145	0.1115	0.2923
0.0231	0.9831	0.0463	0.9565	0.1101	0.8246	0.1354	0.4278
0.0106	0.9936	0.0247	0.9812	0.0783	0.9029	0.1445	0.5722
0.0042	0.9979	0.0116	0.9928	0.0489	0.9519	0.1354	0.7077
0.0015	0.9994	0.0048	0.9975	0.0269	0.9788	0.1115	0.8192
0.0005	0.9998	0.0017	0.9993	0.0129	0.9917	0.0806	0.8998
0.0001	1.0000	0.0005	0.9998	0.0054	0.9971	0.0509	0.9506
0.0000	1.0000	0.0001	1.0000	0.0020	0.9991	0.0280	0.9786
0.0000	1.0000	0.0000	1.0000	0.0006	0.9998	0.0133	0.9919
0.0000	1.0000	0.0000	1.0000	0.0002	1.0000	0.0055	0.9974
0.0000	1.0000	0.0000	1.0000	0.0000	1.0000	0.0019	0.9993
0.0000	1.0000	0.0000	1.0000	0.0000	1.0000	0.0006	0.9998
0.0000	1.0000	0.0000	1.0000	0.0000	1.0000	0.0001	1.0000
0.0000	1.0000	0.0000	1.0000	0.0000	1.0000	0.0000	1.0000
0.0000	1.0000	0.0000	1.0000	0.0000	1.0000	0.0000	1.0000
0.0000	1.0000	0.0000	1.0000	0.0000	1.0000	0.0000	1.0000
0.0000	1.0000	0.0000	1.0000	0.0000	1.0000	0.0000	1.0000
0.0000	1.0000	0.0000	1.0000	0.0000	1.0000	0.0000	1.0000

Table A.3 *continued*

n	x	p = 0.1 f(x)	p = 0.1 F(x)	p = 0.2 f(x)	p = 0.2 F(x)	p = 0.25 f(x)	p = 0.25 F(x)
40	0	0.0148	0.0148	0.0001	0.0001	0.0000	0.0000
	1	0.0657	0.0805	0.0013	0.0015	0.0001	0.0001
	2	0.1423	0.2228	0.0065	0.0079	0.0009	0.0010
	3	0.2003	0.4231	0.0205	0.0285	0.0037	0.0047
	4	0.2059	0.6290	0.0475	0.0759	0.0113	0.0160
	5	0.1647	0.7937	0.0854	0.1613	0.0272	0.0433
	6	0.1068	0.9005	0.1246	0.2859	0.0530	0.0962
	7	0.0576	0.9581	0.1513	0.4371	0.0857	0.1820
	8	0.0264	0.9845	0.1560	0.5931	0.1179	0.2998
	9	0.0104	0.9949	0.1386	0.7318	0.1397	0.4395
	10	0.0036	0.9985	0.1075	0.8392	0.1444	0.5839
	11	0.0011	0.9996	0.0733	0.9125	0.1312	0.7151
	12	0.0003	0.9999	0.0443	0.9568	0.1057	0.8209
	13	0.0001	1.0000	0.0238	0.9806	0.0759	0.8968
	14	0.0000	1.0000	0.0115	0.9921	0.0488	0.9456
	15	0.0000	1.0000	0.0050	0.9971	0.0282	0.9738
	16	0.0000	1.0000	0.0019	0.9990	0.0147	0.9884
	17	0.0000	1.0000	0.0007	0.9997	0.0069	0.9953
	18	0.0000	1.0000	0.0002	0.9999	0.0029	0.9983
	19	0.0000	1.0000	0.0001	1.0000	0.0011	0.9994
	20	0.0000	1.0000	0.0000	1.0000	0.0004	0.9998
	21	0.0000	1.0000	0.0000	1.0000	0.0001	1.0000
	22	0.0000	1.0000	0.0000	1.0000	0.0000	1.0000
	23	0.0000	1.0000	0.0000	1.0000	0.0000	1.0000
	24	0.0000	1.0000	0.0000	1.0000	0.0000	1.0000
	25	0.0000	1.0000	0.0000	1.0000	0.0000	1.0000
	26	0.0000	1.0000	0.0000	1.0000	0.0000	1.0000
	27	0.0000	1.0000	0.0000	1.0000	0.0000	1.0000
	28	0.0000	1.0000	0.0000	1.0000	0.0000	1.0000
	29	0.0000	1.0000	0.0000	1.0000	0.0000	1.0000
	30	0.0000	1.0000	0.0000	1.0000	0.0000	1.0000
	31	0.0000	1.0000	0.0000	1.0000	0.0000	1.0000
	32	0.0000	1.0000	0.0000	1.0000	0.0000	1.0000
	33	0.0000	1.0000	0.0000	1.0000	0.0000	1.0000
	34	0.0000	1.0000	0.0000	1.0000	0.0000	1.0000
	35	0.0000	1.0000	0.0000	1.0000	0.0000	1.0000
	36	0.0000	1.0000	0.0000	1.0000	0.0000	1.0000
	37	0.0000	1.0000	0.0000	1.0000	0.0000	1.0000
	38	0.0000	1.0000	0.0000	1.0000	0.0000	1.0000
	39	0.0000	1.0000	0.0000	1.0000	0.0000	1.0000
	40	0.0000	1.0000	0.0000	1.0000	0.0000	1.0000

p = 0.3		p = 1/3		p = 0.4		p = 0.5	
$f(x)$	$F(x)$	$f(x)$	$F(x)$	$f(x)$	$F(x)$	$f(x)$	$F(x)$
0.0000	0.0000	0.0000	0.0000	0.0000	0.0000	0.0000	0.0000
0.0000	0.0000	0.0000	0.0000	0.0000	0.0000	0.0000	0.0000
0.0001	0.0001	0.0000	0.0000	0.0000	0.0000	0.0000	0.0000
0.0005	0.0006	0.0001	0.0001	0.0000	0.0000	0.0000	0.0000
0.0020	0.0026	0.0005	0.0006	0.0000	0.0000	0.0000	0.0000
0.0061	0.0086	0.0019	0.0025	0.0001	0.0001	0.0000	0.0000
0.0151	0.0238	0.0054	0.0079	0.0005	0.0006	0.0000	0.0000
0.0315	0.0553	0.0132	0.0211	0.0015	0.0021	0.0000	0.0000
0.0557	0.1110	0.0272	0.0483	0.0040	0.0061	0.0001	0.0001
0.0849	0.1959	0.0483	0.0966	0.0095	0.0156	0.0002	0.0003
0.1128	0.3087	0.0749	0.1714	0.0196	0.0352	0.0008	0.0011
0.1319	0.4406	0.1021	0.2735	0.0357	0.0709	0.0021	0.0032
0.1366	0.5772	0.1234	0.3969	0.0576	0.1285	0.0051	0.0083
0.1261	0.7032	0.1328	0.5297	0.0827	0.2112	0.0109	0.0192
0.1042	0.8074	0.1281	0.6578	0.1063	0.3174	0.0211	0.0403
0.0774	0.8849	0.1110	0.7688	0.1228	0.4402	0.0366	0.0769
0.0518	0.9367	0.0867	0.8556	0.1279	0.5681	0.0572	0.1341
0.0314	0.9680	0.0612	0.9168	0.1204	0.6885	0.0807	0.2148
0.0172	0.9852	0.0391	0.9559	0.1026	0.7911	0.1031	0.3179
0.0085	0.9937	0.0226	0.9786	0.0792	0.8702	0.1194	0.4373
0.0038	0.9976	0.0119	0.9904	0.0554	0.9256	0.1254	0.5627
0.0016	0.9991	0.0057	0.9961	0.0352	0.9608	0.1194	0.6821
0.0006	0.9997	0.0024	0.9986	0.0203	0.9811	0.1031	0.7852
0.0002	0.9999	0.0010	0.9995	0.0106	0.9917	0.0807	0.8659
0.0001	1.0000	0.0003	0.9998	0.0050	0.9966	0.0572	0.9231
0.0000	1.0000	0.0001	1.0000	0.0021	0.9988	0.0366	0.9597
0.0000	1.0000	0.0000	1.0000	0.0008	0.9996	0.0211	0.9808
0.0000	1.0000	0.0000	1.0000	0.0003	0.9999	0.0109	0.9917
0.0000	1.0000	0.0000	1.0000	0.0001	1.0000	0.0051	0.9968
0.0000	1.0000	0.0000	1.0000	0.0000	1.0000	0.0021	0.9989
0.0000	1.0000	0.0000	1.0000	0.0000	1.0000	0.0008	0.9997
0.0000	1.0000	0.0000	1.0000	0.0000	1.0000	0.0002	0.9999
0.0000	1.0000	0.0000	1.0000	0.0000	1.0000	0.0001	1.0000
0.0000	1.0000	0.0000	1.0000	0.0000	1.0000	0.0000	1.0000
0.0000	1.0000	0.0000	1.0000	0.0000	1.0000	0.0000	1.0000
0.0000	1.0000	0.0000	1.0000	0.0000	1.0000	0.0000	1.0000
0.0000	1.0000	0.0000	1.0000	0.0000	1.0000	0.0000	1.0000
0.0000	1.0000	0.0000	1.0000	0.0000	1.0000	0.0000	1.0000
0.0000	1.0000	0.0000	1.0000	0.0000	1.0000	0.0000	1.0000
0.0000	1.0000	0.0000	1.0000	0.0000	1.0000	0.0000	1.0000
0.0000	1.0000	0.0000	1.0000	0.0000	1.0000	0.0000	1.0000

Table A.3 *continued*

n	x	p = 0.1 f(x)	p = 0.1 F(x)	p = 0.2 f(x)	p = 0.2 F(x)	p = 0.25 f(x)	p = 0.25 F(x)
50	0	0.0052	0.0052	0.0000	0.0000	0.0000	0.0000
	1	0.0286	0.0338	0.0002	0.0002	0.0000	0.0000
	2	0.0779	0.1117	0.0011	0.0013	0.0001	0.0001
	3	0.1386	0.2503	0.0044	0.0057	0.0004	0.0005
	4	0.1809	0.4312	0.0128	0.0185	0.0016	0.0021
	5	0.1849	0.6161	0.0295	0.0480	0.0049	0.0070
	6	0.1541	0.7702	0.0554	0.1034	0.0123	0.0194
	7	0.1076	0.8779	0.0870	0.1904	0.0259	0.0453
	8	0.0643	0.9421	0.1169	0.3073	0.0463	0.0916
	9	0.0333	0.9755	0.1364	0.4437	0.0721	0.1637
	10	0.0152	0.9906	0.1398	0.5836	0.0985	0.2622
	11	0.0061	0.9968	0.1271	0.7107	0.1194	0.3816
	12	0.0022	0.9990	0.1033	0.8139	0.1294	0.5110
	13	0.0007	0.9997	0.0755	0.8894	0.1261	0.6370
	14	0.0002	0.9999	0.0499	0.9393	0.1110	0.7481
	15	0.0001	1.0000	0.0299	0.9692	0.0888	0.8369
	16	0.0000	1.0000	0.0164	0.9856	0.0648	0.9017
	17	0.0000	1.0000	0.0082	0.9937	0.0432	0.9449
	18	0.0000	1.0000	0.0037	0.9975	0.0264	0.9713
	19	0.0000	1.0000	0.0016	0.9991	0.0148	0.9861
	20	0.0000	1.0000	0.0006	0.9997	0.0077	0.9937
	21	0.0000	1.0000	0.0002	0.9999	0.0036	0.9974
	22	0.0000	1.0000	0.0001	1.0000	0.0016	0.9990
	23	0.0000	1.0000	0.0000	1.0000	0.0006	0.9996
	24	0.0000	1.0000	0.0000	1.0000	0.0002	0.9999
	25	0.0000	1.0000	0.0000	1.0000	0.0001	1.0000
	26	0.0000	1.0000	0.0000	1.0000	0.0000	1.0000
	27	0.0000	1.0000	0.0000	1.0000	0.0000	1.0000
	28	0.0000	1.0000	0.0000	1.0000	0.0000	1.0000
	29	0.0000	1.0000	0.0000	1.0000	0.0000	1.0000
	30	0.0000	1.0000	0.0000	1.0000	0.0000	1.0000
	31	0.0000	1.0000	0.0000	1.0000	0.0000	1.0000
	32	0.0000	1.0000	0.0000	1.0000	0.0000	1.0000
	33	0.0000	1.0000	0.0000	1.0000	0.0000	1.0000
	34	0.0000	1.0000	0.0000	1.0000	0.0000	1.0000
	35	0.0000	1.0000	0.0000	1.0000	0.0000	1.0000
	36	0.0000	1.0000	0.0000	1.0000	0.0000	1.0000
	37	0.0000	1.0000	0.0000	1.0000	0.0000	1.0000
	38	0.0000	1.0000	0.0000	1.0000	0.0000	1.0000
	39	0.0000	1.0000	0.0000	1.0000	0.0000	1.0000
	40	0.0000	1.0000	0.0000	1.0000	0.0000	1.0000
	41	0.0000	1.0000	0.0000	1.0000	0.0000	1.0000
	42	0.0000	1.0000	0.0000	1.0000	0.0000	1.0000
	43	0.0000	1.0000	0.0000	1.0000	0.0000	1.0000
	44	0.0000	1.0000	0.0000	1.0000	0.0000	1.0000
	45	0.0000	1.0000	0.0000	1.0000	0.0000	1.0000
	46	0.0000	1.0000	0.0000	1.0000	0.0000	1.0000
	47	0.0000	1.0000	0.0000	1.0000	0.0000	1.0000
	48	0.0000	1.0000	0.0000	1.0000	0.0000	1.0000
	49	0.0000	1.0000	0.0000	1.0000	0.0000	1.0000
	50	0.0000	1.0000	0.0000	1.0000	0.0000	1.0000

p = 0.3 f(x)	p = 0.3 F(x)	p = 1/3 f(x)	p = 1/3 F(x)	p = 0.4 f(x)	p = 0.4 F(x)	p = 0.5 f(x)	p = 0.5 F(x)
0.0000	0.0000	0.0000	0.0000	0.0000	0.0000	0.0000	0.0000
0.0000	0.0000	0.0000	0.0000	0.0000	0.0000	0.0000	0.0000
0.0000	0.0000	0.0000	0.0000	0.0000	0.0000	0.0000	0.0000
0.0000	0.0000	0.0000	0.0000	0.0000	0.0000	0.0000	0.0000
0.0001	0.0002	0.0000	0.0000	0.0000	0.0000	0.0000	0.0000
0.0006	0.0007	0.0001	0.0001	0.0000	0.0000	0.0000	0.0000
0.0018	0.0025	0.0004	0.0005	0.0000	0.0000	0.0000	0.0000
0.0048	0.0073	0.0012	0.0017	0.0000	0.0001	0.0000	0.0000
0.0110	0.0183	0.0033	0.0050	0.0002	0.0002	0.0000	0.0000
0.0220	0.0402	0.0077	0.0127	0.0005	0.0008	0.0000	0.0000
0.0386	0.0789	0.0157	0.0284	0.0014	0.0022	0.0000	0.0000
0.0602	0.1390	0.0286	0.0570	0.0035	0.0057	0.0000	0.0000
0.0838	0.2229	0.0465	0.1035	0.0076	0.0133	0.0001	0.0002
0.1050	0.3279	0.0679	0.1715	0.0147	0.0280	0.0003	0.0005
0.1189	0.4468	0.0898	0.2612	0.0260	0.0540	0.0008	0.0013
0.1223	0.5692	0.1077	0.3690	0.0415	0.0955	0.0020	0.0033
0.1147	0.6839	0.1178	0.4868	0.0606	0.1561	0.0044	0.0077
0.0983	0.7822	0.1178	0.6046	0.0808	0.2369	0.0087	0.0164
0.0772	0.8594	0.1080	0.7126	0.0987	0.3356	0.0160	0.0325
0.0558	0.9152	0.0910	0.8036	0.1109	0.4465	0.0270	0.0595
0.0370	0.9522	0.0705	0.8741	0.1146	0.5610	0.0419	0.1013
0.0227	0.9749	0.0503	0.9244	0.1091	0.6701	0.0598	0.1611
0.0128	0.9877	0.0332	0.9576	0.0959	0.7660	0.0788	0.2399
0.0067	0.9944	0.0202	0.9778	0.0778	0.8438	0.0960	0.3359
0.0032	0.9976	0.0114	0.9892	0.0584	0.9022	0.1080	0.4439
0.0014	0.9991	0.0059	0.9951	0.0405	0.9427	0.1123	0.5561
0.0006	0.9997	0.0028	0.9979	0.0259	0.9686	0.1080	0.6641
0.0002	0.9999	0.0013	0.9992	0.0154	0.9840	0.0960	0.7601
0.0001	1.0000	0.0005	0.9997	0.0084	0.9924	0.0788	0.8389
0.0000	1.0000	0.0002	0.9999	0.0043	0.9966	0.0598	0.8987
0.0000	1.0000	0.0001	1.0000	0.0020	0.9986	0.0419	0.9405
0.0000	1.0000	0.0000	1.0000	0.0009	0.9995	0.0270	0.9675
0.0000	1.0000	0.0000	1.0000	0.0003	0.9998	0.0160	0.9836
0.0000	1.0000	0.0000	1.0000	0.0001	0.9999	0.0087	0.9923
0.0000	1.0000	0.0000	1.0000	0.0000	1.0000	0.0044	0.9967
0.0000	1.0000	0.0000	1.0000	0.0000	1.0000	0.0020	0.9987
0.0000	1.0000	0.0000	1.0000	0.0000	1.0000	0.0008	0.9995
0.0000	1.0000	0.0000	1.0000	0.0000	1.0000	0.0003	0.9998
0.0000	1.0000	0.0000	1.0000	0.0000	1.0000	0.0001	1.0000
0.0000	1.0000	0.0000	1.0000	0.0000	1.0000	0.0000	1.0000
0.0000	1.0000	0.0000	1.0000	0.0000	1.0000	0.0000	1.0000
0.0000	1.0000	0.0000	1.0000	0.0000	1.0000	0.0000	1.0000
0.0000	1.0000	0.0000	1.0000	0.0000	1.0000	0.0000	1.0000
0.0000	1.0000	0.0000	1.0000	0.0000	1.0000	0.0000	1.0000
0.0000	1.0000	0.0000	1.0000	0.0000	1.0000	0.0000	1.0000
0.0000	1.0000	0.0000	1.0000	0.0000	1.0000	0.0000	1.0000
0.0000	1.0000	0.0000	1.0000	0.0000	1.0000	0.0000	1.0000
0.0000	1.0000	0.0000	1.0000	0.0000	1.0000	0.0000	1.0000
0.0000	1.0000	0.0000	1.0000	0.0000	1.0000	0.0000	1.0000
0.0000	1.0000	0.0000	1.0000	0.0000	1.0000	0.0000	1.0000
0.0000	1.0000	0.0000	1.0000	0.0000	1.0000	0.0000	1.0000

Table A.4 Individual and Cumulative Probabilities of Poisson Distributions†

Let X be a Poisson random variable with parameter λ.

Entries in the $f(x)$ column represent the probability X has the value x:

$$f(x) = P(X = x) = \lambda^x e^{-\lambda}/x!$$

For $\lambda = 0.5$: $f(3) = 0.0126$; and for $\lambda = 4$: $f(8) = 0.0298$.

Entries in the $F(x)$ column represent the probability X has a value *no greater than* x:

$$F(x) = P(X \leqslant x)$$

For $\lambda = 0.5$, $F(3) = 0.9982$; and for $\lambda = 4$, $F(8) = 0.9786$.

† Computed at New Mexico State University.

Table A.4 *continued*

x	$\lambda = 0.5$		$\lambda = 1$		$\lambda = 2$		$\lambda = 3$		$\lambda = 4$		x
	$f(x)$	$F(x)$	$f(x)$	$F(x)$	$f(x)$	$F(x)$	$f(x)$	$F(x)$	$f(x)$	$F(x)$	
0	0.6065	0.6065	0.3679	0.3679	0.1353	0.1353	0.0498	0.0498	0.0183	0.0183	0
1	0.3033	0.9098	0.3679	0.7358	0.2707	0.4060	0.1494	0.1991	0.0733	0.0916	1
2	0.0758	0.9856	0.1839	0.9197	0.2707	0.6767	0.2240	0.4232	0.1465	0.2381	2
3	0.0126	0.9982	0.0613	0.9810	0.1804	0.8571	0.2240	0.6472	0.1954	0.4335	3
4	0.0016	0.9998	0.0153	0.9963	0.0902	0.9473	0.1680	0.8153	0.1954	0.6288	4
5	0.0002	1.0000	0.0031	0.9994	0.0361	0.9834	0.1008	0.9161	0.1563	0.7851	5
6	0.0000	1.0000	0.0005	0.9999	0.0120	0.9955	0.0504	0.9665	0.1042	0.8893	6
7	0.0000	1.0000	0.0001	1.0000	0.0034	0.9989	0.0216	0.9881	0.0595	0.9489	7
8	0.0000	1.0000	0.0000	1.0000	0.0009	0.9998	0.0081	0.9962	0.0298	0.9786	8
9	0.0000	1.0000	0.0000	1.0000	0.0002	1.0000	0.0027	0.9989	0.0132	0.9919	9
10	0.0000	1.0000	0.0000	1.0000	0.0000	1.0000	0.0008	0.9997	0.0053	0.9972	10
11	0.0000	1.0000	0.0000	1.0000	0.0000	1.0000	0.0002	0.9999	0.0019	0.9991	11
12	0.0000	1.0000	0.0000	1.0000	0.0000	1.0000	0.0001	1.0000	0.0006	0.9997	12
13	0.0000	1.0000	0.0000	1.0000	0.0000	1.0000	0.0000	1.0000	0.0002	0.9999	13
14	0.0000	1.0000	0.0000	1.0000	0.0000	1.0000	0.0000	1.0000	0.0001	1.0000	14

x	$\lambda = 5$		$\lambda = 6$		$\lambda = 7$		$\lambda = 8$		$\lambda = 9$		x
	$f(x)$	$F(x)$	$f(x)$	$F(x)$	$f(x)$	$F(x)$	$f(x)$	$F(x)$	$f(x)$	$F(x)$	
0	0.0067	0.0067	0.0025	0.0025	0.0009	0.0009	0.0003	0.0003	0.0001	0.0001	0
1	0.0337	0.0404	0.0149	0.0174	0.0064	0.0073	0.0027	0.0030	0.0011	0.0012	1
2	0.0842	0.1247	0.0446	0.0620	0.0223	0.0296	0.0107	0.0138	0.0050	0.0062	2
3	0.1404	0.2650	0.0892	0.1512	0.0521	0.0818	0.0286	0.0424	0.0150	0.0212	3
4	0.1755	0.4405	0.1339	0.2851	0.0912	0.1730	0.0573	0.0996	0.0337	0.0550	4
5	0.1755	0.6160	0.1606	0.4457	0.1277	0.3007	0.0916	0.1912	0.0607	0.1157	5
6	0.1462	0.7622	0.1606	0.6063	0.1490	0.4497	0.1221	0.3134	0.0911	0.2068	6
7	0.1044	0.8666	0.1377	0.7440	0.1490	0.5987	0.1396	0.4530	0.1171	0.3239	7
8	0.0653	0.9319	0.1033	0.8472	0.1304	0.7291	0.1396	0.5925	0.1318	0.4557	8
9	0.0363	0.9682	0.0688	0.9161	0.1014	0.8305	0.1241	0.7166	0.1318	0.5874	9
10	0.0181	0.9863	0.0413	0.9574	0.0710	0.9015	0.0993	0.8159	0.1186	0.7060	10
11	0.0082	0.9945	0.0225	0.9799	0.0452	0.9467	0.0722	0.8881	0.0970	0.8030	11
12	0.0034	0.9980	0.0113	0.9912	0.0263	0.9730	0.0481	0.9362	0.0728	0.8758	12
13	0.0013	0.9993	0.0052	0.9964	0.0142	0.9872	0.0296	0.9658	0.0504	0.9261	13
14	0.0005	0.9998	0.0022	0.9986	0.0071	0.9943	0.0169	0.9827	0.0324	0.9585	14
15	0.0002	0.9999	0.0009	0.9995	0.0033	0.9976	0.0090	0.9918	0.0194	0.9780	15
16	0.0000	1.0000	0.0003	0.9998	0.0014	0.9990	0.0045	0.9963	0.0109	0.9889	16
17	0.0000	1.0000	0.0001	0.9999	0.0006	0.9996	0.0021	0.9984	0.0058	0.9947	17
18	0.0000	1.0000	0.0000	1.0000	0.0002	0.9999	0.0009	0.9993	0.0029	0.9976	18
19	0.0000	1.0000	0.0000	1.0000	0.0001	1.0000	0.0004	0.9997	0.0014	0.9989	19
20	0.0000	1.0000	0.0000	1.0000	0.0000	1.0000	0.0002	0.9999	0.0006	0.9996	20
21	0.0000	1.0000	0.0000	1.0000	0.0000	1.0000	0.0001	1.0000	0.0003	0.9998	21
22	0.0000	1.0000	0.0000	1.0000	0.0000	1.0000	0.0000	1.0000	0.0001	0.9999	22
23	0.0000	1.0000	0.0000	1.0000	0.0000	1.0000	0.0000	1.0000	0.0000	1.0000	23

Table A.4 *continued*

x	λ = 10 f(x)	λ = 10 F(x)	λ = 12 f(x)	λ = 12 F(x)	λ = 15 f(x)	λ = 15 F(x)	λ = 20 f(x)	λ = 20 F(x)	λ = 25 f(x)	λ = 25 F(x)	x
0	0.0000	0.0000	0.0000	0.0000	0.0000	0.0000	0.0000	0.0000	0.0000	0.0000	0
1	0.0005	0.0005	0.0001	0.0001	0.0000	0.0000	0.0000	0.0000	0.0000	0.0000	1
2	0.0023	0.0028	0.0004	0.0005	0.0000	0.0000	0.0000	0.0000	0.0000	0.0000	2
3	0.0076	0.0103	0.0018	0.0023	0.0002	0.0002	0.0000	0.0000	0.0000	0.0000	3
4	0.0189	0.0293	0.0053	0.0076	0.0006	0.0009	0.0000	0.0000	0.0000	0.0000	4
5	0.0378	0.0671	0.0127	0.0203	0.0019	0.0028	0.0001	0.0001	0.0000	0.0000	5
6	0.0631	0.1301	0.0255	0.0458	0.0048	0.0076	0.0002	0.0003	0.0000	0.0000	6
7	0.0901	0.2202	0.0437	0.0895	0.0104	0.0180	0.0005	0.0008	0.0000	0.0000	7
8	0.1126	0.3328	0.0655	0.1550	0.0194	0.0374	0.0013	0.0021	0.0001	0.0001	8
9	0.1251	0.4579	0.0874	0.2424	0.0324	0.0699	0.0029	0.0050	0.0001	0.0002	9
10	0.1251	0.5830	0.1048	0.3472	0.0486	0.1185	0.0058	0.0108	0.0004	0.0006	10
11	0.1137	0.6968	0.1144	0.4616	0.0663	0.1848	0.0106	0.0214	0.0008	0.0014	11
12	0.0948	0.7916	0.1144	0.5760	0.0829	0.2676	0.0176	0.0390	0.0017	0.0031	12
13	0.0729	0.8645	0.1056	0.6815	0.0956	0.3632	0.0271	0.0661	0.0033	0.0065	13
14	0.0521	0.9165	0.0905	0.7720	0.1024	0.4657	0.0387	0.1049	0.0059	0.0124	14
15	0.0347	0.9513	0.0724	0.8444	0.1024	0.5681	0.0516	0.1565	0.0099	0.0223	15
16	0.0217	0.9730	0.0543	0.8987	0.0960	0.6641	0.0646	0.2211	0.0155	0.0377	16
17	0.0128	0.9857	0.0383	0.9370	0.0847	0.7489	0.0760	0.2970	0.0227	0.0605	17
18	0.0071	0.9928	0.0255	0.9626	0.0706	0.8195	0.0844	0.3814	0.0316	0.0920	18
19	0.0037	0.9965	0.0161	0.9787	0.0557	0.8752	0.0888	0.4703	0.0415	0.1336	19
20	0.0019	0.9984	0.0097	0.9884	0.0418	0.9170	0.0888	0.5591	0.0519	0.1855	20
21	0.0009	0.9993	0.0055	0.9939	0.0299	0.9469	0.0846	0.6437	0.0618	0.2473	21
22	0.0004	0.9997	0.0030	0.9970	0.0204	0.9673	0.0769	0.7206	0.0702	0.3175	22
23	0.0002	0.9999	0.0016	0.9985	0.0133	0.9805	0.0669	0.7875	0.0763	0.3939	23
24	0.0001	1.0000	0.0008	0.9993	0.0083	0.9888	0.0557	0.8432	0.0795	0.4734	24
25	0.0000	1.0000	0.0004	0.9997	0.0050	0.9938	0.0446	0.8878	0.0795	0.5529	25
26	0.0000	1.0000	0.0002	0.9999	0.0029	0.9967	0.0343	0.9221	0.0765	0.6294	26
27	0.0000	1.0000	0.0001	0.9999	0.0016	0.9983	0.0254	0.9475	0.0708	0.7002	27
28	0.0000	1.0000	0.0000	1.0000	0.0009	0.9991	0.0181	0.9657	0.0632	0.7634	28
29	0.0000	1.0000	0.0000	1.0000	0.0004	0.9996	0.0125	0.9782	0.0545	0.8179	29
30	0.0000	1.0000	0.0000	1.0000	0.0002	0.9998	0.0083	0.9865	0.0454	0.8633	30
31	0.0000	1.0000	0.0000	1.0000	0.0001	0.9999	0.0054	0.9919	0.0366	0.8999	31
32	0.0000	1.0000	0.0000	1.0000	0.0001	1.0000	0.0034	0.9953	0.0286	0.9285	32
33	0.0000	1.0000	0.0000	1.0000	0.0000	1.0000	0.0020	0.9973	0.0217	0.9502	33
34	0.0000	1.0000	0.0000	1.0000	0.0000	1.0000	0.0012	0.9985	0.0159	0.9662	34
35	0.0000	1.0000	0.0000	1.0000	0.0000	1.0000	0.0007	0.9992	0.0114	0.9775	35
36	0.0000	1.0000	0.0000	1.0000	0.0000	1.0000	0.0004	0.9996	0.0079	0.9854	36
37	0.0000	1.0000	0.0000	1.0000	0.0000	1.0000	0.0002	0.9998	0.0053	0.9908	37
38	0.0000	1.0000	0.0000	1.0000	0.0000	1.0000	0.0001	0.9999	0.0035	0.9943	38
39	0.0000	1.0000	0.0000	1.0000	0.0000	1.0000	0.0001	0.9999	0.0023	0.9966	39
40	0.0000	1.0000	0.0000	1.0000	0.0000	1.0000	0.0000	1.0000	0.0014	0.9980	40
41	0.0000	1.0000	0.0000	1.0000	0.0000	1.0000	0.0000	1.0000	0.0009	0.9988	41
42	0.0000	1.0000	0.0000	1.0000	0.0000	1.0000	0.0000	1.0000	0.0005	0.9993	42
43	0.0000	1.0000	0.0000	1.0000	0.0000	1.0000	0.0000	1.0000	0.0003	0.9996	43
44	0.0000	1.0000	0.0000	1.0000	0.0000	1.0000	0.0000	1.0000	0.0002	0.9998	44
45	0.0000	1.0000	0.0000	1.0000	0.0000	1.0000	0.0000	1.0000	0.0001	0.9999	45
46	0.0000	1.0000	0.0000	1.0000	0.0000	1.0000	0.0000	1.0000	0.0001	0.9999	46
47	0.0000	1.0000	0.0000	1.0000	0.0000	1.0000	0.0000	1.0000	0.0000	1.0000	47

Table A.5 Cumulative Probabilities of the Standard Normal Distribution[†]

Entries are $F_Z(z_\gamma) = P[Z \leqslant z_\gamma] = \gamma$, where $Z \sim N(0, 1)$.

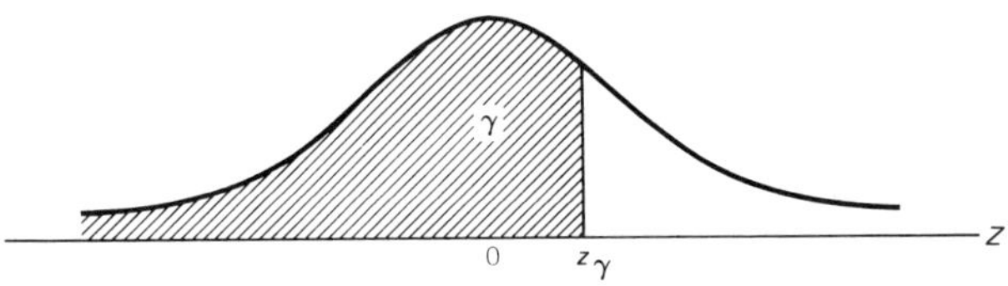

$$F_Z(-2.46) = P(Z \leqslant -2.46) = 0.00695$$

$$F_Z(1.88) = P(Z \leqslant 1.88) = 0.96995$$

[†] Computed at New Mexico State University.

z_γ	.00	.01	.02	.03	.04
-3.9	0.00005	0.00005	0.00004	0.00004	0.00004
-3.8	0.00007	0.00007	0.00007	0.00006	0.00006
-3.7	0.00011	0.00010	0.00010	0.00010	0.00009
-3.6	0.00016	0.00015	0.00015	0.00014	0.00014
-3.5	0.00023	0.00022	0.00022	0.00021	0.00020
-3.4	0.00034	0.00032	0.00031	0.00030	0.00029
-3.3	0.00048	0.00047	0.00045	0.00043	0.00042
-3.2	0.00069	0.00066	0.00064	0.00062	0.00060
-3.1	0.00097	0.00094	0.00090	0.00087	0.00084
-3.0	0.00135	0.00131	0.00126	0.00122	0.00118
-2.9	0.00187	0.00181	0.00175	0.00169	0.00164
-2.8	0.00256	0.00248	0.00240	0.00233	0.00226
-2.7	0.00347	0.00336	0.00326	0.00317	0.00307
-2.6	0.00466	0.00453	0.00440	0.00427	0.00415
-2.5	0.00621	0.00604	0.00587	0.00570	0.00554
-2.4	0.00820	0.00798	0.00776	0.00755	0.00734
-2.3	0.01072	0.01044	0.01017	0.00990	0.00964
-2.2	0.01390	0.01355	0.01321	0.01287	0.01255
-2.1	0.01786	0.01743	0.01700	0.01659	0.01618
-2.0	0.02275	0.02222	0.02169	0.02118	0.02068
-1.9	0.02872	0.02807	0.02743	0.02680	0.02619
-1.8	0.03593	0.03515	0.03438	0.03362	0.03288
-1.7	0.04457	0.04363	0.04272	0.04182	0.04093
-1.6	0.05480	0.05370	0.05262	0.05155	0.05050
-1.5	0.06681	0.06552	0.06426	0.06301	0.06178
-1.4	0.08076	0.07927	0.07780	0.07636	0.07493
-1.3	0.09680	0.09510	0.09342	0.09176	0.09012
-1.2	0.11507	0.11314	0.11123	0.10935	0.10749
-1.1	0.13567	0.13350	0.13136	0.12924	0.12714
-1.0	0.15865	0.15625	0.15386	0.15150	0.14917
-0.9	0.18406	0.18141	0.17879	0.17619	0.17361
-0.8	0.21185	0.20897	0.20611	0.20327	0.20045
-0.7	0.24196	0.23885	0.23576	0.23269	0.22965
-0.6	0.27425	0.27093	0.26763	0.26435	0.26109
-0.5	0.30854	0.30502	0.30153	0.29806	0.29460
-0.4	0.34458	0.34090	0.33724	0.33360	0.32997
-0.3	0.38209	0.37828	0.37448	0.37070	0.36693
-0.2	0.42074	0.41683	0.41293	0.40904	0.40516
-0.1	0.46017	0.45620	0.45224	0.44828	0.44433
-0.0	0.50000	0.49601	0.49202	0.48803	0.48404

Table A.5 *continued*

.05	.06	.07	.08	.09	z_γ
0.00004	0.00004	0.00004	0.00003	0.00003	-3.9
0.00006	0.00006	0.00005	0.00005	0.00005	-3.8
0.00009	0.00008	0.00008	0.00008	0.00008	-3.7
0.00013	0.00013	0.00012	0.00012	0.00011	-3.6
0.00019	0.00019	0.00018	0.00017	0.00017	-3.5
0.00028	0.00027	0.00026	0.00025	0.00024	-3.4
0.00040	0.00039	0.00038	0.00036	0.00035	-3.3
0.00058	0.00056	0.00054	0.00052	0.00050	-3.2
0.00082	0.00079	0.00076	0.00074	0.00071	-3.1
0.00114	0.00111	0.00107	0.00104	0.00100	-3.0
0.00159	0.00154	0.00149	0.00144	0.00139	-2.9
0.00219	0.00212	0.00205	0.00199	0.00193	-2.8
0.00298	0.00289	0.00280	0.00272	0.00264	-2.7
0.00402	0.00391	0.00379	0.00368	0.00357	-2.6
0.00539	0.00523	0.00508	0.00494	0.00480	-2.5
0.00714	0.00695	0.00676	0.00657	0.00639	-2.4
0.00939	0.00914	0.00889	0.00866	0.00842	-2.3
0.01222	0.01191	0.01160	0.01130	0.01101	-2.2
0.01578	0.01539	0.01500	0.01463	0.01426	-2.1
0.02018	0.01970	0.01923	0.01876	0.01831	-2.0
0.02559	0.02500	0.02442	0.02385	0.02330	-1.9
0.03216	0.03144	0.03074	0.03005	0.02938	-1.8
0.04006	0.03920	0.03836	0.03754	0.03673	-1.7
0.04947	0.04846	0.04746	0.04648	0.04551	-1.6
0.06057	0.05938	0.05821	0.05705	0.05592	-1.5
0.07353	0.07214	0.07078	0.06944	0.06811	-1.4
0.08851	0.08691	0.08534	0.08379	0.08226	-1.3
0.10565	0.10383	0.10204	0.10027	0.09853	-1.2
0.12507	0.12302	0.12100	0.11900	0.11702	-1.1
0.14686	0.14457	0.14231	0.14007	0.13786	-1.0
0.17106	0.16853	0.16602	0.16354	0.16109	-0.9
0.19766	0.19489	0.19215	0.18943	0.18673	-0.8
0.22663	0.22363	0.22065	0.21769	0.21476	-0.7
0.25785	0.25463	0.25143	0.24825	0.24510	-0.6
0.29116	0.28774	0.28434	0.28096	0.27759	-0.5
0.32635	0.32276	0.31918	0.31561	0.31207	-0.4
0.36317	0.35942	0.35569	0.35197	0.34827	-0.3
0.40129	0.39743	0.39358	0.38974	0.38591	-0.2
0.44038	0.43644	0.43250	0.42857	0.42465	-0.1
0.48006	0.47608	0.47210	0.46812	0.46414	-0.0

Table A.5 *continued*

z_γ	.00	.01	.02	.03	.04
0.0	0.50000	0.50399	0.50798	0.51197	0.51596
0.1	0.53983	0.54380	0.54776	0.55172	0.55567
0.2	0.57926	0.58317	0.58707	0.59096	0.59484
0.3	0.61791	0.62172	0.62552	0.62930	0.63307
0.4	0.65542	0.65910	0.66276	0.66640	0.67003
0.5	0.69146	0.69498	0.69847	0.70194	0.70540
0.6	0.72575	0.72907	0.73237	0.73565	0.73891
0.7	0.75804	0.76115	0.76424	0.76731	0.77035
0.8	0.78815	0.79103	0.79389	0.79673	0.79955
0.9	0.81594	0.81859	0.82121	0.82381	0.82639
1.0	0.84135	0.84375	0.84614	0.84850	0.85083
1.1	0.86433	0.86650	0.86864	0.87076	0.87286
1.2	0.88493	0.88686	0.88877	0.89065	0.89251
1.3	0.90320	0.90490	0.90658	0.90824	0.90988
1.4	0.91924	0.92073	0.92220	0.92364	0.92507
1.5	0.93319	0.93448	0.93574	0.93699	0.93822
1.6	0.94520	0.94630	0.94738	0.94845	0.94950
1.7	0.95543	0.95637	0.95728	0.95818	0.95907
1.8	0.96407	0.96485	0.96562	0.96638	0.96712
1.9	0.97128	0.97193	0.97257	0.97320	0.97381
2.0	0.97725	0.97778	0.97831	0.97882	0.97932
2.1	0.98214	0.98257	0.98300	0.98341	0.98382
2.2	0.98610	0.98645	0.98679	0.98713	0.98745
2.3	0.98928	0.98956	0.98983	0.99010	0.99036
2.4	0.99180	0.99202	0.99224	0.99245	0.99266
2.5	0.99379	0.99396	0.99413	0.99430	0.99446
2.6	0.99534	0.99547	0.99560	0.99573	0.99585
2.7	0.99653	0.99664	0.99674	0.99683	0.99693
2.8	0.99744	0.99752	0.99760	0.99767	0.99774
2.9	0.99813	0.99819	0.99825	0.99831	0.99836
3.0	0.99865	0.99869	0.99874	0.99878	0.99882
3.1	0.99903	0.99906	0.99910	0.99913	0.99916
3.2	0.99931	0.99934	0.99936	0.99938	0.99940
3.3	0.99952	0.99953	0.99955	0.99957	0.99958
3.4	0.99966	0.99968	0.99969	0.99970	0.99971
3.5	0.99977	0.99978	0.99978	0.99979	0.99980
3.6	0.99984	0.99985	0.99985	0.99986	0.99986
3.7	0.99989	0.99990	0.99990	0.99990	0.99991
3.8	0.99993	0.99993	0.99993	0.99994	0.99994
3.9	0.99995	0.99995	0.99996	0.99996	0.99996

Table A.5 *continued*

.05	.06	.07	.08	.09	z_γ
0.51994	0.52392	0.52790	0.53188	0.53586	0.0
0.55962	0.56356	0.56750	0.57143	0.57535	0.1
0.59871	0.60257	0.60642	0.61026	0.61409	0.2
0.63683	0.64058	0.64431	0.64803	0.65173	0.3
0.67365	0.67724	0.68082	0.68439	0.68793	0.4
0.70884	0.71226	0.71566	0.71904	0.72241	0.5
0.74215	0.74537	0.74857	0.75175	0.75490	0.6
0.77337	0.77637	0.77935	0.78231	0.78524	0.7
0.80234	0.80511	0.80785	0.81057	0.81327	0.8
0.82894	0.83147	0.83398	0.83646	0.83891	0.9
0.85314	0.85543	0.85769	0.85993	0.86214	1.0
0.87493	0.87698	0.87900	0.88100	0.88298	1.1
0.89435	0.89617	0.89796	0.89973	0.90147	1.2
0.91149	0.91309	0.91466	0.91621	0.91774	1.3
0.92647	0.92786	0.92922	0.93056	0.93189	1.4
0.93943	0.94062	0.94179	0.94295	0.94408	1.5
0.95053	0.95154	0.95254	0.95352	0.95449	1.6
0.95994	0.96080	0.96164	0.96246	0.96327	1.7
0.96784	0.96856	0.96926	0.96995	0.97062	1.8
0.97441	0.97500	0.97558	0.97615	0.97670	1.9
0.97982	0.98030	0.98077	0.98124	0.98169	2.0
0.98422	0.98461	0.98500	0.98537	0.98574	2.1
0.98778	0.98809	0.98840	0.98870	0.98899	2.2
0.99061	0.99086	0.99111	0.99134	0.99158	2.3
0.99286	0.99305	0.99324	0.99343	0.99361	2.4
0.99461	0.99477	0.99492	0.99506	0.99520	2.5
0.99598	0.99609	0.99621	0.99632	0.99643	2.6
0.99702	0.99711	0.99720	0.99728	0.99736	2.7
0.99781	0.99788	0.99795	0.99801	0.99807	2.8
0.99841	0.99846	0.99851	0.99856	0.99861	2.9
0.99886	0.99889	0.99893	0.99896	0.99900	3.0
0.99918	0.99921	0.99924	0.99926	0.99929	3.1
0.99942	0.99944	0.99946	0.99948	0.99950	3.2
0.99960	0.99961	0.99962	0.99964	0.99965	3.3
0.99972	0.99973	0.99974	0.99975	0.99976	3.4
0.99981	0.99981	0.99982	0.99983	0.99983	3.5
0.99987	0.99987	0.99988	0.99988	0.99989	3.6
0.99991	0.99992	0.99992	0.99992	0.99992	3.7
0.99994	0.99994	0.99995	0.99995	0.99995	3.8
0.99996	0.99996	0.99996	0.99997	0.99997	3.9

Table A.6 Percentiles of Student T Distributions†

Let T_ν be a Student T variable with ν degrees of freedom. Entries are $t_{\nu;\gamma}$ where $P(T_\nu \leqslant t_{\nu;\gamma}) = \gamma$.

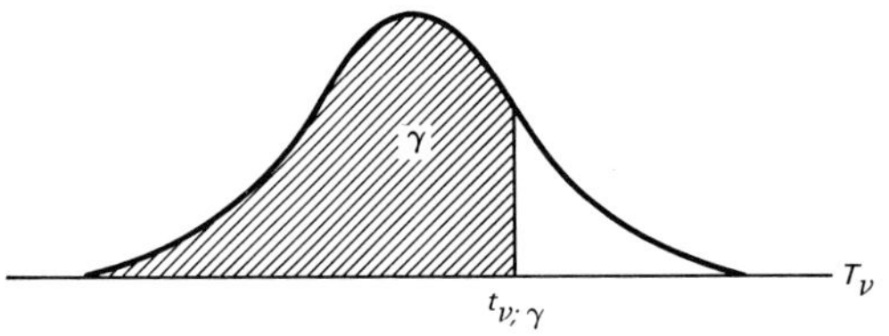

Thus, for $\nu = 16$, the table gives $t_{16;0.1} = -1.3368$ and $t_{16;0.975} = 2.1199$.

† Reproduced with permission from Donald B. Owen, *Handbook of Statistical Tables*, Addison-Wesley, Reading, Mass., 1962. Courtesy of U.S. Atomic Energy Commission.

ν	0.75	0.90	0.95	γ 0.975	0.99	0.995
1	1.0000	3.0777	6.3138	12.7062	31.8207	63.6574
2	0.8165	1.8856	2.9200	4.3027	6.9646	9.9248
3	0.7649	1.6377	2.3534	3.1824	4.5407	5.8409
4	0.7407	1.5332	2.1318	2.7764	3.7469	4.6041
5	0.7267	1.4759	2.0150	2.5706	3.3649	4.0322
6	0.7176	1.4398	1.9432	2.4469	3.1427	3.7074
7	0.7111	1.4149	1.8946	2.3646	2.9980	3.4995
8	0.7064	1.3968	1.8595	2.3060	2.8965	3.3554
9	0.7027	1.3830	1.8331	2.2622	2.8214	3.2498
10	0.6998	1.3722	1.8125	2.2281	2.7638	3.1693
11	0.6974	1.3634	1.7959	2.2010	2.7181	3.1058
12	0.6955	1.3562	1.7823	2.1788	2.6810	3.0545
13	0.6938	1.3502	1.7709	2.1604	2.6503	3.0123
14	0.6924	1.3450	1.7613	2.1448	2.6245	2.9768
15	0.6912	1.3406	1.7531	2.1315	2.6025	2.9467
16	0.6901	1.3368	1.7459	2.1199	2.5835	2.9208
17	0.6892	1.3334	1.7396	2.1098	2.5669	2.8982
18	0.6884	1.3304	1.7341	2.1009	2.5524	2.8784
19	0.6876	1.3277	1.7291	2.0930	2.5395	2.8609
20	0.6870	1.3253	1.7247	2.0860	2.5280	2.8453
21	0.6864	1.3232	1.7207	2.0796	2.5177	2.8314
22	0.6858	1.3212	1.7171	2.0739	2.5083	2.8188
23	0.6853	1.3195	1.7139	2.0687	2.4999	2.8073
24	0.6848	1.3178	1.7109	2.0639	2.4922	2.7969
25	0.6844	1.3163	1.7081	2.0595	2.4851	2.7874
26	0.6840	1.3150	1.7056	2.0555	2.4786	2.7787
27	0.6837	1.3137	1.7033	2.0518	2.4727	2.7707
28	0.6834	1.3125	1.7011	2.0484	2.4671	2.7633
29	0.6830	1.3114	1.6991	2.0452	2.4620	2.7564
30	0.6828	1.3104	1.6973	2.0423	2.4573	2.7500
31	0.6825	1.3095	1.6955	2.0395	2.4528	2.7440
32	0.6822	1.3086	1.6939	2.0369	2.4487	2.7385
33	0.6820	1.3077	1.6924	2.0345	2.4448	2.7333
34	0.6818	1.3070	1.6909	2.0322	2.4411	2.7284
35	0.6816	1.3062	1.6896	2.0301	2.4377	2.7238
36	0.6814	1.3055	1.6883	2.0281	2.4345	2.7195
37	0.6812	1.3049	1.6871	2.0262	2.4314	2.7154
38	0.6810	1.3042	1.6860	2.0244	2.4286	2.7116
39	0.6808	1.3036	1.6849	2.0227	2.4258	2.7079
40	0.6807	1.3031	1.6839	2.0211	2.4233	2.7045
50	0.6794	1.2987	1.6759	2.0086	2.4033	2.6778
60	0.6786	1.2958	1.6706	2.0003	2.3901	2.6603
70	0.6780	1.2938	1.6669	1.9944	2.3808	2.6479
80	0.6776	1.2922	1.6641	1.9901	2.3739	2.6387
90	0.6772	1.2910	1.6620	1.9867	2.3685	2.6316
100	0.6770	1.2901	1.6602	1.9840	2.3642	2.6259
500	0.6750	1.2832	1.6479	1.9647	2.3338	2.5857
1,000	0.6747	1.2824	1.6464	1.9623	2.3301	2.5808
∞	0.6745	1.2816	1.6449	1.9600	2.3263	2.5758
ν	0.25	0.10	0.05	0.025	0.01	0.005

For these values of γ, affix negative sign to the table entry.

Table A.7 Percentiles of Chi-Square Distributions[†]

Let χ_ν^2 denote a chi-square variable with ν degrees of freedom. Entries in the table are $\chi_{\nu;\gamma}^2$ where $P[\chi_\nu^2 \leqslant \chi_{\nu;\gamma}^2] = \gamma$.

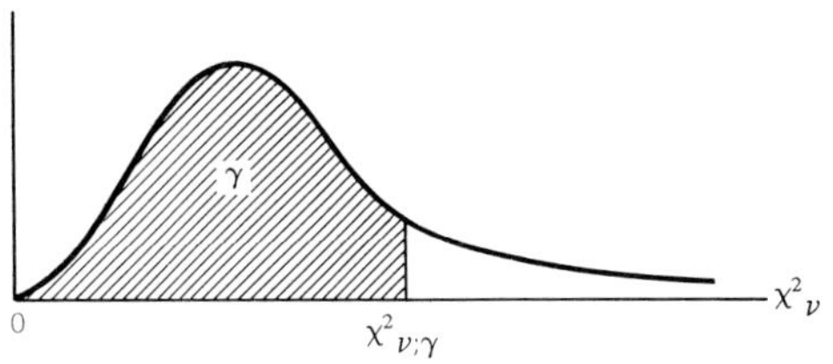

For $\nu = 26$, the table gives $\chi_{26;0.01}^2 = 12.20$; $\chi_{26;0.05}^2 = 15.38$; $\chi_{26;0.975}^2 = 41.92$; and $\chi_{26;0.995}^2 = 48.29$.

[†] Computed at New Mexico State University.

	γ									
ν	0.005	0.010	0.025	0.050	0.100	0.900	0.950	0.975	0.990	0.995
1	0.000	0.000	0.001	0.004	0.016	2.706	3.841	5.024	6.635	7.879
2	0.010	0.020	0.051	0.103	0.211	4.605	5.991	7.378	9.210	10.597
3	0.072	0.115	0.216	0.352	0.584	6.251	7.815	9.348	11.345	12.838
4	0.207	0.297	0.484	0.711	1.064	7.779	9.488	11.143	13.277	14.860
5	0.412	0.554	0.831	1.145	1.610	9.236	11.070	12.833	15.086	16.750
6	0.676	0.872	1.237	1.635	2.204	10.645	12.592	14.449	16.812	18.548
7	0.989	1.239	1.690	2.167	2.833	12.017	14.067	16.013	18.475	20.278
8	1.344	1.646	2.180	2.733	3.490	13.362	15.507	17.534	20.090	21.956
9	1.735	2.088	2.700	3.325	4.168	14.684	16.919	19.023	21.666	23.589
10	2.156	2.558	3.247	3.940	4.865	15.987	18.307	20.483	23.209	25.188
11	2.603	3.053	3.816	4.575	5.578	17.275	19.675	21.920	24.725	26.757
12	3.074	3.570	4.404	5.226	6.304	18.549	21.026	23.337	26.217	28.299
13	3.565	4.107	5.009	5.892	7.041	19.812	22.362	24.735	27.688	29.819
14	4.075	4.660	5.629	6.571	7.789	21.064	23.685	26.119	29.141	31.319
15	4.601	5.229	6.262	7.261	8.547	22.307	24.996	27.488	30.578	32.801
16	5.142	5.812	6.908	7.962	9.312	23.542	26.296	28.845	32.000	34.267
17	5.697	6.408	7.564	8.672	10.085	24.769	27.587	30.191	33.409	35.718
18	6.265	7.015	8.231	9.390	10.865	25.989	28.869	31.526	34.805	37.156
19	6.844	7.633	8.906	10.117	11.651	27.203	30.144	32.852	36.191	38.582
20	7.434	8.260	9.591	10.851	12.443	28.412	31.410	34.170	37.566	39.997
21	8.034	8.897	10.283	11.591	13.239	29.615	32.671	35.479	38.932	41.401
22	8.643	9.542	10.982	12.338	14.041	30.813	33.924	36.781	40.289	42.796
23	9.260	10.196	11.688	13.090	14.848	32.007	35.172	38.076	41.638	44.182
24	9.886	10.856	12.401	13.848	15.659	33.196	36.415	39.364	42.980	45.558
25	10.520	11.524	13.120	14.611	16.474	34.382	37.652	40.646	44.314	46.929
26	11.160	12.198	13.844	15.379	17.298	35.563	38.885	41.923	45.641	48.290
27	11.807	12.878	14.573	16.152	18.162	36.741	40.113	43.194	46.964	49.652
28	12.461	13.565	15.308	16.928	18.939	37.916	41.337	44.461	48.278	50.993
29	13.121	14.256	16.047	17.708	19.768	39.087	42.557	45.722	49.591	52.376
30	13.787	14.953	16.791	18.493	20.599	40.256	43.773	46.979	50.892	53.672
31	14.457	15.655	17.539	19.280	21.433	41.412	44.979	48.235	52.213	55.044
32	15.134	16.362	18.291	20.072	22.270	42.585	46.194	49.480	53.486	56.328
33	15.815	17.073	19.047	20.866	23.110	43.735	47.394	50.728	54.797	57.688
34	16.501	17.789	19.806	21.664	23.952	44.903	48.602	51.966	56.061	58.964
35	17.192	18.509	20.569	22.465	24.796	46.049	49.796	53.206	57.363	60.313
36	17.887	19.233	21.336	23.269	25.643	47.212	50.998	54.437	58.619	61.581
37	18.585	19.960	22.105	24.075	26.492	48.354	52.187	55.671	59.913	62.921
38	19.289	20.691	22.878	24.884	27.343	49.504	53.378	56.898	61.182	64.219
39	19.996	21.426	23.654	25.695	28.199	50.651	54.567	58.123	62.448	65.512
40	20.706	22.164	24.433	26.509	29.073	51.796	55.753	59.345	63.710	66.802

Table A.7 *continued*

	γ									
ν	0.005	0.010	0.025	0.050	0.100	0.900	0.950	0.975	0.990	0.995
41	21.38	22.88	25.20	27.32	29.91	52.94	56.94	60.56	64.97	68.09
42	22.10	23.63	25.99	28.14	30.77	54.08	58.12	61.78	66.23	69.37
43	22.82	24.37	26.78	28.96	31.63	55.22	59.30	62.99	67.48	70.65
44	23.55	25.12	27.57	29.79	32.49	56.36	60.48	64.20	68.73	71.93
45	24.28	25.88	28.36	30.61	33.36	57.50	61.65	65.41	69.98	73.20
46	25.01	26.63	29.15	31.44	34.22	58.63	62.83	66.62	71.22	74.47
47	25.74	27.39	29.95	32.27	35.09	59.77	64.00	67.82	72.46	75.74
48	26.48	28.15	30.75	33.10	35.95	60.90	65.17	69.03	73.70	77.00
49	27.22	28.92	31.55	33.93	36.82	62.03	66.33	70.23	74.94	78.26
50	27.96	29.68	32.35	34.76	37.69	63.16	67.50	71.42	76.17	79.52
51	28.70	30.45	33.15	35.60	38.57	64.29	68.67	72.62	77.40	80.78
52	29.45	31.22	33.96	36.44	39.44	65.41	69.83	73.81	78.63	82.03
53	30.20	32.00	34.77	37.28	40.31	66.54	70.99	75.00	79.86	83.28
54	30.95	32.77	35.58	38.12	41.19	67.67	72.15	76.20	81.09	84.53
55	31.70	33.55	36.39	38.96	42.06	68.79	73.31	77.38	82.31	85.78
56	32.46	34.33	37.20	39.80	42.94	69.91	74.46	78.57	83.53	87.03
57	33.22	35.11	38.02	40.65	43.82	71.03	75.62	79.76	84.75	88.27
58	33.98	35.89	38.84	41.49	44.70	72.15	76.77	80.94	85.97	89.51
59	34.74	36.68	39.65	42.34	45.58	73.27	77.93	82.12	87.18	90.75
60	35.50	37.46	40.47	43.19	46.46	74.39	79.08	83.30	88.40	91.98
61	36.27	38.25	41.30	44.04	47.35	75.51	80.23	84.48	89.61	93.22
62	37.04	39.04	42.12	44.89	48.23	76.62	81.38	85.66	90.82	94.45
63	37.81	39.84	42.94	45.74	49.11	77.74	82.53	86.83	92.03	95.68
64	38.58	40.63	43.77	46.59	50.00	78.85	83.67	88.01	93.23	96.91
65	39.35	41.42	44.60	47.45	50.89	79.97	84.82	89.18	94.44	98.13
66	40.13	42.22	45.42	48.30	51.77	81.08	85.96	90.35	95.64	99.36
67	40.91	43.02	46.25	49.16	52.66	82.19	87.10	91.52	96.84	100.58
68	41.68	43.82	47.08	50.02	53.55	83.30	88.25	92.69	98.04	101.80
69	42.46	44.62	47.92	50.88	54.44	84.41	89.39	93.86	99.24	103.02
70	43.25	45.42	48.75	51.74	55.33	85.52	90.53	95.03	100.44	104.24
71	44.03	46.23	49.59	52.60	56.22	86.63	91.67	96.19	101.64	105.46
72	44.81	47.03	50.42	53.46	57.12	87.74	92.81	97.36	102.83	106.68
73	45.60	47.84	51.26	54.32	58.01	88.84	93.94	98.52	104.03	107.89
74	46.39	48.65	52.10	55.19	58.90	89.95	95.08	99.68	105.22	109.10
75	47.18	49.46	52.94	56.05	59.80	91.06	96.21	100.84	106.41	110.31
76	47.97	50.27	53.78	56.92	60.69	92.16	97.35	102.00	107.60	111.52
77	48.76	51.08	54.62	57.79	61.59	93.26	98.48	103.16	108.79	112.73
78	49.55	51.89	55.46	58.65	62.49	94.37	99.61	104.32	109.97	113.94
79	50.35	52.71	56.30	59.52	63.38	95.47	100.75	105.48	111.16	115.14
80	51.14	53.52	57.15	60.39	64.28	96.57	101.88	106.63	112.34	116.35

Table A.8 Factor Related to Range Estimation of σ for Normal Populations[†]

Entries d_n are multipliers of the sample range to use in estimating σ; that is,

$$\hat{\sigma} = d_n \cdot r$$

where r is the sample range.
The column n_r/n_s gives the ratio of sample sizes needed—when using the range estimator and standard deviation—to have the same efficiency of estimation. The entries in the relative efficiency column are reciprocals of the ratios n_r/n_s.

[†] Compilation based on entries extracted with permission of E. S. Pearson for the Biometrika Trustees from E. S. Pearson, "The Percentage Limits for the Distribution of Range in Samples from a Normal Population ($n \leqslant 100$)," *Biometrika*, **24** (1932). Calculations performed by the computing staff of the Department of Experimental Statistics, New Mexico State University.

n	d_n	n_r/n_s	Relative Efficiency
2	0.886	1.000	1.000
3	0.591	1.008	0.992
4	0.486	1.025	0.975
5	0.430	1.047	0.955
6	0.395	1.072	0.933
7	0.370	1.097	0.912
8	0.351	1.124	0.890
9	0.337	1.151	0.869
10	0.325	1.177	0.850
11	0.315	1.202	0.832
12	0.307	1.228	0.815
13	0.300	1.254	0.798
14	0.294	1.277	0.783
15	0.288	1.302	0.768
16	0.283	1.328	0.753
17	0.279	1.352	0.740
18	0.275	1.378	0.726
19	0.271	1.402	0.713
20	0.268	1.430	0.699
25	0.254	1.546	0.647
30	0.245	1.655	0.604
35	0.237	1.764	0.567
40	0.231	1.863	0.537
45	0.226	1.955	0.511
50	0.222	2.047	0.488
55	0.219	2.141	0.467
60	0.216	2.231	0.448
65	0.213	2.314	0.432
70	0.210	2.399	0.417
75	0.208	2.489	0.402
80	0.206	2.563	0.390
85	0.204	2.642	0.378
90	0.202	2.724	0.367
95	0.201	2.798	0.357
100	0.199	2.875	0.348

Table A.9 Percentiles of T^* Distributions†

Let T_n^* denote a T^* variable for sample size n. Entries are $t_{n;\gamma}^*$ where $P[T_n^* \leqslant t_{n;\gamma}^*] = \gamma$.

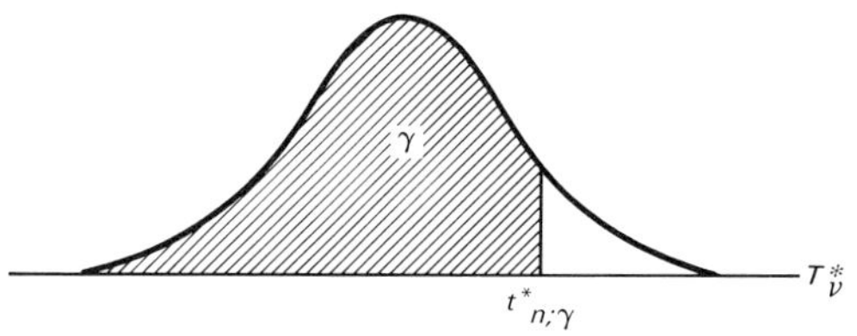

For $n = 12$, the table gives $t_{12;0.025}^* = -0.194$ and $t_{12;0.99}^* = 0.241$.

† Extracted with permission of E. S. Pearson for the Biometrika Trustees from E. Lord, "The Use of Range in Place of Standard Deviation in the t-test," *Biometrika*, **34** (1947).

n	γ 0.95	0.975	0.99	0.995	0.999	0.9995
2	3.157	6.353	15.910	31.828	159.16	318.31
3	0.885	1.304	2.111	3.008	6.77	9.58
4	.529	0.717	1.023	1.316	2.29	2.85
5	.388	.507	0.685	0.843	1.32	1.58
6	0.312	0.399	0.523	0.628	0.92	1.07
7	.263	.333	.429	.507	.71	0.82
8	.230	.288	.366	.429	.59	.67
9	.205	.255	.322	.374	.50	.57
10	.186	.230	.288	.333	.44	.50
11	0.170	0.210	0.262	0.302	0.40	0.44
12	.158	.194	.241	.277	.36	.40
13	.147	.181	.224	.256	.33	.37
14	.138	.170	.209	.239	.31	.34
15	.131	.160	.197	.224	.29	.32
16	0.124	0.151	0.186	0.212	0.27	0.30
17	.118	.144	.177	.201	.26	.28
18	.113	.137	.168	.191	.24	.26
19	.108	.131	.161	.182	.23	.25
20	.104	.126	.154	.175	.22	.24

Entries below are for two samples of size $n_1 = n_2 = n$.

n	0.95	0.975	0.99	0.995	0.999	0.9995
2	1.161	1.713	2.776	3.958	8.90	12.62
3	.487	.636	.857	1.046	1.63	2.09
4	.322	.406	.523	.618	.87	.99
5	.246	.306	.386	.448	.60	.67
6	.202	.249	.310	.357	.47	.51
7	.173	.213	.262	.300	.38	.42
8	.153	.186	.229	.260	.33	.36
9	.137	.167	.204	.232	.29	.32
10	.125	.152	.185	.209	.26	.29
11	.116	.140	.170	.192	.24	.26
12	.107	.130	.157	.177	.22	.24
13	.100	.121	.147	.165	.20	.22
14	.094	.114	.138	.155	.19	.21
15	.089	.108	.130	.146	.18	.19
16	.085	.102	.123	.139	.17	.18
17	.081	.097	.118	.132	.16	.17
18	.077	.093	.112	.126	.15	.17
19	.074	.089	.108	.121	.15	.16
20	.071	.086	.103	.116	.14	.15

Table A.10 Percentiles of F Distributions[†]

Let F_{ν_1,ν_2} denote an F variable with ν_1 numerator degrees of freedom and ν_2 denominator degrees of freedom. Entries are $F_{\nu_1,\nu_2;\gamma}$ where $P[F_{\nu_1,\nu_2} \leqslant F_{\nu_1,\nu_2;\gamma}] = \gamma$.

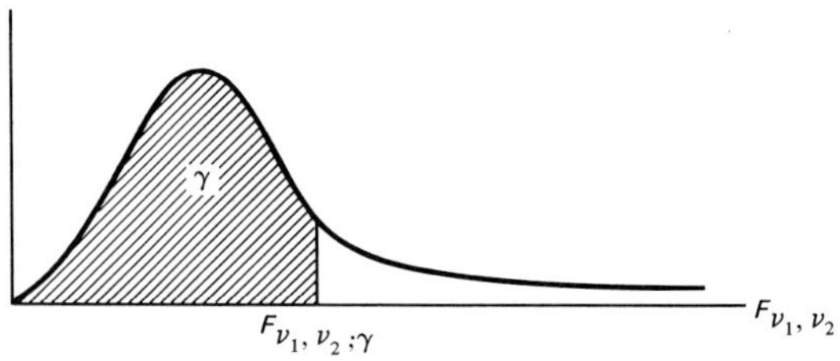

For $\nu_1 = 10$ and $\nu_2 = 17$, the table gives:

$$F_{10,17;0.005} = 0.186\,; \qquad F_{10,17;0.10} = 0.450$$

$$F_{10,17;0.975} = 2.92\,; \qquad F_{10,17;0.99} = 3.59$$

[†] Computed at New Mexico State University.

Table A.10 *continued*

ν_2	γ \ ν_1	1	2	3	4	5	6	7	8	9	10	ν_1 / γ
1	0.005	0.000	0.005	0.018	0.032	0.044	0.054	0.062	0.068	0.073	0.078	0.005
	0.010	0.000	0.010	0.029	0.047	0.062	0.073	0.082	0.089	0.095	0.100	0.010
	0.025	0.002	0.026	0.057	0.082	0.100	0.113	0.124	0.132	0.139	0.144	0.025
	0.050	0.006	0.054	0.099	0.130	0.151	0.167	0.179	0.188	0.195	0.201	0.050
	0.100	0.025	0.117	0.181	0.220	0.246	0.265	0.279	0.289	0.298	0.304	0.100
	0.900	39.86	49.50	53.59	55.83	57.24	58.20	58.91	59.44	59.86	60.19	0.900
	0.950	161.4	199.5	215.7	224.6	230.2	234.0	236.8	238.9	240.5	241.9	0.950
	0.975	647.8	799.5	864.2	899.6	921.8	937.1	948.2	956.7	963.3	968.6	0.975
	0.990	4052.	5000.	5403.	5625.	5764.	5859.	5928.	5981.	6022.	6056.	0.990
	0.995	16211	20000	21615	22500	23056	23437	23715	23926	24091	24225	0.995
2	0.005	0.000	0.005	0.020	0.038	0.055	0.069	0.081	0.091	0.099	0.106	0.005
	0.010	0.000	0.010	0.032	0.056	0.075	0.092	0.105	0.116	0.125	0.132	0.010
	0.025	0.001	0.026	0.062	0.094	0.119	0.138	0.153	0.165	0.175	0.183	0.025
	0.050	0.005	0.053	0.105	0.144	0.173	0.194	0.211	0.224	0.235	0.244	0.050
	0.100	0.020	0.111	0.183	0.231	0.265	0.289	0.307	0.321	0.333	0.342	0.100
	0.900	8.53	9.00	9.16	9.24	9.29	9.33	9.35	9.37	9.38	9.39	0.900
	0.950	18.51	19.00	19.16	19.25	19.30	19.33	19.35	19.37	19.38	19.40	0.950
	0.975	38.51	39.00	39.17	39.25	39.30	39.33	39.36	39.37	39.39	39.40	0.975
	0.990	98.50	99.00	99.17	99.25	99.30	99.33	99.36	99.37	99.39	99.40	0.990
	0.995	198.5	199.0	199.2	199.2	199.3	199.3	199.4	199.4	199.4	199.4	0.995
3	0.005	0.000	0.005	0.021	0.041	0.060	0.077	0.092	0.104	0.115	0.124	0.005
	0.010	0.000	0.010	0.034	0.060	0.083	0.102	0.118	0.132	0.143	0.153	0.010
	0.025	0.001	0.026	0.065	0.100	0.129	0.152	0.170	0.185	0.197	0.207	0.025
	0.050	0.005	0.052	0.108	0.152	0.185	0.210	0.230	0.246	0.259	0.270	0.050
	0.100	0.019	0.109	0.185	0.239	0.276	0.304	0.325	0.342	0.356	0.367	0.100
	0.900	5.54	5.46	5.39	5.34	5.31	5.29	5.27	5.25	5.24	5.23	0.900
	0.950	10.13	9.55	9.28	9.12	9.01	8.94	8.89	8.85	8.81	8.79	0.950
	0.975	17.44	16.04	15.44	15.10	14.89	14.74	14.63	14.55	14.48	14.43	0.975
	0.990	34.12	30.82	29.47	28.72	28.25	27.93	27.70	27.52	27.38	27.27	0.990
	0.995	55.55	49.80	47.50	46.24	45.45	44.91	44.52	44.23	44.00	43.82	0.995
4	0.005	0.000	0.005	0.022	0.043	0.064	0.083	0.099	0.114	0.126	0.136	0.005
	0.010	0.000	0.010	0.035	0.063	0.088	0.109	0.127	0.143	0.156	0.167	0.010
	0.025	0.001	0.025	0.066	0.104	0.135	0.161	0.181	0.198	0.212	0.224	0.025
	0.050	0.004	0.052	0.110	0.157	0.193	0.221	0.243	0.261	0.275	0.288	0.050
	0.100	0.018	0.108	0.187	0.243	0.284	0.314	0.338	0.356	0.371	0.384	0.100
	0.900	4.54	4.32	4.19	4.11	4.05	4.01	3.98	3.95	3.94	3.92	0.900
	0.950	7.71	6.94	6.59	6.39	6.26	6.16	6.09	6.04	6.00	5.96	0.950
	0.975	12.22	10.65	9.98	9.60	9.36	9.20	9.07	8.98	8.90	8.84	0.975
	0.990	21.20	18.00	16.69	15.98	15.52	15.21	14.98	14.80	14.66	14.55	0.990
	0.995	31.33	26.28	24.26	23.15	22.46	21.97	21.62	21.35	21.14	20.97	0.995
5	0.005	0.000	0.005	0.022	0.045	0.067	0.087	0.105	0.120	0.134	0.146	0.005
	0.010	0.000	0.010	0.035	0.064	0.091	0.114	0.134	0.151	0.165	0.177	0.010
	0.025	0.001	0.025	0.067	0.107	0.140	0.167	0.189	0.208	0.223	0.236	0.025
	0.050	0.004	0.052	0.111	0.160	0.198	0.228	0.252	0.271	0.287	0.301	0.050
	0.100	0.017	0.108	0.188	0.247	0.290	0.322	0.347	0.367	0.383	0.397	0.100
	0.900	4.06	3.78	3.62	3.52	3.45	3.40	3.37	3.34	3.32	3.30	0.900
	0.950	6.61	5.79	5.41	5.19	5.05	4.95	4.88	4.82	4.77	4.73	0.950
	0.975	10.01	8.43	7.76	7.39	7.15	6.98	6.85	6.76	6.68	6.62	0.975
	0.990	16.26	13.27	12.06	11.39	10.97	10.67	10.46	10.29	10.16	10.05	0.990
	0.995	22.78	18.31	16.53	15.56	14.94	14.51	14.20	13.96	13.77	13.62	0.995

Table A.10 *continued*

ν_2	ν_1 / γ	1	2	3	4	5	6	7	8	9	10	ν_1 / γ
6	0.005	0.000	0.005	0.022	0.046	0.069	0.090	0.109	0.126	0.140	0.153	0.005
	0.010	0.000	0.010	0.036	0.066	0.094	0.118	0.139	0.157	0.172	0.186	0.010
	0.025	0.001	0.025	0.068	0.109	0.143	0.172	0.195	0.215	0.231	0.246	0.025
	0.050	0.004	0.052	0.112	0.162	0.202	0.233	0.259	0.279	0.296	0.311	0.050
	0.100	0.017	0.107	0.189	0.249	0.294	0.327	0.354	0.375	0.392	0.406	0.100
	0.900	3.78	3.46	3.29	3.18	3.11	3.05	3.01	2.98	2.96	2.94	0.900
	0.950	5.99	5.14	4.76	4.53	4.39	4.28	4.21	4.15	4.10	4.06	0.950
	0.975	8.81	7.26	6.60	6.23	5.99	5.82	5.70	5.60	5.52	5.46	0.975
	0.990	13.75	10.92	9.78	9.15	8.75	8.47	8.26	8.10	7.98	7.87	0.990
	0.995	18.63	14.54	12.92	12.03	11.46	11.07	10.79	10.57	10.39	10.25	0.995
7	0.005	0.000	0.005	0.022	0.046	0.070	0.093	0.113	0.130	0.145	0.159	0.005
	0.010	0.000	0.010	0.036	0.067	0.096	0.121	0.143	0.162	0.178	0.192	0.010
	0.025	0.001	0.025	0.068	0.110	0.146	0.176	0.200	0.221	0.238	0.253	0.025
	0.050	0.004	0.052	0.113	0.164	0.205	0.238	0.264	0.286	0.304	0.319	0.050
	0.100	0.017	0.107	0.190	0.251	0.297	0.332	0.359	0.381	0.399	0.414	0.100
	0.900	3.59	3.26	3.07	2.96	2.88	2.83	2.78	2.75	2.72	2.70	0.900
	0.950	5.59	4.74	4.35	4.12	3.97	3.87	3.79	3.73	3.68	3.64	0.950
	0.975	8.07	6.54	5.89	5.52	5.29	5.12	4.99	4.90	4.82	4.76	0.975
	0.990	12.25	9.55	8.45	7.85	7.46	7.19	6.99	6.84	6.72	6.62	0.990
	0.995	16.24	12.40	10.88	10.05	9.52	9.16	8.89	8.68	8.51	8.38	0.995
8	0.005	0.000	0.005	0.023	0.047	0.072	0.095	0.115	0.133	0.149	0.164	0.005
	0.010	0.000	0.010	0.036	0.068	0.097	0.123	0.146	0.166	0.183	0.198	0.010
	0.025	0.001	0.025	0.069	0.111	0.148	0.179	0.204	0.226	0.244	0.259	0.025
	0.050	0.004	0.052	0.113	0.166	0.208	0.241	0.268	0.291	0.310	0.326	0.050
	0.100	0.017	0.107	0.190	0.253	0.299	0.335	0.363	0.386	0.405	0.421	0.100
	0.900	3.46	3.11	2.92	2.81	2.73	2.67	2.62	2.59	2.56	2.54	0.900
	0.950	5.32	4.46	4.07	3.84	3.69	3.58	3.50	3.44	3.39	3.35	0.950
	0.975	7.57	6.06	5.42	5.05	4.82	4.65	4.53	4.43	4.36	4.30	0.975
	0.990	11.26	8.65	7.59	7.01	6.63	6.37	6.18	6.03	5.91	5.81	0.990
	0.995	14.69	11.04	9.60	8.81	8.30	7.95	7.69	7.50	7.34	7.21	0.995
9	0.005	0.000	0.005	0.023	0.047	0.073	0.096	0.117	0.136	0.153	0.168	0.005
	0.010	0.000	0.010	0.037	0.068	0.098	0.125	0.149	0.169	0.187	0.202	0.010
	0.025	0.001	0.025	0.069	0.112	0.150	0.181	0.207	0.230	0.248	0.265	0.025
	0.050	0.004	0.052	0.113	0.167	0.210	0.244	0.272	0.295	0.315	0.331	0.050
	0.100	0.017	0.107	0.191	0.254	0.302	0.338	0.367	0.390	0.410	0.426	0.100
	0.900	3.36	3.01	2.81	2.69	2.61	2.55	2.51	2.47	2.44	2.42	0.900
	0.950	5.12	4.26	3.86	3.63	3.48	3.37	3.29	3.23	3.18	3.14	0.950
	0.975	7.21	5.71	5.08	4.72	4.48	4.32	4.20	4.10	4.03	3.96	0.975
	0.990	10.56	8.02	6.99	6.42	6.06	5.80	5.61	5.47	5.35	5.26	0.990
	0.995	13.61	10.11	8.72	7.96	7.47	7.13	6.88	6.69	6.54	6.42	0.995
10	0.005	0.000	0.005	0.023	0.048	0.073	0.098	0.119	0.139	0.156	0.171	0.005
	0.010	0.000	0.010	0.037	0.069	0.099	0.127	0.151	0.172	0.190	0.206	0.010
	0.025	0.001	0.025	0.069	0.113	0.151	0.183	0.210	0.233	0.252	0.269	0.025
	0.050	0.004	0.052	0.114	0.168	0.211	0.246	0.275	0.299	0.319	0.336	0.050
	0.100	0.017	0.106	0.191	0.255	0.303	0.340	0.370	0.394	0.414	0.431	0.100
	0.900	3.29	2.92	2.73	2.61	2.52	2.46	2.41	2.38	2.35	2.32	0.900
	0.950	4.96	4.10	3.71	3.48	3.33	3.22	3.14	3.07	3.02	2.98	0.950
	0.975	6.94	5.46	4.83	4.47	4.24	4.07	3.95	3.85	3.78	3.72	0.975
	0.990	10.04	7.56	6.55	5.99	5.64	5.39	5.20	5.06	4.94	4.85	0.990
	0.995	12.83	9.43	8.08	7.34	6.87	6.54	6.30	6.12	5.97	5.85	0.995

ν_2	γ \ ν_1	1	2	3	4	5	6	7	8	9	10	ν_1 / γ
11	0.005	0.000	0.005	0.023	0.048	0.074	0.099	0.121	0.141	0.158	0.174	0.005
	0.010	0.000	0.010	0.037	0.069	0.100	0.128	0.153	0.174	0.193	0.210	0.010
	0.025	0.001	0.025	0.070	0.114	0.152	0.185	0.212	0.236	0.256	0.273	0.025
	0.050	0.004	0.052	0.114	0.168	0.213	0.248	0.278	0.302	0.322	0.340	0.050
	0.100	0.017	0.106	0.191	0.256	0.305	0.343	0.373	0.397	0.417	0.434	0.100
	0.900	3.23	2.86	2.66	2.54	2.45	2.39	2.34	2.30	2.27	2.25	0.900
	0.950	4.84	3.98	3.59	3.36	3.20	3.09	3.01	2.95	2.90	2.85	0.950
	0.975	6.72	5.26	4.63	4.28	4.04	3.88	3.76	3.66	3.59	3.53	0.975
	0.990	9.65	7.21	6.22	5.67	5.32	5.07	4.89	4.74	4.63	4.54	0.990
	0.995	12.23	8.91	7.60	6.88	6.42	6.10	5.86	5.68	5.54	5.42	0.995
12	0.005	0.000	0.005	0.023	0.048	0.075	0.100	0.122	0.143	0.161	0.177	0.005
	0.010	0.000	0.010	0.037	0.070	0.101	0.130	0.155	0.176	0.196	0.213	0.010
	0.025	0.001	0.025	0.070	0.114	0.153	0.186	0.214	0.238	0.259	0.276	0.025
	0.050	0.004	0.052	0.114	0.169	0.214	0.250	0.280	0.305	0.325	0.343	0.050
	0.100	0.016	0.106	0.192	0.257	0.306	0.344	0.375	0.400	0.420	0.438	0.100
	0.900	3.18	2.81	2.61	2.48	2.39	2.33	2.28	2.24	2.21	2.19	0.900
	0.950	4.75	3.89	3.49	3.26	3.11	3.00	2.91	2.85	2.80	2.75	0.950
	0.975	6.55	5.10	4.47	4.12	3.89	3.73	3.61	3.51	3.44	3.37	0.975
	0.990	9.33	6.93	5.95	5.41	5.06	4.82	4.64	4.50	4.39	4.30	0.990
	0.995	11.75	8.51	7.23	6.52	6.07	5.76	5.52	5.35	5.20	5.09	0.995
13	0.005	0.000	0.005	0.023	0.049	0.075	0.101	0.124	0.144	0.163	0.179	0.005
	0.010	0.000	0.010	0.037	0.070	0.102	0.131	0.156	0.178	0.198	0.215	0.010
	0.025	0.001	0.025	0.070	0.115	0.154	0.188	0.216	0.240	0.261	0.279	0.025
	0.050	0.004	0.051	0.115	0.170	0.215	0.251	0.282	0.307	0.328	0.346	0.050
	0.100	0.016	0.106	0.192	0.257	0.307	0.346	0.377	0.402	0.423	0.441	0.100
	0.900	3.14	2.76	2.56	2.43	2.35	2.28	2.23	2.20	2.16	2.14	0.900
	0.950	4.67	3.81	3.41	3.18	3.03	2.92	2.83	2.77	2.71	2.67	0.950
	0.975	6.41	4.97	4.35	4.00	3.77	3.60	3.48	3.39	3.31	3.25	0.975
	0.990	9.07	6.70	5.74	5.21	4.86	4.62	4.44	4.30	4.19	4.10	0.990
	0.995	11.37	8.19	6.93	6.23	5.79	5.48	5.25	5.08	4.94	4.82	0.995
14	0.005	0.000	0.005	0.023	0.049	0.076	0.101	0.125	0.146	0.164	0.181	0.005
	0.010	0.000	0.010	0.037	0.070	0.102	0.131	0.157	0.180	0.200	0.217	0.010
	0.025	0.001	0.025	0.070	0.115	0.155	0.189	0.218	0.242	0.263	0.282	0.025
	0.050	0.004	0.051	0.115	0.170	0.216	0.253	0.283	0.309	0.331	0.349	0.050
	0.100	0.016	0.106	0.192	0.258	0.308	0.347	0.378	0.404	0.425	0.443	0.100
	0.900	3.10	2.73	2.52	2.39	2.31	2.24	2.19	2.15	2.12	2.10	0.900
	0.950	4.60	3.74	3.34	3.11	2.96	2.85	2.76	2.70	2.65	2.60	0.950
	0.975	6.30	4.86	4.24	3.89	3.66	3.50	3.38	3.29	3.21	3.15	0.975
	0.990	8.86	6.51	5.56	5.04	4.69	4.46	4.28	4.14	4.03	3.94	0.990
	0.995	11.06	7.92	6.68	6.00	5.56	5.26	5.03	4.86	4.72	4.60	0.995
15	0.005	0.000	0.005	0.023	0.049	0.076	0.102	0.126	0.147	0.166	0.183	0.005
	0.010	0.000	0.010	0.037	0.070	0.103	0.132	0.158	0.181	0.202	0.219	0.010
	0.025	0.001	0.025	0.070	0.116	0.156	0.190	0.219	0.244	0.265	0.284	0.025
	0.050	0.004	0.051	0.115	0.171	0.217	0.254	0.285	0.311	0.333	0.351	0.050
	0.100	0.016	0.106	0.192	0.258	0.309	0.348	0.380	0.406	0.427	0.446	0.100
	0.900	3.07	2.70	2.49	2.36	2.27	2.21	2.16	2.12	2.09	2.06	0.900
	0.950	4.54	3.68	3.29	3.06	2.90	2.79	2.71	2.64	2.59	2.54	0.950
	0.975	6.20	4.77	4.15	3.80	3.58	3.41	3.29	3.20	3.12	3.06	0.975
	0.990	8.68	6.36	5.42	4.89	4.56	4.32	4.14	4.00	3.89	3.80	0.990
	0.995	10.80	7.70	6.48	5.80	5.37	5.07	4.85	4.67	4.54	4.42	0.995

Table A.10 *continued*

ν_2	γ \ ν_1	1	2	3	4	5	6	7	8	9	10	ν_1 / γ
16	0.005	0.000	0.005	0.023	0.049	0.076	0.102	0.126	0.148	0.167	0.184	0.005
	0.010	0.000	0.010	0.037	0.071	0.103	0.133	0.159	0.183	0.203	0.221	0.010
	0.025	0.001	0.025	0.070	0.116	0.156	0.191	0.220	0.245	0.267	0.286	0.025
	0.050	0.004	0.051	0.115	0.171	0.217	0.255	0.286	0.312	0.335	0.354	0.050
	0.100	0.016	0.106	0.192	0.259	0.310	0.349	0.381	0.407	0.429	0.448	0.100
	0.900	3.05	2.67	2.46	2.33	2.24	2.18	2.13	2.09	2.06	2.03	0.900
	0.950	4.49	3.63	3.24	3.01	2.85	2.74	2.66	2.59	2.54	2.49	0.950
	0.975	6.12	4.69	4.08	3.73	3.50	3.34	3.22	3.12	3.05	2.99	0.975
	0.990	8.53	6.23	5.29	4.77	4.44	4.20	4.03	3.89	3.78	3.69	0.990
	0.995	10.58	7.51	6.30	5.64	5.21	4.91	4.69	4.52	4.38	4.27	0.995
17	0.005	0.000	0.005	0.023	0.049	0.077	0.103	0.127	0.149	0.168	0.186	0.005
	0.010	0.000	0.010	0.037	0.071	0.104	0.134	0.160	0.184	0.204	0.223	0.010
	0.025	0.001	0.025	0.070	0.116	0.157	0.192	0.221	0.247	0.269	0.288	0.025
	0.050	0.004	0.051	0.115	0.171	0.218	0.256	0.287	0.314	0.336	0.356	0.050
	0.100	0.016	0.106	0.192	0.259	0.310	0.350	0.382	0.409	0.431	0.450	0.100
	0.900	3.03	2.64	2.44	2.31	2.22	2.15	2.10	2.06	2.03	2.00	0.900
	0.950	4.45	3.59	3.20	2.96	2.81	2.70	2.61	2.55	2.49	2.45	0.950
	0.975	6.04	4.62	4.01	3.66	3.44	3.28	3.16	3.06	2.98	2.92	0.975
	0.990	8.40	6.11	5.18	4.67	4.34	4.10	3.93	3.79	3.68	3.59	0.990
	0.995	10.38	7.35	6.16	5.50	5.07	4.78	4.56	4.39	4.25	4.14	0.995
18	0.005	0.000	0.005	0.023	0.049	0.077	0.103	0.128	0.150	0.170	0.187	0.005
	0.010	0.000	0.010	0.037	0.071	0.104	0.134	0.161	0.185	0.206	0.224	0.010
	0.025	0.001	0.025	0.070	0.116	0.157	0.192	0.222	0.248	0.270	0.290	0.025
	0.050	0.004	0.051	0.115	0.172	0.218	0.257	0.288	0.315	0.338	0.357	0.050
	0.100	0.016	0.106	0.193	0.260	0.311	0.351	0.384	0.410	0.432	0.451	0.100
	0.900	3.01	2.62	2.42	2.29	2.20	2.13	2.08	2.04	2.00	1.98	0.900
	0.950	4.41	3.55	3.16	2.93	2.77	2.66	2.58	2.51	2.46	2.41	0.950
	0.975	5.98	4.56	3.95	3.61	3.38	3.22	3.10	3.01	2.93	2.87	0.975
	0.990	8.29	6.01	5.09	4.58	4.25	4.01	3.84	3.71	3.60	3.51	0.990
	0.995	10.22	7.21	6.03	5.37	4.96	4.66	4.44	4.28	4.14	4.03	0.995
19	0.005	0.000	0.005	0.023	0.049	0.077	0.104	0.128	0.151	0.171	0.188	0.005
	0.010	0.000	0.010	0.037	0.071	0.104	0.135	0.162	0.186	0.207	0.226	0.010
	0.025	0.001	0.025	0.070	0.117	0.158	0.193	0.223	0.249	0.271	0.291	0.025
	0.050	0.004	0.051	0.115	0.172	0.219	0.257	0.289	0.316	0.339	0.359	0.050
	0.100	0.016	0.106	0.193	0.260	0.311	0.352	0.385	0.411	0.434	0.453	0.100
	0.900	2.99	2.61	2.40	2.27	2.18	2.11	2.06	2.02	1.98	1.96	0.900
	0.950	4.38	3.52	3.13	2.90	2.74	2.63	2.54	2.48	2.42	2.38	0.950
	0.975	5.92	4.51	3.90	3.56	3.33	3.17	3.05	2.96	2.88	2.82	0.975
	0.990	8.18	5.93	5.01	4.50	4.17	3.94	3.77	3.63	3.52	3.43	0.990
	0.995	10.07	7.09	5.92	5.27	4.85	4.56	4.34	4.18	4.04	3.93	0.995
20	0.005	0.000	0.005	0.023	0.050	0.078	0.104	0.129	0.151	0.171	0.190	0.005
	0.010	0.000	0.010	0.037	0.071	0.105	0.135	0.162	0.187	0.208	0.227	0.010
	0.025	0.001	0.025	0.070	0.117	0.158	0.193	0.224	0.250	0.273	0.293	0.025
	0.050	0.004	0.051	0.115	0.172	0.219	0.258	0.290	0.317	0.341	0.360	0.050
	0.100	0.016	0.106	0.193	0.260	0.312	0.353	0.385	0.412	0.435	0.454	0.100
	0.900	2.97	2.59	2.38	2.25	2.16	2.09	2.04	2.00	1.96	1.94	0.900
	0.950	4.35	3.49	3.10	2.87	2.71	2.60	2.51	2.45	2.39	2.35	0.950
	0.975	5.87	4.46	3.86	3.51	3.29	3.13	3.01	2.91	2.84	2.77	0.975
	0.990	8.10	5.85	4.94	4.43	4.10	3.87	3.70	3.56	3.46	3.37	0.990
	0.995	9.94	6.99	5.82	5.17	4.76	4.47	4.26	4.09	3.96	3.85	0.995

ν_2	γ \ ν_1	1	2	3	4	5	6	7	8	9	10	ν_1 / γ
21	0.005	0.000	0.005	0.023	0.050	0.078	0.105	0.129	0.152	0.172	0.191	0.005
	0.010	0.000	0.010	0.037	0.071	0.105	0.136	0.163	0.187	0.209	0.228	0.010
	0.025	0.001	0.025	0.071	0.117	0.158	0.194	0.225	0.251	0.274	0.294	0.025
	0.050	0.004	0.051	0.115	0.173	0.220	0.259	0.291	0.318	0.342	0.362	0.050
	0.100	0.016	0.106	0.193	0.260	0.312	0.353	0.386	0.413	0.436	0.456	0.100
	0.900	2.96	2.57	2.36	2.23	2.14	2.08	2.02	1.98	1.95	1.92	0.900
	0.950	4.32	3.47	3.07	2.84	2.68	2.57	2.49	2.42	2.37	2.32	0.950
	0.975	5.83	4.42	3.82	3.48	3.25	3.09	2.97	2.87	2.80	2.73	0.975
	0.990	8.02	5.78	4.87	4.37	4.04	3.81	3.64	3.51	3.40	3.31	0.990
	0.995	9.83	6.89	5.73	5.09	4.68	4.39	4.18	4.01	3.88	3.77	0.995
22	0.005	0.000	0.005	0.023	0.050	0.078	0.105	0.130	0.153	0.173	0.192	0.005
	0.010	0.000	0.010	0.037	0.072	0.105	0.136	0.164	0.188	0.210	0.229	0.010
	0.025	0.001	0.025	0.071	0.117	0.159	0.195	0.225	0.252	0.275	0.295	0.025
	0.050	0.004	0.051	0.116	0.173	0.220	0.259	0.292	0.319	0.343	0.363	0.050
	0.100	0.016	0.106	0.193	0.261	0.313	0.354	0.387	0.414	0.437	0.457	0.100
	0.900	2.95	2.56	2.35	2.22	2.13	2.06	2.01	1.97	1.93	1.90	0.900
	0.950	4.30	3.44	3.05	2.82	2.66	2.55	2.46	2.40	2.34	2.30	0.950
	0.975	5.79	4.38	3.78	3.44	3.22	3.05	2.93	2.84	2.76	2.70	0.975
	0.990	7.95	5.72	4.82	4.31	3.99	3.76	3.59	3.45	3.35	3.26	0.990
	0.995	9.73	6.81	5.65	5.02	4.61	4.32	4.11	3.94	3.81	3.70	0.995
23	0.005	0.000	0.005	0.023	0.050	0.078	0.105	0.130	0.153	0.174	0.192	0.005
	0.010	0.000	0.010	0.037	0.072	0.105	0.136	0.164	0.189	0.211	0.230	0.010
	0.025	0.001	0.025	0.071	0.117	0.159	0.195	0.226	0.253	0.276	0.296	0.025
	0.050	0.004	0.051	0.116	0.173	0.221	0.260	0.293	0.320	0.344	0.364	0.050
	0.100	0.016	0.106	0.193	0.261	0.313	0.354	0.388	0.415	0.438	0.458	0.100
	0.900	2.94	2.55	2.34	2.21	2.11	2.05	1.99	1.95	1.92	1.89	0.900
	0.950	4.28	3.42	3.03	2.80	2.64	2.53	2.44	2.37	2.32	2.27	0.950
	0.975	5.75	4.35	3.75	3.41	3.18	3.02	2.90	2.81	2.73	2.67	0.975
	0.990	7.88	5.66	4.76	4.26	3.94	3.71	3.54	3.41	3.30	3.21	0.990
	0.995	9.63	6.73	5.58	4.95	4.54	4.26	4.05	3.88	3.75	3.64	0.995
24	0.005	0.000	0.005	0.023	0.050	0.078	0.106	0.131	0.154	0.175	0.193	0.005
	0.010	0.000	0.010	0.037	0.072	0.106	0.137	0.165	0.189	0.211	0.231	0.010
	0.025	0.001	0.025	0.071	0.117	0.159	0.195	0.227	0.253	0.277	0.297	0.025
	0.050	0.004	0.051	0.116	0.173	0.221	0.260	0.293	0.321	0.345	0.365	0.050
	0.100	0.016	0.106	0.193	0.261	0.313	0.355	0.388	0.416	0.439	0.459	0.100
	0.900	2.93	2.54	2.33	2.19	2.10	2.04	1.98	1.94	1.91	1.88	0.900
	0.950	4.26	3.40	3.01	2.78	2.62	2.51	2.42	2.36	2.30	2.25	0.950
	0.975	5.72	4.32	3.72	3.38	3.15	2.99	2.87	2.78	2.70	2.64	0.975
	0.990	7.82	5.61	4.72	4.22	3.90	3.67	3.50	3.36	3.26	3.17	0.990
	0.995	9.55	6.66	5.52	4.89	4.49	4.20	3.99	3.83	3.69	3.59	0.995
25	0.005	0.000	0.005	0.023	0.050	0.078	0.106	0.131	0.154	0.175	0.194	0.005
	0.010	0.000	0.010	0.037	0.072	0.106	0.137	0.165	0.190	0.212	0.232	0.010
	0.025	0.001	0.025	0.071	0.118	0.160	0.196	0.227	0.254	0.278	0.298	0.025
	0.050	0.004	0.051	0.116	0.173	0.221	0.261	0.294	0.322	0.346	0.366	0.050
	0.100	0.016	0.106	0.193	0.261	0.314	0.355	0.389	0.417	0.440	0.460	0.100
	0.900	2.92	2.53	2.32	2.18	2.09	2.02	1.97	1.93	1.89	1.87	0.900
	0.950	4.24	3.39	2.99	2.76	2.60	2.49	2.40	2.34	2.28	2.24	0.950
	0.975	5.69	4.29	3.69	3.35	3.13	2.97	2.85	2.75	2.68	2.61	0.975
	0.990	7.77	5.57	4.68	4.18	3.85	3.63	3.46	3.32	3.22	3.13	0.990
	0.995	9.48	6.60	5.46	4.84	4.43	4.15	3.94	3.78	3.64	3.54	0.995

Table A.10 *continued*

ν_2	ν_1 / γ	1	2	3	4	5	6	7	8	9	10	ν_1 / γ
30	0.005	0.000	0.005	0.023	0.050	0.079	0.107	0.133	0.156	0.178	0.197	0.005
	0.010	0.000	0.010	0.037	0.072	0.107	0.138	0.167	0.192	0.215	0.235	0.010
	0.025	0.001	0.025	0.071	0.118	0.161	0.197	0.229	0.257	0.281	0.302	0.025
	0.050	0.004	0.051	0.116	0.174	0.222	0.263	0.296	0.325	0.349	0.370	0.050
	0.100	0.016	0.106	0.193	0.262	0.315	0.357	0.391	0.420	0.444	0.464	0.100
	0.900	2.88	2.49	2.28	2.14	2.05	1.98	1.93	1.88	1.85	1.82	0.900
	0.950	4.17	3.32	2.92	2.69	2.53	2.42	2.33	2.27	2.21	2.16	0.950
	0.975	5.57	4.18	3.59	3.25	3.03	2.87	2.75	2.65	2.57	2.51	0.975
	0.990	7.56	5.39	4.51	4.02	3.70	3.47	3.30	3.17	3.07	2.98	0.990
	0.995	9.18	6.35	5.24	4.62	4.23	3.95	3.74	3.58	3.45	3.34	0.995
40	0.005	0.000	0.005	0.023	0.051	0.080	0.108	0.135	0.159	0.181	0.201	0.005
	0.010	0.000	0.010	0.037	0.073	0.108	0.140	0.169	0.195	0.219	0.240	0.010
	0.025	0.001	0.025	0.071	0.119	0.162	0.200	0.232	0.260	0.285	0.307	0.025
	0.050	0.004	0.051	0.116	0.175	0.224	0.265	0.299	0.329	0.354	0.376	0.050
	0.100	0.016	0.106	0.194	0.263	0.317	0.360	0.394	0.423	0.448	0.469	0.100
	0.900	2.84	2.44	2.23	2.09	2.00	1.93	1.87	1.83	1.79	1.76	0.900
	0.950	4.08	3.23	2.84	2.61	2.45	2.34	2.25	2.18	2.12	2.08	0.950
	0.975	5.42	4.05	3.46	3.13	2.90	2.74	2.62	2.53	2.45	2.39	0.975
	0.990	7.31	5.18	4.31	3.83	3.51	3.29	3.12	2.99	2.89	2.80	0.990
	0.995	8.83	6.07	4.98	4.37	3.99	3.71	3.51	3.35	3.22	3.12	0.995
60	0.005	0.000	0.005	0.023	0.051	0.081	0.110	0.137	0.162	0.185	0.206	0.005
	0.010	0.000	0.010	0.037	0.073	0.109	0.142	0.172	0.199	0.223	0.245	0.010
	0.025	0.001	0.025	0.071	0.120	0.163	0.202	0.235	0.264	0.290	0.313	0.025
	0.050	0.004	0.051	0.116	0.176	0.226	0.267	0.303	0.333	0.359	0.382	0.050
	0.100	0.016	0.106	0.194	0.264	0.318	0.362	0.398	0.428	0.453	0.475	0.100
	0.900	2.79	2.39	2.18	2.04	1.95	1.87	1.82	1.77	1.74	1.71	0.900
	0.950	4.00	3.15	2.76	2.53	2.37	2.25	2.17	2.10	2.04	1.99	0.950
	0.975	5.29	3.93	3.34	3.01	2.79	2.63	2.51	2.41	2.33	2.27	0.975
	0.990	7.08	4.98	4.13	3.65	3.34	3.12	2.95	2.82	2.72	2.63	0.990
	0.995	8.49	5.79	4.73	4.14	3.76	3.49	3.29	3.13	3.01	2.90	0.995
120	0.005	0.000	0.005	0.024	0.051	0.082	0.111	0.139	0.165	0.189	0.211	0.005
	0.010	0.000	0.010	0.038	0.074	0.110	0.144	0.174	0.202	0.227	0.250	0.010
	0.025	0.001	0.025	0.072	0.120	0.165	0.204	0.238	0.268	0.295	0.319	0.025
	0.050	0.004	0.051	0.117	0.177	0.227	0.270	0.306	0.337	0.364	0.388	0.050
	0.100	0.016	0.105	0.194	0.265	0.320	0.365	0.401	0.432	0.458	0.480	0.100
	0.900	2.75	2.35	2.13	1.99	1.90	1.82	1.77	1.72	1.68	1.65	0.900
	0.950	3.92	3.07	2.68	2.45	2.29	2.17	2.09	2.02	1.96	1.91	0.950
	0.975	5.15	3.80	3.23	2.89	2.67	2.52	2.39	2.30	2.22	2.16	0.975
	0.990	6.85	4.79	3.95	3.48	3.17	2.96	2.79	2.66	2.56	2.47	0.990
	0.995	8.18	5.54	4.50	3.92	3.55	3.28	3.09	2.93	2.81	2.71	0.995
∞	0.005	0.000	0.005	0.024	0.052	0.082	0.113	0.141	0.168	0.193	0.216	0.005
	0.010	0.000	0.010	0.038	0.074	0.111	0.145	0.177	0.206	0.232	0.256	0.010
	0.025	0.001	0.025	0.072	0.121	0.166	0.206	0.241	0.272	0.300	0.325	0.025
	0.050	0.004	0.051	0.117	0.178	0.229	0.273	0.310	0.342	0.369	0.394	0.050
	0.100	0.016	0.105	0.195	0.266	0.322	0.367	0.405	0.436	0.463	0.487	0.100
	0.900	2.71	2.30	2.08	1.94	1.85	1.77	1.72	1.67	1.63	1.60	0.900
	0.950	3.84	3.00	2.60	2.37	2.21	2.10	2.01	1.94	1.88	1.83	0.950
	0.975	5.02	3.69	3.12	2.79	2.57	2.41	2.29	2.19	2.11	2.05	0.975
	0.990	6.63	4.61	3.78	3.32	3.02	2.80	2.64	2.51	2.41	2.32	0.990
	0.995	7.88	5.30	4.28	3.72	3.35	3.09	2.90	2.74	2.63	2.52	0.995

ν_2	γ \ ν_1	11	12	13	14	15	16	17	18	19	20	ν_1 / γ
1	0.005	0.082	0.085	0.088	0.090	0.093	0.095	0.096	0.098	0.099	0.101	0.005
	0.010	0.104	0.107	0.110	0.113	0.115	0.117	0.119	0.121	0.122	0.124	0.010
	0.025	0.149	0.153	0.156	0.159	0.161	0.164	0.166	0.167	0.169	0.170	0.025
	0.050	0.206	0.211	0.214	0.217	0.220	0.223	0.225	0.227	0.228	0.230	0.050
	0.100	0.310	0.315	0.319	0.322	0.325	0.328	0.330	0.333	0.334	0.336	0.100
	0.900	60.47	60.71	60.90	61.07	61.22	61.35	61.46	61.57	61.66	61.74	0.900
	0.950	243.0	243.9	244.7	245.4	246.0	246.5	246.9	247.3	247.7	248.0	0.950
	0.975	973.0	976.7	979.8	982.5	984.9	986.9	988.7	990.3	991.8	993.1	0.975
	0.990	6083.	6106.	6126.	6143.	6157.	6170.	6181.	6192.	6201.	6209.	0.990
	0.995	24334	24426	24505	24572	24631	24682	24727	24767	24804	24836	0.995
2	0.005	0.112	0.118	0.122	0.126	0.130	0.133	0.136	0.139	0.141	0.143	0.005
	0.010	0.139	0.144	0.149	0.153	0.157	0.161	0.164	0.166	0.169	0.171	0.010
	0.025	0.190	0.196	0.201	0.206	0.210	0.213	0.217	0.219	0.222	0.224	0.025
	0.050	0.251	0.257	0.263	0.267	0.272	0.275	0.278	0.281	0.284	0.286	0.050
	0.100	0.350	0.356	0.362	0.367	0.371	0.375	0.378	0.381	0.384	0.386	0.100
	0.900	9.40	9.41	9.41	9.42	9.42	9.43	9.43	9.44	9.44	9.44	0.900
	0.950	19.40	19.41	19.42	19.42	19.43	19.43	19.44	19.44	19.44	19.45	0.950
	0.975	39.41	39.41	39.42	39.43	39.43	39.44	39.44	39.44	39.45	39.45	0.975
	0.990	99.41	99.42	99.42	99.43	99.43	99.44	99.44	99.44	99.45	99.45	0.990
	0.995	199.4	199.4	199.4	199.4	199.4	199.4	199.4	199.4	199.4	199.4	0.995
3	0.005	0.132	0.138	0.144	0.150	0.154	0.159	0.162	0.166	0.169	0.172	0.005
	0.010	0.161	0.168	0.174	0.180	0.185	0.189	0.193	0.196	0.200	0.203	0.010
	0.025	0.216	0.224	0.230	0.236	0.241	0.245	0.249	0.253	0.256	0.259	0.025
	0.050	0.279	0.287	0.293	0.299	0.304	0.309	0.313	0.316	0.320	0.323	0.050
	0.100	0.376	0.384	0.391	0.396	0.402	0.406	0.410	0.414	0.417	0.420	0.100
	0.900	5.22	5.22	5.21	5.21	5.20	5.20	5.19	5.19	5.19	5.19	0.900
	0.950	8.77	8.75	8.73	8.72	8.71	8.70	8.69	8.68	8.67	8.67	0.950
	0.975	14.38	14.35	14.32	14.29	14.27	14.25	14.23	14.22	14.20	14.19	0.975
	0.990	27.18	27.11	27.04	26.99	26.94	26.91	26.88	26.85	26.82	26.80	0.990
	0.995	43.68	43.56	43.46	43.38	43.31	43.26	43.22	43.17	43.14	43.12	0.995
4	0.005	0.145	0.153	0.160	0.167	0.172	0.177	0.182	0.186	0.190	0.193	0.005
	0.010	0.176	0.185	0.192	0.199	0.204	0.210	0.214	0.218	0.222	0.226	0.010
	0.025	0.234	0.243	0.250	0.257	0.263	0.268	0.273	0.277	0.281	0.285	0.025
	0.050	0.298	0.307	0.315	0.321	0.327	0.333	0.337	0.342	0.345	0.349	0.050
	0.100	0.394	0.403	0.411	0.418	0.423	0.429	0.433	0.437	0.441	0.445	0.100
	0.900	3.91	3.90	3.89	3.88	3.87	3.86	3.86	3.85	3.85	3.84	0.900
	0.950	5.94	5.91	5.89	5.87	5.86	5.84	5.83	5.82	5.81	5.80	0.950
	0.975	8.79	8.75	8.71	8.68	8.66	8.63	8.61	8.59	8.58	8.56	0.975
	0.990	14.45	14.37	14.31	14.25	14.20	14.15	14.11	14.08	14.05	14.02	0.990
	0.995	20.82	20.71	20.60	20.52	20.44	20.37	20.31	20.26	20.21	20.17	0.995
5	0.005	0.156	0.165	0.173	0.180	0.186	0.192	0.197	0.202	0.206	0.210	0.005
	0.010	0.188	0.197	0.206	0.213	0.220	0.225	0.231	0.235	0.240	0.244	0.010
	0.025	0.247	0.257	0.265	0.273	0.280	0.286	0.291	0.296	0.300	0.304	0.025
	0.050	0.312	0.322	0.331	0.338	0.345	0.351	0.356	0.361	0.365	0.369	0.050
	0.100	0.408	0.418	0.426	0.433	0.440	0.446	0.451	0.455	0.460	0.463	0.100
	0.900	3.28	3.27	3.26	3.25	3.24	3.23	3.22	3.22	3.21	3.21	0.900
	0.950	4.70	4.68	4.66	4.64	4.62	4.60	4.59	4.58	4.57	4.56	0.950
	0.975	6.57	6.52	6.49	6.46	6.43	6.40	6.38	6.36	6.34	6.33	0.975
	0.990	9.96	9.89	9.82	9.77	9.72	9.68	9.64	9.61	9.58	9.55	0.990
	0.995	13.49	13.38	13.29	13.21	13.15	13.08	13.03	12.98	12.94	12.90	0.995

Table A.10 *continued*

ν_2	γ \ ν_1	11	12	13	14	15	16	17	18	19	20	ν_1 / γ
6	0.005	0.164	0.174	0.182	0.190	0.197	0.204	0.209	0.214	0.219	0.224	0.005
	0.010	0.197	0.207	0.216	0.224	0.232	0.238	0.244	0.249	0.254	0.258	0.010
	0.025	0.258	0.268	0.277	0.286	0.293	0.299	0.305	0.310	0.315	0.320	0.025
	0.050	0.323	0.334	0.343	0.351	0.358	0.365	0.371	0.376	0.380	0.385	0.050
	0.100	0.419	0.429	0.438	0.446	0.453	0.459	0.465	0.470	0.474	0.478	0.100
	0.900	2.92	2.90	2.89	2.88	2.87	2.86	2.85	2.85	2.84	2.84	0.900
	0.950	4.03	4.00	3.98	3.96	3.94	3.92	3.91	3.90	3.88	3.87	0.950
	0.975	5.41	5.37	5.33	5.30	5.27	5.24	5.22	5.20	5.18	5.17	0.975
	0.990	7.79	7.72	7.66	7.60	7.56	7.52	7.48	7.45	7.42	7.40	0.990
	0.995	10.13	10.03	9.95	9.88	9.81	9.76	9.71	9.66	9.62	9.59	0.995
7	0.005	0.171	0.181	0.190	0.199	0.206	0.213	0.219	0.225	0.230	0.235	0.005
	0.010	0.205	0.216	0.225	0.234	0.241	0.248	0.255	0.260	0.266	0.270	0.010
	0.025	0.266	0.277	0.287	0.296	0.304	0.311	0.317	0.323	0.328	0.333	0.025
	0.050	0.332	0.343	0.353	0.362	0.369	0.376	0.383	0.388	0.393	0.398	0.050
	0.100	0.427	0.438	0.448	0.456	0.463	0.470	0.476	0.481	0.486	0.490	0.100
	0.900	2.68	2.67	2.65	2.64	2.63	2.62	2.61	2.61	2.60	2.59	0.900
	0.950	3.60	3.57	3.55	3.53	3.51	3.49	3.48	3.47	3.46	3.44	0.950
	0.975	4.71	4.67	4.63	4.60	4.57	4.54	4.52	4.50	4.48	4.47	0.975
	0.990	6.54	6.47	6.41	6.36	6.31	6.27	6.24	6.21	6.18	6.16	0.990
	0.995	8.27	8.18	8.10	8.03	7.97	7.91	7.87	7.83	7.79	7.75	0.995
8	0.005	0.176	0.187	0.197	0.206	0.214	0.221	0.228	0.234	0.239	0.245	0.005
	0.010	0.211	0.222	0.232	0.242	0.250	0.257	0.264	0.270	0.275	0.281	0.010
	0.025	0.273	0.285	0.295	0.304	0.313	0.320	0.327	0.333	0.338	0.343	0.025
	0.050	0.339	0.351	0.361	0.371	0.379	0.386	0.392	0.398	0.404	0.409	0.050
	0.100	0.434	0.446	0.456	0.464	0.472	0.479	0.485	0.491	0.496	0.500	0.100
	0.900	2.52	2.50	2.49	2.48	2.46	2.45	2.45	2.44	2.43	2.42	0.900
	0.950	3.31	3.28	3.26	3.24	3.22	3.20	3.19	3.17	3.16	3.15	0.950
	0.975	4.24	4.20	4.16	4.13	4.10	4.08	4.05	4.03	4.02	4.00	0.975
	0.990	5.73	5.67	5.61	5.56	5.52	5.48	5.44	5.41	5.38	5.36	0.990
	0.995	7.10	7.01	6.94	6.87	6.81	6.76	6.72	6.68	6.64	6.61	0.995
9	0.005	0.181	0.192	0.203	0.212	0.220	0.228	0.235	0.241	0.247	0.253	0.005
	0.010	0.216	0.228	0.239	0.248	0.257	0.265	0.272	0.278	0.284	0.289	0.010
	0.025	0.279	0.291	0.302	0.312	0.320	0.328	0.335	0.341	0.347	0.353	0.025
	0.050	0.345	0.358	0.368	0.378	0.386	0.394	0.401	0.407	0.413	0.418	0.050
	0.100	0.440	0.452	0.462	0.471	0.479	0.487	0.493	0.499	0.504	0.509	0.100
	0.900	2.40	2.38	2.36	2.35	2.34	2.33	2.32	2.31	2.30	2.30	0.900
	0.950	3.10	3.07	3.05	3.03	3.01	2.99	2.97	2.96	2.95	2.94	0.950
	0.975	3.91	3.87	3.83	3.80	3.77	3.74	3.72	3.70	3.68	3.67	0.975
	0.990	5.18	5.11	5.05	5.01	4.96	4.92	4.89	4.86	4.83	4.81	0.990
	0.995	6.31	6.23	6.15	6.09	6.03	5.98	5.94	5.90	5.86	5.83	0.995
10	0.005	0.185	0.197	0.207	0.217	0.226	0.234	0.241	0.248	0.254	0.260	0.005
	0.010	0.220	0.233	0.244	0.254	0.263	0.271	0.278	0.285	0.291	0.297	0.010
	0.025	0.284	0.296	0.308	0.318	0.327	0.335	0.342	0.349	0.355	0.361	0.025
	0.050	0.350	0.363	0.374	0.384	0.393	0.401	0.408	0.415	0.421	0.426	0.050
	0.100	0.445	0.457	0.468	0.477	0.486	0.493	0.500	0.506	0.511	0.516	0.100
	0.900	2.30	2.28	2.27	2.26	2.24	2.23	2.22	2.22	2.21	2.20	0.900
	0.950	2.94	2.91	2.89	2.86	2.84	2.83	2.81	2.80	2.79	2.77	0.950
	0.975	3.66	3.62	3.58	3.55	3.52	3.50	3.47	3.45	3.44	3.42	0.975
	0.990	4.77	4.71	4.65	4.60	4.56	4.52	4.49	4.46	4.43	4.41	0.990
	0.995	5.75	5.66	5.59	5.53	5.47	5.42	5.38	5.34	5.31	5.27	0.995

ν_2	γ \ ν_1	11	12	13	14	15	16	17	18	19	20	ν_1 / γ
11	0.005	0.188	0.200	0.212	0.222	0.231	0.239	0.247	0.254	0.260	0.266	0.005
	0.010	0.224	0.237	0.248	0.259	0.268	0.277	0.284	0.291	0.298	0.304	0.010
	0.025	0.288	0.301	0.313	0.323	0.332	0.341	0.348	0.355	0.362	0.368	0.025
	0.050	0.355	0.368	0.380	0.390	0.399	0.407	0.414	0.421	0.427	0.433	0.050
	0.100	0.449	0.462	0.473	0.482	0.491	0.499	0.506	0.512	0.518	0.523	0.100
	0.900	2.23	2.21	2.19	2.18	2.17	2.16	2.15	2.14	2.13	2.12	0.900
	0.950	2.82	2.79	2.76	2.74	2.72	2.70	2.69	2.67	2.66	2.65	0.950
	0.975	3.47	3.43	3.39	3.36	3.33	3.30	3.28	3.26	3.24	3.23	0.975
	0.990	4.46	4.40	4.34	4.29	4.25	4.21	4.18	4.15	4.12	4.10	0.990
	0.995	5.32	5.24	5.16	5.10	5.05	5.00	4.96	4.92	4.89	4.86	0.995
12	0.005	0.191	0.204	0.215	0.226	0.235	0.244	0.252	0.259	0.266	0.272	0.005
	0.010	0.227	0.241	0.253	0.263	0.273	0.281	0.289	0.297	0.303	0.309	0.010
	0.025	0.292	0.305	0.317	0.328	0.337	0.346	0.354	0.361	0.368	0.374	0.025
	0.050	0.359	0.372	0.384	0.395	0.404	0.412	0.420	0.427	0.433	0.439	0.050
	0.100	0.453	0.466	0.477	0.487	0.496	0.504	0.511	0.517	0.523	0.528	0.100
	0.900	2.17	2.15	2.13	2.12	2.10	2.09	2.08	2.08	2.07	2.06	0.900
	0.950	2.72	2.69	2.66	2.64	2.62	2.60	2.58	2.57	2.56	2.54	0.950
	0.975	3.32	3.28	3.24	3.21	3.18	3.15	3.13	3.11	3.09	3.07	0.975
	0.990	4.22	4.16	4.10	4.05	4.01	3.97	3.94	3.91	3.88	3.86	0.990
	0.995	4.99	4.91	4.84	4.77	4.72	4.67	4.63	4.59	4.56	4.53	0.995
13	0.005	0.194	0.207	0.219	0.229	0.239	0.248	0.256	0.264	0.271	0.277	0.005
	0.010	0.230	0.244	0.256	0.267	0.277	0.286	0.294	0.302	0.308	0.315	0.010
	0.025	0.295	0.309	0.321	0.332	0.342	0.351	0.359	0.366	0.373	0.379	0.025
	0.050	0.362	0.376	0.388	0.399	0.408	0.417	0.425	0.432	0.439	0.445	0.050
	0.100	0.456	0.469	0.481	0.491	0.500	0.508	0.515	0.522	0.528	0.533	0.100
	0.900	2.12	2.10	2.08	2.07	2.05	2.04	2.03	2.02	2.01	2.01	0.900
	0.950	2.63	2.60	2.58	2.55	2.53	2.51	2.50	2.48	2.47	2.46	0.950
	0.975	3.20	3.15	3.12	3.08	3.05	3.03	3.00	2.98	2.96	2.95	0.975
	0.990	4.02	3.96	3.91	3.86	3.82	3.78	3.75	3.72	3.69	3.66	0.990
	0.995	4.72	4.64	4.57	4.51	4.46	4.41	4.37	4.33	4.30	4.27	0.995
14	0.005	0.196	0.209	0.222	0.233	0.243	0.252	0.260	0.268	0.275	0.281	0.005
	0.010	0.233	0.247	0.259	0.270	0.281	0.290	0.298	0.306	0.313	0.320	0.010
	0.025	0.298	0.312	0.324	0.336	0.346	0.355	0.363	0.371	0.378	0.384	0.025
	0.050	0.365	0.379	0.392	0.403	0.412	0.421	0.429	0.437	0.443	0.449	0.050
	0.100	0.459	0.472	0.484	0.494	0.504	0.512	0.519	0.526	0.532	0.538	0.100
	0.900	2.07	2.05	2.04	2.02	2.01	2.00	1.99	1.98	1.97	1.96	0.900
	0.950	2.57	2.53	2.51	2.48	2.46	2.44	2.43	2.41	2.40	2.39	0.950
	0.975	3.09	3.05	3.01	2.98	2.95	2.92	2.90	2.88	2.86	2.84	0.975
	0.990	3.86	3.80	3.75	3.70	3.66	3.62	3.59	3.56	3.53	3.51	0.990
	0.995	4.51	4.43	4.36	4.30	4.25	4.20	4.16	4.12	4.09	4.06	0.995
15	0.005	0.198	0.212	0.224	0.235	0.246	0.255	0.264	0.272	0.279	0.286	0.005
	0.010	0.235	0.249	0.262	0.274	0.284	0.293	0.302	0.310	0.317	0.324	0.010
	0.025	0.300	0.315	0.328	0.339	0.349	0.359	0.367	0.375	0.382	0.389	0.025
	0.050	0.368	0.382	0.395	0.406	0.416	0.425	0.433	0.441	0.448	0.454	0.050
	0.100	0.461	0.475	0.487	0.498	0.507	0.515	0.523	0.530	0.536	0.542	0.100
	0.900	2.04	2.02	2.00	1.99	1.97	1.96	1.95	1.94	1.93	1.92	0.900
	0.950	2.51	2.48	2.45	2.42	2.40	2.38	2.37	2.35	2.34	2.33	0.950
	0.975	3.01	2.96	2.92	2.89	2.86	2.84	2.81	2.79	2.77	2.76	0.975
	0.990	3.73	3.67	3.61	3.56	3.52	3.49	3.45	3.42	3.40	3.37	0.990
	0.995	4.33	4.25	4.18	4.12	4.07	4.02	3.98	3.95	3.91	3.88	0.995

Table A.10 *continued*

ν_2	γ \ ν_1	11	12	13	14	15	16	17	18	19	20	ν_1 / γ
16	0.005	0.200	0.214	0.227	0.238	0.249	0.258	0.267	0.275	0.282	0.289	0.005
	0.010	0.237	0.252	0.265	0.276	0.287	0.297	0.305	0.313	0.321	0.328	0.010
	0.025	0.303	0.317	0.330	0.342	0.353	0.362	0.371	0.379	0.386	0.393	0.025
	0.050	0.370	0.385	0.398	0.409	0.419	0.429	0.437	0.445	0.451	0.458	0.050
	0.100	0.464	0.478	0.490	0.500	0.510	0.519	0.526	0.533	0.540	0.546	0.100
	0.900	2.01	1.99	1.97	1.95	1.94	1.93	1.92	1.91	1.90	1.89	0.900
	0.950	2.46	2.42	2.40	2.37	2.35	2.33	2.32	2.30	2.29	2.28	0.950
	0.975	2.93	2.89	2.85	2.82	2.79	2.76	2.74	2.72	2.70	2.68	0.975
	0.990	3.62	3.55	3.50	3.45	3.41	3.37	3.34	3.31	3.28	3.26	0.990
	0.995	4.18	4.10	4.03	3.97	3.92	3.87	3.83	3.80	3.76	3.73	0.995
17	0.005	0.202	0.216	0.229	0.240	0.251	0.261	0.270	0.278	0.286	0.293	0.005
	0.010	0.239	0.254	0.267	0.279	0.290	0.299	0.308	0.317	0.324	0.331	0.010
	0.025	0.305	0.320	0.333	0.345	0.356	0.365	0.374	0.382	0.390	0.396	0.025
	0.050	0.372	0.387	0.400	0.412	0.422	0.432	0.440	0.448	0.455	0.462	0.050
	0.100	0.466	0.480	0.492	0.503	0.513	0.522	0.529	0.537	0.543	0.549	0.100
	0.900	1.98	1.96	1.94	1.93	1.91	1.90	1.89	1.88	1.87	1.86	0.900
	0.950	2.41	2.38	2.35	2.33	2.31	2.29	2.27	2.26	2.24	2.23	0.950
	0.975	2.87	2.82	2.79	2.75	2.72	2.70	2.67	2.65	2.63	2.62	0.975
	0.990	3.52	3.46	3.40	3.35	3.31	3.27	3.24	3.21	3.19	3.16	0.990
	0.995	4.05	3.97	3.90	3.84	3.79	3.75	3.71	3.67	3.64	3.61	0.995
18	0.005	0.203	0.218	0.231	0.243	0.253	0.263	0.272	0.281	0.289	0.296	0.005
	0.010	0.241	0.256	0.269	0.281	0.292	0.302	0.311	0.320	0.327	0.335	0.010
	0.025	0.307	0.322	0.335	0.347	0.358	0.368	0.377	0.385	0.393	0.400	0.025
	0.050	0.374	0.389	0.403	0.414	0.425	0.434	0.443	0.451	0.458	0.465	0.050
	0.100	0.468	0.482	0.494	0.505	0.515	0.524	0.532	0.539	0.546	0.552	0.100
	0.900	1.95	1.93	1.92	1.90	1.89	1.87	1.86	1.85	1.84	1.84	0.900
	0.950	2.37	2.34	2.31	2.29	2.27	2.25	2.23	2.22	2.20	2.19	0.950
	0.975	2.81	2.77	2.73	2.70	2.67	2.64	2.62	2.60	2.58	2.56	0.975
	0.990	3.43	3.37	3.32	3.27	3.23	3.19	3.16	3.13	3.10	3.08	0.990
	0.995	3.94	3.86	3.79	3.73	3.68	3.64	3.60	3.56	3.53	3.50	0.995
19	0.005	0.205	0.219	0.233	0.245	0.256	0.266	0.275	0.283	0.291	0.299	0.005
	0.010	0.243	0.258	0.271	0.283	0.294	0.305	0.314	0.322	0.330	0.338	0.010
	0.025	0.308	0.324	0.337	0.350	0.361	0.371	0.380	0.388	0.396	0.403	0.025
	0.050	0.376	0.391	0.405	0.417	0.427	0.437	0.446	0.454	0.461	0.468	0.050
	0.100	0.469	0.484	0.496	0.508	0.518	0.527	0.535	0.542	0.549	0.555	0.100
	0.900	1.93	1.91	1.89	1.88	1.86	1.85	1.84	1.83	1.82	1.81	0.900
	0.950	2.34	2.31	2.28	2.26	2.23	2.21	2.20	2.18	2.17	2.16	0.950
	0.975	2.76	2.72	2.68	2.65	2.62	2.59	2.57	2.55	2.53	2.51	0.975
	0.990	3.36	3.30	3.24	3.19	3.15	3.12	3.08	3.05	3.03	3.00	0.990
	0.995	3.84	3.76	3.70	3.64	3.59	3.54	3.50	3.46	3.43	3.40	0.995
20	0.005	0.206	0.221	0.234	0.246	0.258	0.268	0.277	0.286	0.294	0.301	0.005
	0.010	0.244	0.259	0.273	0.285	0.297	0.307	0.316	0.325	0.333	0.340	0.010
	0.025	0.310	0.325	0.339	0.352	0.363	0.373	0.382	0.391	0.399	0.406	0.025
	0.050	0.378	0.393	0.407	0.419	0.430	0.439	0.448	0.456	0.464	0.471	0.050
	0.100	0.471	0.486	0.498	0.510	0.520	0.529	0.537	0.544	0.551	0.557	0.100
	0.900	1.91	1.89	1.87	1.86	1.84	1.83	1.82	1.81	1.80	1.79	0.900
	0.950	2.31	2.28	2.25	2.22	2.20	2.18	2.17	2.15	2.14	2.12	0.950
	0.975	2.72	2.68	2.64	2.60	2.57	2.55	2.52	2.50	2.48	2.46	0.975
	0.990	3.29	3.23	3.18	3.13	3.09	3.05	3.02	2.99	2.96	2.94	0.990
	0.995	3.76	3.68	3.61	3.55	3.50	3.46	3.42	3.38	3.35	3.32	0.995

ν_2	γ \ ν_1	11	12	13	14	15	16	17	18	19	20	ν_1 / γ
21	0.005	0.207	0.222	0.236	0.248	0.259	0.270	0.279	0.288	0.296	0.304	0.005
	0.010	0.245	0.261	0.275	0.287	0.299	0.309	0.319	0.327	0.335	0.343	0.010
	0.025	0.311	0.327	0.341	0.354	0.365	0.375	0.385	0.393	0.401	0.408	0.025
	0.050	0.379	0.395	0.409	0.421	0.432	0.442	0.451	0.459	0.466	0.473	0.050
	0.100	0.472	0.487	0.500	0.511	0.522	0.531	0.539	0.547	0.554	0.560	0.100
	0.900	1.90	1.87	1.86	1.84	1.83	1.81	1.80	1.79	1.78	1.78	0.900
	0.950	2.28	2.25	2.22	2.20	2.18	2.16	2.14	2.12	2.11	2.10	0.950
	0.975	2.68	2.64	2.60	2.56	2.53	2.51	2.48	2.46	2.44	2.42	0.975
	0.990	3.24	3.17	3.12	3.07	3.03	2.99	2.96	2.93	2.90	2.88	0.990
	0.995	3.68	3.60	3.54	3.48	3.43	3.38	3.34	3.31	3.27	3.24	0.995
22	0.005	0.208	0.223	0.237	0.250	0.261	0.272	0.281	0.290	0.299	0.306	0.005
	0.010	0.247	0.262	0.276	0.289	0.300	0.311	0.321	0.330	0.338	0.345	0.010
	0.025	0.313	0.329	0.343	0.355	0.367	0.377	0.387	0.395	0.404	0.411	0.025
	0.050	0.381	0.396	0.410	0.423	0.434	0.444	0.453	0.461	0.469	0.476	0.050
	0.100	0.474	0.489	0.502	0.513	0.523	0.533	0.541	0.549	0.556	0.562	0.100
	0.900	1.88	1.86	1.84	1.83	1.81	1.80	1.79	1.78	1.77	1.76	0.900
	0.950	2.26	2.23	2.20	2.17	2.15	2.13	2.11	2.10	2.08	2.07	0.950
	0.975	2.65	2.60	2.56	2.53	2.50	2.47	2.45	2.43	2.41	2.39	0.975
	0.990	3.18	3.12	3.07	3.02	2.98	2.94	2.91	2.88	2.85	2.83	0.990
	0.995	3.61	3.54	3.47	3.41	3.36	3.31	3.27	3.24	3.21	3.18	0.995
23	0.005	0.209	0.225	0.238	0.251	0.263	0.273	0.283	0.292	0.301	0.308	0.005
	0.010	0.248	0.263	0.277	0.290	0.302	0.313	0.323	0.332	0.340	0.348	0.010
	0.025	0.314	0.330	0.344	0.357	0.369	0.379	0.389	0.398	0.406	0.413	0.025
	0.050	0.382	0.398	0.412	0.424	0.435	0.446	0.455	0.463	0.471	0.478	0.050
	0.100	0.475	0.490	0.503	0.515	0.525	0.534	0.543	0.551	0.558	0.564	0.100
	0.900	1.87	1.84	1.83	1.81	1.80	1.78	1.77	1.76	1.75	1.74	0.900
	0.950	2.24	2.20	2.18	2.15	2.13	2.11	2.09	2.08	2.06	2.05	0.950
	0.975	2.62	2.57	2.53	2.50	2.47	2.44	2.42	2.39	2.37	2.36	0.975
	0.990	3.14	3.07	3.02	2.97	2.93	2.89	2.86	2.83	2.80	2.78	0.990
	0.995	3.55	3.47	3.41	3.35	3.30	3.25	3.21	3.18	3.15	3.12	0.995
24	0.005	0.210	0.226	0.240	0.252	0.264	0.275	0.285	0.294	0.302	0.310	0.005
	0.010	0.249	0.265	0.279	0.292	0.304	0.314	0.324	0.333	0.342	0.350	0.010
	0.025	0.315	0.331	0.346	0.359	0.370	0.381	0.391	0.400	0.408	0.415	0.025
	0.050	0.383	0.399	0.413	0.426	0.437	0.447	0.457	0.465	0.473	0.480	0.050
	0.100	0.476	0.491	0.504	0.516	0.527	0.536	0.545	0.552	0.560	0.566	0.100
	0.900	1.85	1.83	1.81	1.80	1.78	1.77	1.76	1.75	1.74	1.73	0.900
	0.950	2.22	2.18	2.15	2.13	2.11	2.09	2.07	2.05	2.04	2.03	0.950
	0.975	2.59	2.54	2.50	2.47	2.44	2.41	2.39	2.36	2.35	2.33	0.975
	0.990	3.09	3.03	2.98	2.93	2.89	2.85	2.82	2.79	2.76	2.74	0.990
	0.995	3.50	3.42	3.35	3.30	3.25	3.20	3.16	3.12	3.09	3.06	0.995
25	0.005	0.211	0.227	0.241	0.254	0.266	0.276	0.286	0.296	0.304	0.312	0.005
	0.010	0.250	0.266	0.280	0.293	0.305	0.316	0.326	0.335	0.344	0.352	0.010
	0.025	0.316	0.332	0.347	0.360	0.372	0.383	0.392	0.401	0.410	0.417	0.025
	0.050	0.384	0.400	0.415	0.427	0.439	0.449	0.458	0.467	0.475	0.482	0.050
	0.100	0.477	0.492	0.506	0.517	0.528	0.538	0.546	0.554	0.561	0.568	0.100
	0.900	1.84	1.82	1.80	1.79	1.77	1.76	1.75	1.74	1.73	1.72	0.900
	0.950	2.20	2.16	2.14	2.11	2.09	2.07	2.05	2.04	2.02	2.01	0.950
	0.975	2.56	2.51	2.48	2.44	2.41	2.38	2.36	2.34	2.32	2.30	0.975
	0.990	3.06	2.99	2.94	2.89	2.85	2.81	2.78	2.75	2.72	2.70	0.990
	0.995	3.45	3.37	3.30	3.25	3.20	3.15	3.11	3.08	3.04	3.01	0.995

Table A.10 *continued*

ν_2	γ \ ν_1	11	12	13	14	15	16	17	18	19	20	ν_1 / γ
30	0.005	0.215	0.231	0.246	0.259	0.271	0.283	0.293	0.303	0.312	0.320	0.005
	0.010	0.254	0.270	0.285	0.299	0.311	0.323	0.333	0.343	0.352	0.360	0.010
	0.025	0.321	0.337	0.352	0.366	0.378	0.389	0.400	0.409	0.418	0.426	0.025
	0.050	0.389	0.405	0.420	0.433	0.445	0.456	0.466	0.475	0.483	0.490	0.050
	0.100	0.482	0.497	0.511	0.523	0.534	0.544	0.553	0.561	0.568	0.575	0.100
	0.900	1.79	1.77	1.75	1.74	1.72	1.71	1.70	1.69	1.68	1.67	0.900
	0.950	2.13	2.09	2.06	2.04	2.01	1.99	1.98	1.96	1.95	1.93	0.950
	0.975	2.46	2.41	2.37	2.34	2.31	2.28	2.26	2.23	2.21	2.20	0.975
	0.990	2.91	2.84	2.79	2.74	2.70	2.66	2.63	2.60	2.57	2.55	0.990
	0.995	3.25	3.18	3.11	3.06	3.01	2.96	2.92	2.89	2.85	2.82	0.995
40	0.005	0.220	0.237	0.252	0.266	0.279	0.291	0.302	0.312	0.322	0.331	0.005
	0.010	0.259	0.276	0.292	0.306	0.319	0.331	0.342	0.353	0.362	0.371	0.010
	0.025	0.327	0.344	0.360	0.374	0.387	0.399	0.409	0.419	0.429	0.437	0.025
	0.050	0.395	0.412	0.428	0.441	0.454	0.465	0.475	0.485	0.493	0.502	0.050
	0.100	0.487	0.504	0.518	0.530	0.542	0.552	0.562	0.570	0.578	0.585	0.100
	0.900	1.74	1.71	1.70	1.68	1.66	1.65	1.64	1.62	1.61	1.61	0.900
	0.950	2.04	2.00	1.97	1.95	1.92	1.90	1.89	1.87	1.85	1.84	0.950
	0.975	2.33	2.29	2.25	2.21	2.18	2.15	2.13	2.11	2.09	2.07	0.975
	0.990	2.73	2.66	2.61	2.56	2.52	2.48	2.45	2.42	2.39	2.37	0.990
	0.995	3.03	2.95	2.89	2.83	2.78	2.74	2.70	2.66	2.63	2.60	0.995
60	0.005	0.225	0.243	0.259	0.274	0.287	0.300	0.312	0.323	0.333	0.343	0.005
	0.010	0.265	0.283	0.299	0.314	0.328	0.341	0.353	0.364	0.374	0.383	0.010
	0.025	0.333	0.351	0.368	0.383	0.396	0.409	0.420	0.431	0.441	0.450	0.025
	0.050	0.402	0.419	0.435	0.450	0.463	0.475	0.486	0.496	0.505	0.514	0.050
	0.100	0.494	0.510	0.525	0.538	0.550	0.561	0.571	0.580	0.589	0.596	0.100
	0.900	1.68	1.66	1.64	1.62	1.60	1.59	1.58	1.56	1.55	1.54	0.900
	0.950	1.95	1.92	1.89	1.86	1.84	1.82	1.80	1.78	1.76	1.75	0.950
	0.975	2.22	2.17	2.13	2.09	2.06	2.03	2.01	1.98	1.96	1.94	0.975
	0.990	2.56	2.50	2.44	2.39	2.35	2.31	2.28	2.25	2.22	2.20	0.990
	0.995	2.82	2.74	2.68	2.62	2.57	2.53	2.49	2.45	2.42	2.39	0.995
120	0.005	0.231	0.249	0.266	0.282	0.297	0.310	0.323	0.335	0.346	0.356	0.005
	0.010	0.271	0.290	0.307	0.323	0.338	0.352	0.364	0.376	0.387	0.397	0.010
	0.025	0.340	0.359	0.376	0.392	0.406	0.420	0.432	0.443	0.454	0.464	0.025
	0.050	0.409	0.427	0.444	0.459	0.473	0.486	0.497	0.508	0.518	0.527	0.050
	0.100	0.500	0.518	0.533	0.547	0.560	0.571	0.582	0.591	0.600	0.609	0.100
	0.900	1.62	1.60	1.58	1.56	1.54	1.53	1.52	1.50	1.49	1.48	0.900
	0.950	1.87	1.83	1.80	1.78	1.75	1.73	1.71	1.69	1.67	1.66	0.950
	0.975	2.10	2.05	2.01	1.98	1.94	1.92	1.89	1.87	1.84	1.82	0.975
	0.990	2.40	2.34	2.28	2.23	2.19	2.15	2.12	2.09	2.06	2.03	0.990
	0.995	2.62	2.54	2.48	2.42	2.37	2.33	2.29	2.25	2.22	2.19	0.995
∞	0.005	0.237	0.256	0.274	0.291	0.307	0.321	0.335	0.348	0.360	0.372	0.005
	0.010	0.278	0.298	0.316	0.333	0.349	0.363	0.377	0.390	0.402	0.413	0.010
	0.025	0.347	0.367	0.385	0.402	0.417	0.432	0.445	0.457	0.469	0.480	0.025
	0.050	0.416	0.436	0.453	0.469	0.484	0.498	0.510	0.522	0.532	0.543	0.050
	0.100	0.507	0.525	0.542	0.556	0.570	0.582	0.593	0.604	0.613	0.622	0.100
	0.900	1.57	1.55	1.52	1.50	1.49	1.47	1.46	1.44	1.43	1.42	0.900
	0.950	1.79	1.75	1.72	1.69	1.67	1.64	1.62	1.60	1.59	1.57	0.950
	0.975	1.99	1.94	1.90	1.87	1.83	1.80	1.78	1.75	1.73	1.71	0.975
	0.990	2.25	2.18	2.13	2.08	2.04	2.00	1.97	1.93	1.90	1.88	0.990
	0.995	2.43	2.36	2.29	2.24	2.19	2.14	2.10	2.06	2.03	2.00	0.995

ν_2	γ \ ν_1	21	22	23	24	25	30	40	60	120	∞	ν_1 / γ
1	0.005	0.102	0.103	0.104	0.105	0.106	0.109	0.113	0.118	0.122	0.127	0.005
	0.010	0.125	0.126	0.127	0.128	0.129	0.132	0.137	0.141	0.146	0.151	0.010
	0.025	0.172	0.173	0.174	0.175	0.176	0.180	0.184	0.189	0.194	0.199	0.025
	0.050	0.231	0.233	0.234	0.235	0.236	0.240	0.245	0.250	0.255	0.260	0.050
	0.100	0.338	0.339	0.340	0.342	0.343	0.347	0.353	0.358	0.364	0.370	0.100
	0.900	61.81	61.88	61.95	62.00	62.05	62.26	62.53	62.79	63.06	63.33	0.900
	0.950	248.3	248.6	248.8	249.1	249.3	250.1	251.1	252.2	253.3	254.3	0.950
	0.975	994.3	995.4	996.3	997.3	998.1	1001.	1006.	1010.	1014.	1018.	0.975
	0.990	6216.	6223.	6229.	6235.	6240.	6261.	6287.	6313.	6340.	6366.	0.990
	0.995	24866	24893	24917	24940	24961	25044	25148	25256	25360	25465	0.995
2	0.005	0.145	0.147	0.149	0.150	0.152	0.157	0.165	0.173	0.181	0.189	0.005
	0.010	0.173	0.175	0.177	0.178	0.180	0.186	0.193	0.201	0.209	0.217	0.010
	0.025	0.226	0.228	0.230	0.232	0.233	0.239	0.247	0.255	0.263	0.271	0.025
	0.050	0.288	0.290	0.292	0.294	0.295	0.302	0.309	0.317	0.326	0.334	0.050
	0.100	0.388	0.390	0.392	0.394	0.396	0.402	0.410	0.418	0.426	0.434	0.100
	0.900	9.44	9.45	9.45	9.45	9.45	9.46	9.47	9.47	9.48	9.49	0.900
	0.950	19.45	19.45	19.45	19.45	19.46	19.46	19.47	19.48	19.49	19.50	0.950
	0.975	39.45	39.45	39.45	39.46	39.46	39.46	39.47	39.48	39.49	39.50	0.975
	0.990	99.45	99.45	99.46	99.46	99.46	99.47	99.47	99.48	99.49	99.50	0.990
	0.995	199.5	199.5	199.5	199.5	199.5	199.5	199.5	199.5	199.5	199.5	0.995
3	0.005	0.175	0.177	0.179	0.181	0.183	0.191	0.201	0.211	0.222	0.234	0.005
	0.010	0.205	0.208	0.210	0.212	0.214	0.222	0.232	0.242	0.253	0.264	0.010
	0.025	0.262	0.264	0.267	0.269	0.271	0.279	0.289	0.299	0.310	0.321	0.025
	0.050	0.325	0.328	0.330	0.332	0.334	0.342	0.352	0.363	0.373	0.384	0.050
	0.100	0.423	0.425	0.428	0.430	0.432	0.439	0.449	0.459	0.470	0.480	0.100
	0.900	5.18	5.18	5.18	5.18	5.18	5.17	5.17	5.16	5.15	5.13	0.900
	0.950	8.66	8.66	8.65	8.65	8.64	8.63	8.61	8.61	8.56	8.53	0.950
	0.975	14.18	14.17	14.16	14.16	14.15	14.12	14.10	14.10	13.96	13.90	0.975
	0.990	26.78	26.77	26.75	26.73	26.72	26.70	26.67	26.76	26.27	26.13	0.990
	0.995	43.10	43.08	43.08	43.05	43.04	43.05	43.09	43.64	42.15	41.84	0.995
4	0.005	0.196	0.199	0.202	0.205	0.207	0.216	0.229	0.242	0.255	0.269	0.005
	0.010	0.229	0.232	0.235	0.237	0.239	0.249	0.261	0.274	0.287	0.301	0.010
	0.025	0.288	0.291	0.293	0.296	0.298	0.308	0.320	0.332	0.346	0.359	0.025
	0.050	0.352	0.355	0.358	0.360	0.362	0.372	0.384	0.396	0.409	0.422	0.050
	0.100	0.448	0.451	0.453	0.456	0.458	0.467	0.478	0.490	0.502	0.514	0.100
	0.900	3.84	3.84	3.83	3.83	3.83	3.82	3.80	3.79	3.78	3.76	0.900
	0.950	5.79	5.79	5.78	5.77	5.77	5.75	5.72	5.69	5.66	5.63	0.950
	0.975	8.55	8.53	8.52	8.51	8.50	8.46	8.41	8.36	8.31	8.26	0.975
	0.990	13.99	13.97	13.95	13.93	13.91	13.84	13.75	13.66	13.57	13.46	0.990
	0.995	20.13	20.09	20.06	20.03	20.01	19.89	19.76	19.63	19.50	19.33	0.995
5	0.005	0.214	0.217	0.220	0.223	0.226	0.237	0.251	0.266	0.282	0.299	0.005
	0.010	0.247	0.251	0.254	0.257	0.259	0.270	0.285	0.300	0.315	0.331	0.010
	0.025	0.308	0.311	0.314	0.317	0.320	0.330	0.344	0.359	0.374	0.390	0.025
	0.050	0.372	0.376	0.379	0.382	0.384	0.395	0.408	0.422	0.437	0.452	0.050
	0.100	0.467	0.470	0.473	0.476	0.478	0.488	0.501	0.514	0.527	0.541	0.100
	0.900	3.20	3.20	3.19	3.19	3.19	3.17	3.16	3.14	3.12	3.11	0.900
	0.950	4.55	4.54	4.53	4.53	4.52	4.50	4.46	4.43	4.40	4.37	0.950
	0.975	6.31	6.30	6.29	6.28	6.27	6.23	6.17	6.12	6.06	6.02	0.975
	0.990	9.53	9.50	9.48	9.47	9.45	9.38	9.29	9.20	9.09	9.02	0.990
	0.995	12.87	12.83	12.81	12.78	12.75	12.65	12.52	12.39	12.23	12.14	0.995

Table A.10 *continued*

ν_2	γ \ ν_1	21	22	23	24	25	30	40	60	120	∞	ν_1 / γ
6	0.005	0.228	0.231	0.235	0.238	0.241	0.253	0.269	0.286	0.304	0.323	0.005
	0.010	0.262	0.266	0.270	0.273	0.276	0.288	0.304	0.321	0.338	0.357	0.010
	0.025	0.324	0.327	0.331	0.334	0.337	0.349	0.364	0.381	0.398	0.415	0.025
	0.050	0.389	0.392	0.396	0.399	0.402	0.413	0.428	0.444	0.460	0.477	0.050
	0.100	0.482	0.485	0.488	0.491	0.494	0.505	0.519	0.533	0.548	0.564	0.100
	0.900	2.83	2.83	2.82	2.82	2.81	2.80	2.78	2.76	2.74	2.72	0.900
	0.950	3.86	3.86	3.85	3.84	3.83	3.81	3.77	3.74	3.70	3.67	0.950
	0.975	5.15	5.14	5.13	5.12	5.11	5.06	5.01	4.96	4.90	4.85	0.975
	0.990	7.37	7.35	7.33	7.31	7.30	7.23	7.14	7.05	6.96	6.88	0.990
	0.995	9.56	9.53	9.50	9.47	9.45	9.36	9.24	9.12	8.98	8.88	0.995
7	0.005	0.239	0.243	0.247	0.251	0.254	0.267	0.285	0.304	0.324	0.345	0.005
	0.010	0.275	0.279	0.283	0.286	0.289	0.303	0.320	0.339	0.358	0.379	0.010
	0.025	0.337	0.341	0.345	0.348	0.351	0.364	0.381	0.399	0.418	0.437	0.025
	0.050	0.402	0.406	0.409	0.413	0.416	0.428	0.445	0.462	0.479	0.498	0.050
	0.100	0.494	0.498	0.501	0.504	0.507	0.519	0.534	0.550	0.566	0.583	0.100
	0.900	2.59	2.58	2.58	2.58	2.57	2.56	2.54	2.51	2.49	2.47	0.900
	0.950	3.43	3.43	3.42	3.41	3.40	3.38	3.34	3.30	3.27	3.23	0.950
	0.975	4.45	4.44	4.43	4.41	4.40	4.36	4.31	4.25	4.20	4.14	0.975
	0.990	6.13	6.11	6.09	6.07	6.06	5.99	5.91	5.82	5.73	5.65	0.990
	0.995	7.72	7.69	7.67	7.64	7.62	7.53	7.42	7.31	7.18	7.08	0.995
8	0.005	0.249	0.254	0.258	0.261	0.265	0.279	0.299	0.319	0.341	0.364	0.005
	0.010	0.285	0.290	0.294	0.297	0.301	0.315	0.334	0.354	0.376	0.398	0.010
	0.025	0.348	0.352	0.356	0.360	0.363	0.377	0.395	0.415	0.435	0.456	0.025
	0.050	0.413	0.417	0.421	0.425	0.428	0.441	0.459	0.477	0.496	0.516	0.050
	0.100	0.505	0.508	0.512	0.515	0.518	0.531	0.547	0.563	0.581	0.599	0.100
	0.900	2.42	2.41	2.41	2.40	2.40	2.38	2.36	2.34	2.32	2.29	0.900
	0.950	3.14	3.13	3.12	3.12	3.11	3.08	3.04	3.01	2.97	2.93	0.950
	0.975	3.98	3.97	3.96	3.95	3.94	3.89	3.84	3.78	3.73	3.67	0.975
	0.990	5.34	5.32	5.30	5.28	5.26	5.20	5.12	5.03	4.94	4.86	0.990
	0.995	6.58	6.55	6.53	6.50	6.48	6.40	6.29	6.18	6.06	5.95	0.995
9	0.005	0.258	0.262	0.267	0.271	0.274	0.290	0.310	0.332	0.356	0.381	0.005
	0.010	0.294	0.299	0.303	0.307	0.311	0.326	0.346	0.368	0.391	0.415	0.010
	0.025	0.357	0.362	0.366	0.370	0.374	0.388	0.408	0.428	0.450	0.473	0.025
	0.050	0.423	0.427	0.431	0.435	0.438	0.452	0.471	0.490	0.511	0.532	0.050
	0.100	0.513	0.517	0.521	0.525	0.528	0.541	0.558	0.575	0.594	0.613	0.100
	0.900	2.29	2.29	2.28	2.28	2.27	2.25	2.23	2.21	2.18	2.16	0.900
	0.950	2.93	2.92	2.91	2.90	2.89	2.86	2.83	2.79	2.75	2.71	0.950
	0.975	3.65	3.64	3.63	3.61	3.60	3.56	3.51	3.45	3.39	3.33	0.975
	0.990	4.79	4.76	4.75	4.73	4.71	4.65	4.57	4.48	4.40	4.31	0.990
	0.995	5.80	5.78	5.75	5.73	5.71	5.62	5.52	5.41	5.30	5.19	0.995
10	0.005	0.265	0.270	0.275	0.279	0.283	0.299	0.321	0.344	0.370	0.397	0.005
	0.010	0.302	0.307	0.311	0.316	0.320	0.336	0.357	0.380	0.405	0.431	0.010
	0.025	0.366	0.370	0.375	0.379	0.383	0.398	0.419	0.440	0.464	0.488	0.025
	0.050	0.431	0.435	0.440	0.444	0.447	0.462	0.481	0.502	0.523	0.546	0.050
	0.100	0.521	0.525	0.529	0.533	0.536	0.550	0.567	0.586	0.605	0.625	0.100
	0.900	2.19	2.19	2.18	2.18	2.17	2.16	2.13	2.11	2.08	2.06	0.900
	0.950	2.76	2.75	2.75	2.74	2.73	2.70	2.66	2.62	2.58	2.54	0.950
	0.975	3.40	3.39	3.38	3.37	3.35	3.31	3.26	3.20	3.14	3.08	0.975
	0.990	4.38	4.36	4.34	4.33	4.31	4.25	4.17	4.08	4.00	3.91	0.990
	0.995	5.25	5.22	5.20	5.17	5.15	5.07	4.97	4.86	4.75	4.64	0.995

ν_2	ν_1 / γ	21	22	23	24	25	30	40	60	120	∞	ν_1 / γ
11	0.005	0.272	0.277	0.282	0.286	0.290	0.307	0.330	0.355	0.382	0.411	0.005
	0.010	0.309	0.314	0.319	0.323	0.327	0.344	0.367	0.391	0.417	0.445	0.010
	0.025	0.373	0.378	0.382	0.387	0.391	0.407	0.428	0.451	0.476	0.502	0.025
	0.050	0.438	0.443	0.447	0.451	0.455	0.470	0.491	0.512	0.535	0.559	0.050
	0.100	0.528	0.532	0.536	0.540	0.543	0.557	0.576	0.595	0.615	0.637	0.100
	0.900	2.12	2.11	2.11	2.10	2.10	2.08	2.05	2.03	2.00	1.97	0.900
	0.950	2.64	2.63	2.62	2.61	2.60	2.57	2.53	2.49	2.45	2.40	0.950
	0.975	3.21	3.20	3.18	3.17	3.16	3.12	3.06	3.00	2.94	2.88	0.975
	0.990	4.08	4.06	4.04	4.02	4.01	3.94	3.86	3.78	3.69	3.60	0.990
	0.995	4.83	4.80	4.78	4.76	4.74	4.65	4.55	4.44	4.34	4.23	0.995
12	0.005	0.278	0.283	0.288	0.292	0.297	0.315	0.339	0.365	0.393	0.424	0.005
	0.010	0.315	0.320	0.325	0.330	0.334	0.352	0.375	0.401	0.428	0.458	0.010
	0.025	0.379	0.384	0.389	0.394	0.398	0.415	0.437	0.461	0.487	0.514	0.025
	0.050	0.444	0.449	0.454	0.458	0.462	0.478	0.499	0.522	0.545	0.571	0.050
	0.100	0.533	0.538	0.542	0.546	0.549	0.564	0.583	0.603	0.625	0.647	0.100
	0.900	2.05	2.05	2.04	2.04	2.03	2.01	1.99	1.96	1.93	1.90	0.900
	0.950	2.53	2.52	2.51	2.51	2.50	2.47	2.43	2.38	2.34	2.30	0.950
	0.975	3.06	3.04	3.03	3.02	3.01	2.96	2.91	2.85	2.79	2.72	0.975
	0.990	3.84	3.82	3.80	3.78	3.76	3.70	3.62	3.54	3.45	3.36	0.990
	0.995	4.50	4.48	4.45	4.43	4.41	4.33	4.23	4.12	4.01	3.90	0.995
13	0.005	0.283	0.288	0.293	0.298	0.303	0.321	0.346	0.374	0.403	0.436	0.005
	0.010	0.321	0.326	0.331	0.336	0.340	0.359	0.383	0.410	0.438	0.470	0.010
	0.025	0.385	0.390	0.395	0.400	0.404	0.422	0.445	0.470	0.497	0.526	0.025
	0.050	0.450	0.455	0.460	0.464	0.468	0.485	0.507	0.530	0.555	0.581	0.050
	0.100	0.539	0.543	0.547	0.551	0.555	0.570	0.590	0.611	0.633	0.656	0.100
	0.900	2.00	1.99	1.99	1.98	1.98	1.96	1.93	1.90	1.88	1.85	0.900
	0.950	2.45	2.44	2.43	2.42	2.41	2.38	2.34	2.30	2.25	2.21	0.950
	0.975	2.93	2.92	2.91	2.89	2.88	2.84	2.78	2.72	2.66	2.60	0.975
	0.990	3.64	3.62	3.60	3.59	3.57	3.51	3.43	3.34	3.25	3.17	0.990
	0.995	4.24	4.22	4.19	4.17	4.15	4.07	3.97	3.87	3.76	3.65	0.995
14	0.005	0.288	0.293	0.298	0.303	0.308	0.327	0.353	0.382	0.413	0.447	0.005
	0.010	0.326	0.331	0.336	0.341	0.346	0.365	0.390	0.418	0.448	0.480	0.010
	0.025	0.390	0.395	0.401	0.405	0.410	0.428	0.452	0.478	0.506	0.536	0.025
	0.050	0.455	0.460	0.465	0.470	0.474	0.491	0.513	0.538	0.563	0.591	0.050
	0.100	0.543	0.548	0.552	0.556	0.560	0.576	0.596	0.618	0.640	0.665	0.100
	0.900	1.96	1.95	1.94	1.94	1.93	1.91	1.89	1.86	1.83	1.80	0.900
	0.950	2.38	2.37	2.36	2.35	2.34	2.31	2.27	2.22	2.18	2.13	0.950
	0.975	2.83	2.81	2.80	2.79	2.78	2.73	2.67	2.61	2.55	2.49	0.975
	0.990	3.48	3.46	3.44	3.43	3.41	3.35	3.27	3.18	3.09	3.00	0.990
	0.995	4.03	4.01	3.98	3.96	3.94	3.86	3.76	3.66	3.55	3.44	0.995
15	0.005	0.292	0.298	0.303	0.308	0.313	0.333	0.360	0.389	0.421	0.457	0.005
	0.010	0.330	0.336	0.341	0.346	0.351	0.370	0.397	0.425	0.456	0.491	0.010
	0.025	0.395	0.400	0.405	0.410	0.415	0.433	0.458	0.485	0.514	0.546	0.025
	0.050	0.460	0.465	0.470	0.474	0.479	0.496	0.520	0.545	0.571	0.600	0.050
	0.100	0.547	0.552	0.557	0.561	0.565	0.581	0.602	0.624	0.647	0.672	0.100
	0.900	1.92	1.91	1.90	1.90	1.89	1.87	1.85	1.82	1.79	1.76	0.900
	0.950	2.32	2.31	2.30	2.29	2.28	2.25	2.20	2.16	2.11	2.07	0.950
	0.975	2.74	2.73	2.71	2.70	2.69	2.64	2.58	2.52	2.46	2.40	0.975
	0.990	3.35	3.33	3.31	3.29	3.28	3.21	3.13	3.05	2.96	2.87	0.990
	0.995	3.86	3.83	3.81	3.79	3.77	3.69	3.58	3.48	3.37	3.26	0.995

Table A.10 *continued*

ν_2	γ \ ν_1	21	22	23	24	25	30	40	60	120	∞	ν_1 / γ
16	0.005	0.296	0.302	0.307	0.312	0.317	0.338	0.365	0.396	0.430	0.467	0.005
	0.010	0.334	0.340	0.346	0.351	0.355	0.375	0.403	0.432	0.464	0.500	0.010
	0.025	0.399	0.405	0.410	0.415	0.419	0.439	0.464	0.492	0.522	0.555	0.025
	0.050	0.464	0.469	0.474	0.479	0.483	0.501	0.525	0.551	0.579	0.608	0.050
	0.100	0.551	0.556	0.561	0.565	0.569	0.585	0.607	0.629	0.654	0.680	0.100
	0.900	1.88	1.88	1.87	1.87	1.86	1.84	1.81	1.78	1.75	1.72	0.900
	0.950	2.26	2.25	2.24	2.24	2.23	2.19	2.15	2.11	2.06	2.01	0.950
	0.975	2.67	2.65	2.64	2.63	2.61	2.57	2.51	2.45	2.38	2.32	0.975
	0.990	3.24	3.22	3.20	3.18	3.16	3.10	3.02	2.93	2.84	2.75	0.990
	0.995	3.71	3.68	3.66	3.64	3.62	3.54	3.44	3.33	3.22	3.11	0.995
17	0.005	0.299	0.305	0.311	0.316	0.321	0.342	0.371	0.402	0.437	0.476	0.005
	0.010	0.338	0.344	0.349	0.355	0.360	0.380	0.408	0.438	0.472	0.509	0.010
	0.025	0.403	0.409	0.414	0.419	0.424	0.443	0.470	0.498	0.529	0.563	0.025
	0.050	0.468	0.473	0.478	0.483	0.487	0.506	0.530	0.557	0.585	0.616	0.050
	0.100	0.555	0.560	0.564	0.569	0.573	0.589	0.611	0.635	0.659	0.686	0.100
	0.900	1.85	1.85	1.84	1.84	1.83	1.81	1.78	1.75	1.72	1.69	0.900
	0.950	2.22	2.21	2.20	2.19	2.18	2.15	2.10	2.06	2.01	1.96	0.950
	0.975	2.60	2.59	2.57	2.56	2.55	2.50	2.44	2.38	2.32	2.25	0.975
	0.990	3.14	3.12	3.10	3.08	3.07	3.00	2.92	2.83	2.75	2.65	0.990
	0.995	3.58	3.56	3.53	3.51	3.49	3.41	3.31	3.21	3.10	2.98	0.995
18	0.005	0.303	0.309	0.315	0.320	0.325	0.347	0.376	0.408	0.444	0.484	0.005
	0.010	0.341	0.347	0.353	0.359	0.364	0.385	0.413	0.444	0.479	0.517	0.010
	0.025	0.406	0.412	0.418	0.423	0.428	0.448	0.475	0.504	0.536	0.571	0.025
	0.050	0.471	0.477	0.482	0.487	0.491	0.510	0.535	0.562	0.592	0.624	0.050
	0.100	0.558	0.563	0.568	0.572	0.576	0.593	0.615	0.639	0.665	0.693	0.100
	0.900	1.83	1.82	1.82	1.81	1.80	1.78	1.75	1.72	1.69	1.66	0.900
	0.950	2.18	2.17	2.16	2.15	2.14	2.11	2.06	2.02	1.97	1.92	0.950
	0.975	2.54	2.53	2.52	2.50	2.49	2.44	2.38	2.32	2.26	2.19	0.975
	0.990	3.05	3.03	3.02	3.00	2.98	2.92	2.84	2.75	2.66	2.57	0.990
	0.995	3.47	3.45	3.42	3.40	3.38	3.30	3.20	3.10	2.99	2.87	0.995
19	0.005	0.306	0.312	0.318	0.323	0.329	0.351	0.381	0.414	0.451	0.492	0.005
	0.010	0.344	0.351	0.357	0.362	0.367	0.389	0.418	0.450	0.485	0.525	0.010
	0.025	0.409	0.416	0.421	0.426	0.431	0.452	0.479	0.509	0.542	0.578	0.025
	0.050	0.474	0.480	0.485	0.490	0.495	0.514	0.540	0.567	0.597	0.630	0.050
	0.100	0.561	0.566	0.571	0.575	0.579	0.597	0.619	0.644	0.670	0.698	0.100
	0.900	1.81	1.80	1.79	1.79	1.78	1.76	1.73	1.70	1.67	1.63	0.900
	0.950	2.14	2.13	2.12	2.11	2.11	2.07	2.03	1.98	1.93	1.88	0.950
	0.975	2.49	2.48	2.46	2.45	2.44	2.39	2.33	2.27	2.20	2.13	0.975
	0.990	2.98	2.96	2.94	2.92	2.91	2.84	2.76	2.67	2.58	2.49	0.990
	0.995	3.37	3.35	3.33	3.31	3.29	3.21	3.11	3.00	2.89	2.78	0.995
20	0.005	0.308	0.315	0.321	0.327	0.332	0.354	0.385	0.419	0.457	0.500	0.005
	0.010	0.347	0.354	0.360	0.365	0.370	0.392	0.422	0.455	0.492	0.532	0.010
	0.025	0.412	0.419	0.424	0.430	0.435	0.456	0.484	0.514	0.548	0.585	0.025
	0.050	0.477	0.483	0.488	0.493	0.498	0.518	0.544	0.572	0.603	0.637	0.050
	0.100	0.563	0.569	0.573	0.578	0.582	0.600	0.623	0.648	0.675	0.704	0.100
	0.900	1.79	1.78	1.77	1.77	1.76	1.74	1.71	1.68	1.64	1.61	0.900
	0.950	2.11	2.10	2.09	2.08	2.07	2.04	1.99	1.95	1.90	1.84	0.950
	0.975	2.45	2.43	2.42	2.41	2.40	2.35	2.29	2.22	2.16	2.09	0.975
	0.990	2.92	2.90	2.88	2.86	2.84	2.78	2.69	2.61	2.52	2.42	0.990
	0.995	3.29	3.27	3.24	3.22	3.20	3.12	3.02	2.92	2.81	2.69	0.995

ν_2	γ \ ν_1	21	22	23	24	25	30	40	60	120	∞	ν_1 / γ
21	0.005	0.311	0.318	0.324	0.329	0.335	0.358	0.389	0.424	0.463	0.507	0.005
	0.010	0.350	0.356	0.363	0.368	0.374	0.396	0.426	0.460	0.497	0.539	0.010
	0.025	0.415	0.421	0.427	0.433	0.438	0.459	0.488	0.519	0.554	0.592	0.025
	0.050	0.480	0.486	0.491	0.496	0.501	0.521	0.548	0.577	0.608	0.643	0.050
	0.100	0.566	0.571	0.576	0.581	0.585	0.603	0.626	0.652	0.679	0.709	0.100
	0.900	1.77	1.76	1.75	1.75	1.74	1.72	1.69	1.66	1.62	1.59	0.900
	0.950	2.08	2.07	2.06	2.05	2.05	2.01	1.96	1.92	1.87	1.81	0.950
	0.975	2.41	2.39	2.38	2.37	2.36	2.31	2.25	2.18	2.11	2.04	0.975
	0.990	2.86	2.84	2.82	2.80	2.79	2.72	2.64	2.55	2.46	2.36	0.990
	0.995	3.22	3.19	3.17	3.15	3.13	3.05	2.95	2.84	2.73	2.61	0.995
22	0.005	0.313	0.320	0.326	0.332	0.338	0.361	0.393	0.428	0.468	0.514	0.005
	0.010	0.352	0.359	0.365	0.371	0.376	0.399	0.430	0.464	0.503	0.546	0.010
	0.025	0.418	0.424	0.430	0.436	0.441	0.462	0.491	0.523	0.559	0.598	0.025
	0.050	0.482	0.488	0.494	0.499	0.504	0.524	0.551	0.581	0.613	0.649	0.050
	0.100	0.568	0.573	0.578	0.583	0.588	0.606	0.630	0.655	0.683	0.714	0.100
	0.900	1.75	1.74	1.74	1.73	1.73	1.70	1.67	1.64	1.60	1.57	0.900
	0.950	2.06	2.05	2.04	2.03	2.02	1.98	1.94	1.89	1.84	1.78	0.950
	0.975	2.37	2.36	2.34	2.33	2.32	2.27	2.21	2.14	2.08	2.00	0.975
	0.990	2.81	2.78	2.77	2.75	2.73	2.67	2.58	2.50	2.40	2.31	0.990
	0.995	3.15	3.12	3.10	3.08	3.06	2.98	2.88	2.77	2.66	2.55	0.995
23	0.005	0.316	0.322	0.329	0.335	0.340	0.364	0.396	0.433	0.474	0.521	0.005
	0.010	0.355	0.362	0.368	0.374	0.379	0.402	0.434	0.469	0.508	0.552	0.010
	0.025	0.420	0.427	0.433	0.438	0.443	0.465	0.495	0.528	0.564	0.604	0.025
	0.050	0.485	0.491	0.496	0.502	0.507	0.527	0.555	0.585	0.617	0.654	0.050
	0.100	0.570	0.576	0.581	0.585	0.590	0.608	0.633	0.659	0.687	0.719	0.100
	0.900	1.74	1.73	1.72	1.72	1.71	1.69	1.66	1.62	1.59	1.55	0.900
	0.950	2.04	2.02	2.01	2.01	2.00	1.96	1.91	1.86	1.81	1.76	0.950
	0.975	2.34	2.33	2.31	2.30	2.29	2.24	2.18	2.11	2.04	1.97	0.975
	0.990	2.76	2.74	2.72	2.70	2.69	2.62	2.54	2.45	2.35	2.26	0.990
	0.995	3.09	3.06	3.04	3.02	3.00	2.92	2.82	2.71	2.60	2.48	0.995
24	0.005	0.318	0.325	0.331	0.337	0.343	0.367	0.400	0.437	0.479	0.527	0.005
	0.010	0.357	0.364	0.370	0.376	0.382	0.405	0.437	0.473	0.513	0.558	0.010
	0.025	0.422	0.429	0.435	0.441	0.446	0.468	0.498	0.531	0.568	0.610	0.025
	0.050	0.487	0.493	0.499	0.504	0.509	0.530	0.558	0.588	0.622	0.659	0.050
	0.100	0.572	0.578	0.583	0.588	0.592	0.611	0.635	0.662	0.691	0.723	0.100
	0.900	1.72	1.71	1.71	1.70	1.70	1.67	1.64	1.61	1.57	1.53	0.900
	0.950	2.01	2.00	1.99	1.98	1.97	1.94	1.89	1.84	1.79	1.73	0.950
	0.975	2.31	2.30	2.28	2.27	2.26	2.21	2.15	2.08	2.01	1.94	0.975
	0.990	2.72	2.70	2.68	2.66	2.64	2.58	2.49	2.40	2.31	2.21	0.990
	0.995	3.04	3.01	2.99	2.97	2.95	2.87	2.77	2.66	2.55	2.43	0.995
25	0.005	0.320	0.327	0.333	0.339	0.345	0.369	0.403	0.441	0.483	0.533	0.005
	0.010	0.359	0.366	0.372	0.378	0.384	0.408	0.440	0.477	0.517	0.564	0.010
	0.025	0.424	0.431	0.437	0.443	0.448	0.471	0.501	0.535	0.573	0.615	0.025
	0.050	0.489	0.495	0.501	0.506	0.511	0.532	0.561	0.592	0.626	0.664	0.050
	0.100	0.574	0.580	0.585	0.590	0.594	0.613	0.638	0.665	0.694	0.727	0.100
	0.900	1.71	1.70	1.70	1.69	1.68	1.66	1.63	1.59	1.56	1.52	0.900
	0.950	2.00	1.98	1.97	1.96	1.96	1.92	1.87	1.82	1.77	1.71	0.950
	0.975	2.28	2.27	2.26	2.24	2.23	2.18	2.12	2.05	1.98	1.91	0.975
	0.990	2.68	2.66	2.64	2.62	2.60	2.54	2.45	2.36	2.27	2.17	0.990
	0.995	2.99	2.96	2.94	2.92	2.90	2.82	2.72	2.61	2.50	2.38	0.995

Table A.10 *continued*

ν_2	ν_1 / γ	21	22	23	24	25	30	40	60	120	∞	ν_1 / γ
30	0.005	0.328	0.335	0.342	0.349	0.355	0.381	0.416	0.457	0.504	0.559	0.005
	0.010	0.368	0.375	0.382	0.388	0.394	0.419	0.454	0.493	0.538	0.589	0.010
	0.025	0.433	0.440	0.447	0.453	0.458	0.482	0.515	0.551	0.592	0.639	0.025
	0.050	0.497	0.504	0.510	0.516	0.521	0.543	0.573	0.606	0.643	0.685	0.050
	0.100	0.582	0.588	0.593	0.598	0.603	0.622	0.649	0.678	0.710	0.745	0.100
	0.900	1.66	1.65	1.64	1.64	1.63	1.61	1.57	1.54	1.50	1.46	0.900
	0.950	1.92	1.91	1.90	1.89	1.88	1.84	1.79	1.74	1.68	1.62	0.950
	0.975	2.18	2.16	2.15	2.14	2.12	2.07	2.01	1.94	1.87	1.79	0.975
	0.990	2.53	2.51	2.49	2.47	2.45	2.39	2.30	2.21	2.11	2.01	0.990
	0.995	2.80	2.77	2.75	2.73	2.71	2.63	2.52	2.42	2.30	2.18	0.995
40	0.005	0.339	0.347	0.355	0.362	0.368	0.396	0.436	0.481	0.534	0.599	0.005
	0.010	0.379	0.387	0.394	0.401	0.408	0.435	0.473	0.517	0.567	0.628	0.010
	0.025	0.445	0.453	0.459	0.466	0.472	0.498	0.533	0.573	0.620	0.674	0.025
	0.050	0.509	0.516	0.522	0.529	0.534	0.558	0.591	0.627	0.669	0.717	0.050
	0.100	0.592	0.598	0.604	0.610	0.615	0.636	0.664	0.696	0.731	0.772	0.100
	0.900	1.60	1.59	1.58	1.57	1.57	1.54	1.51	1.47	1.42	1.38	0.900
	0.950	1.83	1.81	1.80	1.79	1.78	1.74	1.69	1.64	1.58	1.51	0.950
	0.975	2.05	2.03	2.02	2.01	1.99	1.94	1.88	1.80	1.72	1.64	0.975
	0.990	2.35	2.33	2.31	2.29	2.27	2.20	2.11	2.02	1.92	1.80	0.990
	0.995	2.57	2.55	2.52	2.50	2.48	2.40	2.30	2.18	2.06	1.93	0.995
60	0.005	0.352	0.361	0.369	0.376	0.383	0.414	0.458	0.510	0.572	0.652	0.005
	0.010	0.392	0.401	0.409	0.416	0.423	0.453	0.495	0.545	0.604	0.679	0.010
	0.025	0.458	0.466	0.474	0.481	0.487	0.515	0.555	0.600	0.654	0.720	0.025
	0.050	0.522	0.529	0.536	0.543	0.549	0.575	0.611	0.652	0.700	0.759	0.050
	0.100	0.604	0.610	0.616	0.622	0.628	0.650	0.682	0.717	0.757	0.807	0.100
	0.900	1.53	1.53	1.52	1.51	1.50	1.48	1.44	1.40	1.35	1.29	0.900
	0.950	1.73	1.72	1.71	1.70	1.69	1.65	1.59	1.53	1.47	1.39	0.950
	0.975	1.93	1.91	1.90	1.88	1.87	1.82	1.74	1.67	1.58	1.48	0.975
	0.990	2.17	2.15	2.13	2.12	2.10	2.03	1.94	1.84	1.73	1.60	0.990
	0.995	2.36	2.33	2.31	2.29	2.27	2.19	2.08	1.96	1.83	1.69	0.995
120	0.005	0.366	0.376	0.384	0.393	0.401	0.435	0.485	0.545	0.623	0.733	0.005
	0.010	0.407	0.416	0.425	0.433	0.441	0.474	0.522	0.579	0.652	0.755	0.010
	0.025	0.473	0.482	0.490	0.498	0.505	0.536	0.580	0.633	0.698	0.788	0.025
	0.050	0.536	0.544	0.552	0.559	0.565	0.594	0.634	0.682	0.740	0.819	0.050
	0.100	0.616	0.623	0.630	0.636	0.642	0.667	0.702	0.742	0.791	0.856	0.100
	0.900	1.47	1.46	1.46	1.45	1.44	1.41	1.37	1.32	1.26	1.19	0.900
	0.950	1.64	1.63	1.62	1.61	1.60	1.55	1.50	1.43	1.35	1.25	0.950
	0.975	1.81	1.79	1.77	1.76	1.75	1.69	1.61	1.53	1.43	1.31	0.975
	0.990	2.01	1.99	1.97	1.95	1.93	1.86	1.76	1.66	1.53	1.38	0.990
	0.995	2.16	2.13	2.11	2.09	2.07	1.98	1.87	1.75	1.61	1.43	0.995
∞	0.005	0.383	0.393	0.403	0.412	0.421	0.460	0.518	0.592	0.699	1.000	0.005
	0.010	0.424	0.434	0.443	0.452	0.461	0.498	0.554	0.624	0.724	1.000	0.010
	0.025	0.490	0.499	0.508	0.517	0.525	0.560	0.611	0.675	0.763	1.000	0.025
	0.050	0.552	0.561	0.569	0.577	0.584	0.616	0.663	0.720	0.798	1.000	0.050
	0.100	0.630	0.638	0.646	0.652	0.659	0.687	0.727	0.774	0.839	1.000	0.100
	0.900	1.41	1.40	1.39	1.38	1.38	1.34	1.29	1.24	1.17	1.00	0.900
	0.950	1.56	1.54	1.53	1.52	1.51	1.46	1.39	1.32	1.22	1.00	0.950
	0.975	1.69	1.67	1.66	1.64	1.63	1.57	1.48	1.39	1.27	1.00	0.975
	0.990	1.85	1.83	1.81	1.79	1.77	1.70	1.59	1.47	1.32	1.00	0.990
	0.995	1.97	1.95	1.92	1.90	1.88	1.79	1.67	1.53	1.36	1.00	0.995

Table A.11 Values of the Correlation Transformation[†]

$$Z(r) = 0.5 \log_e [(1 + r)/(1 - r)]$$

$$Z(-r) = -Z(r)$$

$$Z(0.535) = 0.59712$$

$$Z(-0.535) = -0.59712$$

[†] Based on calculations performed by the computing staff of the Department of Experimental Statistics, New Mexico State University.

Table A.11 *continued*

r	$Z(r)$	r	$Z(r)$	r	$Z(r)$	r	$Z(r)$
0.00000	0.00000	0.40000	0.42365	0.65000	0.77530	0.85000	1.25615
0.01000	0.01000	0.41000	0.43561	0.65500	0.78401	0.85500	1.27445
0.02000	0.02000	0.42000	0.44769	0.66000	0.79281	0.86000	1.29334
0.03000	0.03001	0.43000	0.45990	0.66500	0.80172	0.86500	1.31287
0.04000	0.04002	0.44000	0.47223	0.67000	0.81074	0.87000	1.33308
0.05000	0.05004	0.45000	0.48470	0.67500	0.81987	0.87500	1.35403
0.06000	0.06007	0.46000	0.49731	0.68000	0.82911	0.88000	1.37577
0.07000	0.07011	0.47000	0.51007	0.68500	0.83847	0.88500	1.39838
0.08000	0.08017	0.48000	0.52298	0.69000	0.84796	0.89000	1.42193
0.09000	0.09024	0.49000	0.53606	0.69500	0.85756	0.89500	1.44651
0.10000	0.10034	0.50000	0.54931	0.70000	0.86730	0.90000	1.47222
0.11000	0.11045	0.50500	0.55600	0.70500	0.87717	0.90500	1.49918
0.12000	0.12058	0.51000	0.56273	0.71000	0.88718	0.91000	1.52752
0.13000	0.13074	0.51500	0.56951	0.71500	0.89734	0.91500	1.55741
0.14000	0.14093	0.52000	0.57634	0.72000	0.90764	0.92000	1.58903
0.15000	0.15114	0.52500	0.58322	0.72500	0.91811	0.92500	1.62260
0.16000	0.16139	0.53000	0.59015	0.73000	0.92873	0.93000	1.65839
0.17000	0.17167	0.53500	0.59712	0.73500	0.93952	0.93500	1.69674
0.18000	0.18198	0.54000	0.60416	0.74000	0.95048	0.94000	1.73805
0.19000	0.19234	0.54500	0.61124	0.74500	0.96162	0.94500	1.78284
0.20000	0.20273	0.55000	0.61838	0.75000	0.97296	0.95000	1.83178
0.21000	0.21317	0.55500	0.62558	0.75500	0.98448	0.95500	1.88574
0.22000	0.22366	0.56000	0.63283	0.76000	0.99622	0.96000	1.94591
0.23000	0.23419	0.56500	0.64015	0.76500	1.00816	0.96500	2.01395
0.24000	0.24477	0.57000	0.64752	0.77000	1.02033	0.97000	2.09230
0.25000	0.25541	0.57500	0.65496	0.77500	1.03273	0.97500	2.18472
0.26000	0.26611	0.58000	0.66246	0.78000	1.04537	0.98000	2.29756
0.27000	0.27686	0.58500	0.67003	0.78500	1.05827	0.98500	2.44266
0.28000	0.28768	0.59000	0.67767	0.79000	1.07143	0.99000	2.64665
0.29000	0.29857	0.59500	0.68537	0.79500	1.08488	0.99100	2.69958
0.30000	0.30952	0.60000	0.69315	0.80000	1.09861	0.99200	2.75873
0.31000	0.32055	0.60500	0.70100	0.80500	1.11266	0.99300	2.82574
0.32000	0.33165	0.61000	0.70892	0.81000	1.12703	0.99400	2.90307
0.33000	0.34283	0.61500	0.71692	0.81500	1.14174	0.99500	2.99448
0.34000	0.35409	0.62000	0.72501	0.82000	1.15682	0.99600	3.10630
0.35000	0.36544	0.62500	0.73317	0.82500	1.17227	0.99700	3.25039
0.36000	0.37689	0.63000	0.74142	0.83000	1.18814	0.99800	3.45338
0.37000	0.38842	0.63500	0.74975	0.83500	1.20443	0.99900	3.80020
0.38000	0.40006	0.64000	0.75817	0.84000	1.22117	0.99950	4.14690
0.39000	0.41180	0.64500	0.76669	0.84500	1.23840	0.99990	4.95172
ρ	$Z(\rho)$	ρ	$Z(\rho)$	ρ	$Z(\rho)$	ρ	$Z(\rho)$

Table A.12 Critical Points for Testing Zero Correlation in a Bivariate Normal Distribution†

Let R_{XY} denote the correlation coefficient in a random sample of n pairs from a bivariate normal.

Entries are $r_{\nu;\gamma}$ where $\nu = n - 2$ and $P[R_{XY} \leqslant r_{\nu;\gamma}] = \gamma$.

For a sample of size $n = 60$, $\nu = 58$ and $r_{58;0.95} = 0.2144$ and $r_{58;0.995} = 0.3301$.

† Based on calculations performed by the computing staff of the Department of Experimental Statistics, New Mexico State University.

Table A.12 *continued*

	γ			
ν	0.95	0.975	0.99	0.995
1	0.9877	0.9969	0.9995	0.9999
2	0.9000	0.9500	0.9800	0.9900
3	0.8054	0.8783	0.9343	0.9587
4	0.7293	0.8114	0.8822	0.9172
5	0.6694	0.7545	0.8329	0.8745
6	0.6215	0.7067	0.7887	0.8343
7	0.5822	0.6664	0.7498	0.7977
8	0.5493	0.6319	0.7155	0.7646
9	0.5214	0.6021	0.6851	0.7348
10	0.4973	0.5760	0.6581	0.7079
11	0.4762	0.5529	0.6339	0.6835
12	0.4575	0.5324	0.6120	0.6614
13	0.4409	0.5140	0.5923	0.6411
14	0.4259	0.4973	0.5742	0.6226
15	0.4124	0.4822	0.5577	0.6055
16	0.4000	0.4683	0.5426	0.5897
17	0.3887	0.4555	0.5285	0.5751
18	0.3783	0.4438	0.5155	0.5614
19	0.3687	0.4328	0.5034	0.5487
20	0.3598	0.4227	0.4921	0.5368
21	0.3515	0.4132	0.4815	0.5256
22	0.3438	0.4043	0.4716	0.5151
23	0.3365	0.3961	0.4622	0.5052
24	0.3297	0.3882	0.4534	0.4958
25	0.3233	0.3809	0.4451	0.4869
26	0.3172	0.3739	0.4372	0.4785
27	0.3115	0.3673	0.4297	0.4705
28	0.3061	0.3610	0.4226	0.4629
29	0.3009	0.3550	0.4158	0.4556
30	0.2960	0.3494	0.4093	0.4487
31	0.2913	0.3440	0.4031	0.4421
32	0.2869	0.3388	0.3973	0.4357
33	0.2826	0.3338	0.3916	0.4297
34	0.2785	0.3291	0.3862	0.4238
35	0.2746	0.3246	0.3810	0.4182
36	0.2709	0.3202	0.3760	0.4128
37	0.2673	0.3160	0.3712	0.4076
38	0.2638	0.3120	0.3665	0.4026
39	0.2605	0.3081	0.3621	0.3978
40	0.2573	0.3044	0.3578	0.3932
48	0.2353	0.2787	0.3281	0.3610
58	0.2144	0.2542	0.2997	0.3301
68	0.1982	0.2352	0.2776	0.3060
78	0.1852	0.2199	0.2597	0.2864
88	0.1745	0.2072	0.2449	0.2702
98	0.1654	0.1966	0.2324	0.2565
108	0.1577	0.1874	0.2216	0.2446
118	0.1509	0.1793	0.2122	0.2343
128	0.1449	0.1723	0.2039	0.2252
138	0.1396	0.1660	0.1965	0.2170
148	0.1348	0.1603	0.1898	0.2097
200	0.1161	0.1381	0.1636	0.1809
300	0.0948	0.1129	0.1338	0.1480
400	0.0822	0.0978	0.1160	0.1283
500	0.0735	0.0875	0.1038	0.1149
1000	0.0520	0.0619	0.0735	0.0813

APPENDIX B

Appendix B.1 Graphs of the Power Function of Student T Tests[†]

Charts are for *one-sided* tests of size α. Enter with δ on horizontal scale, project upward to intersection with curve for the degrees of freedom on the T variable then project horizontally to left scale for power.

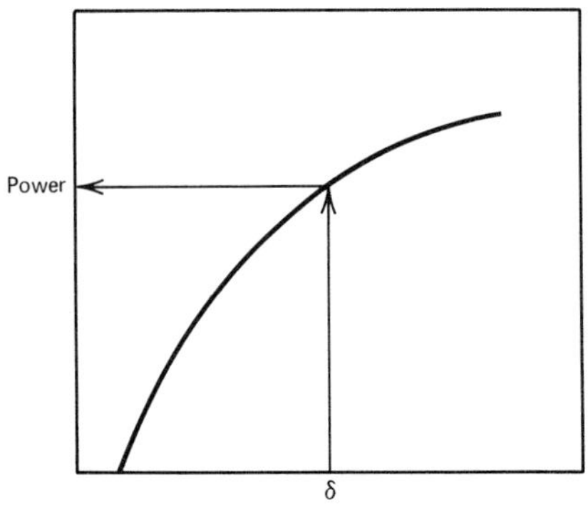

The chart for $\alpha/2$ may be used for *two-sided* tests of size α; a maximum error of $\alpha/2$ occurs as δ approaches 0 and diminishes for δ greater than unity.

[†] Reproduced with permission from Donald B. Owen, *Handbook of Statistical Tables*, Addison-Wesley, Reading, Mass., 1962. Courtesy of U.S. Atomic Energy Commission.

$\delta = |\mu - \mu_0|\sqrt{n}/\sigma$

ν = degrees of freedom

$\alpha = 0.005$

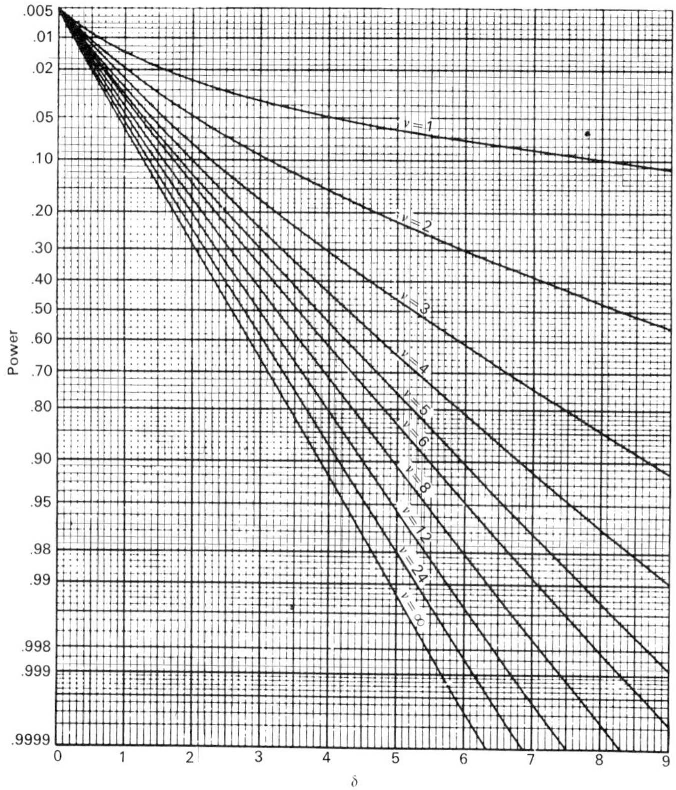

$$\delta = |\mu - \mu_0|\sqrt{n}/\sigma$$

ν = degrees of freedom

$\alpha = 0.01$

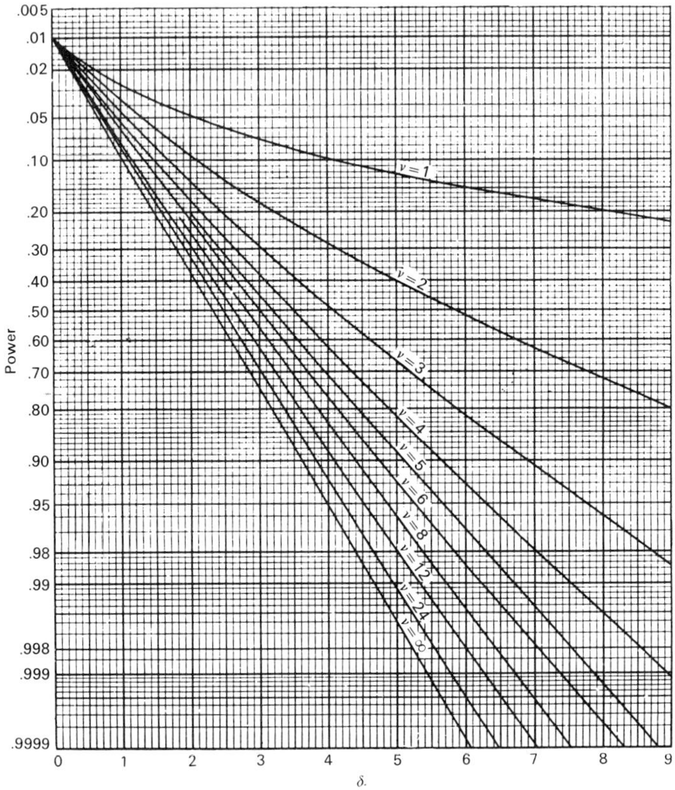

$$\delta = |\mu - \mu_0|\sqrt{n}/\sigma$$

ν = degrees of freedom

$\alpha = 0.025$

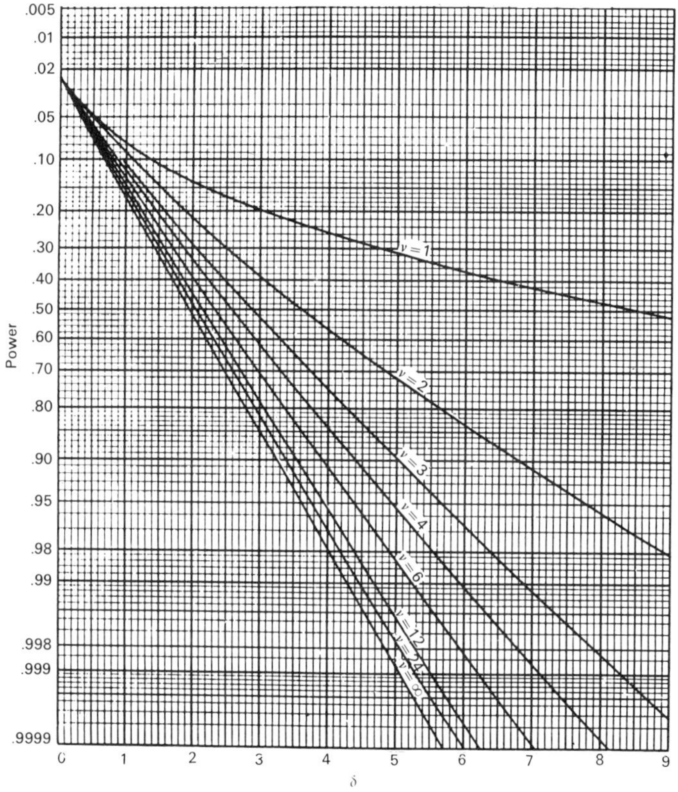

Appendix B.1 *continued*

$\delta = |\mu - \mu_0|\sqrt{n}/\sigma$

ν = degrees of freedom

$\alpha = 0.05$

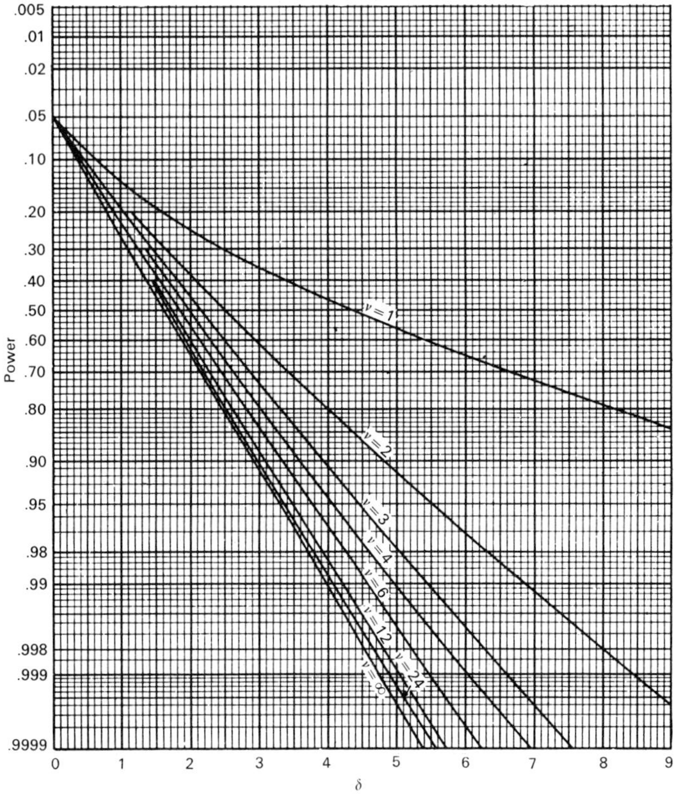

Appendix B.2 Graphs of the Power Function of Tests about a Single Normal Variance†

Charts are for *one-sided* tests of size α. Those for $R_a \leqslant 1$ correspond to the alternative $H_1 : \sigma^2 < \sigma_0^2$ while those for $R_a > 1$ correspond to the alternative $H_1 : \sigma^2 > \sigma_0^2$.
Enter on the horizontal scale of the graphs with $R_a = \sigma_a/\sigma_0$ where σ_0 is the null value and σ_a is a value of σ specified by the alternative, project vertically to the curve for the appropriate degrees of freedom, and then project horizontally to the left scale for the power.

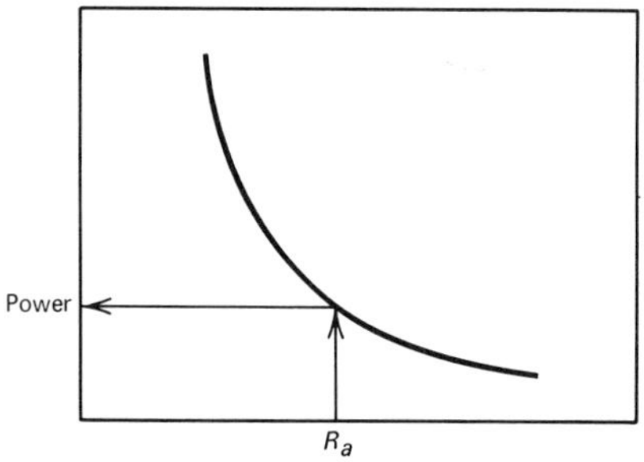

The chart for $\alpha/2$ may be used for *two-sided* tests of size α; a maximum error of $\alpha/2$ occurs as R_a approaches unity, and diminishes as the power exceeds 0.30.

† From *Concepts of Statistical Inference*, by W. C. Guenther. Copyright © 1965 by McGraw-Hill, Inc. Used by permission of McGraw-Hill Book Company.

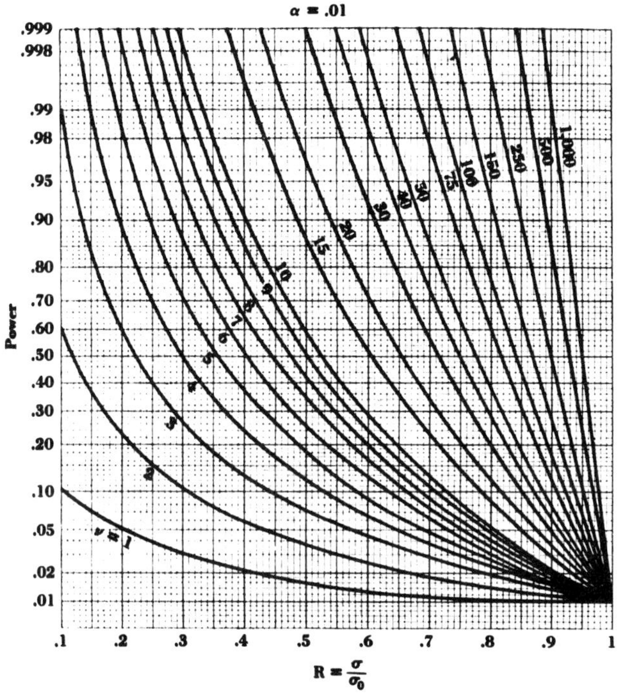
α = .01
.999
.998
.99
.98
.95
.90
.80
.70
.60
.50
.40
.30
.20
.10
.05
.02
.01
Power
ν = 1
2
3
4
5
6
7
8
9
10
15
20
30
40
50
75
100
150
250
500
1,000
.1
.2
.3
.4
.5
.6
.7
.8
.9
1
R = σ/σ0

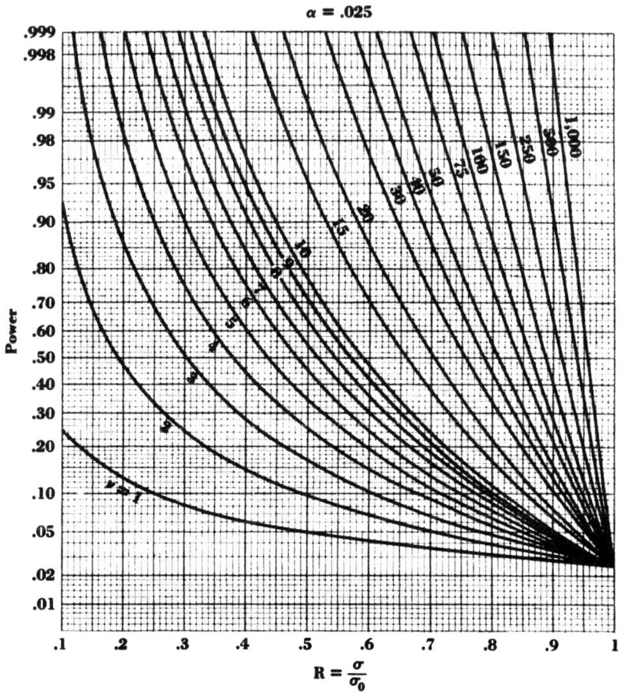
α = .025
.999
.998
.99
.98
.95
.90
.80
.70
.60
.50
.40
.30
.20
.10
.05
.02
.01
Power
ν = 1
2
3
4
5
6
7
8
9
10
15
20
30
40
50
75
100
150
250
500
1,000
.1
.2
.3
.4
.5
.6
.7
.8
.9
1
R = σ/σ0

Appendix B.2 *continued*

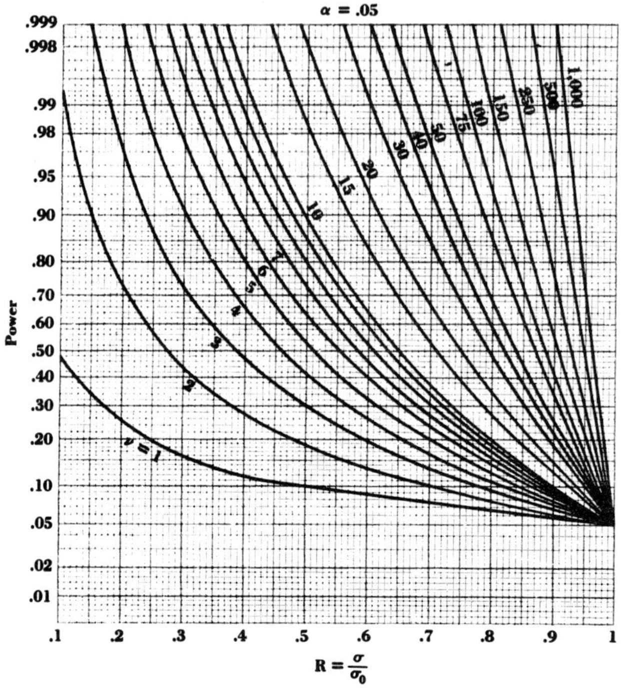

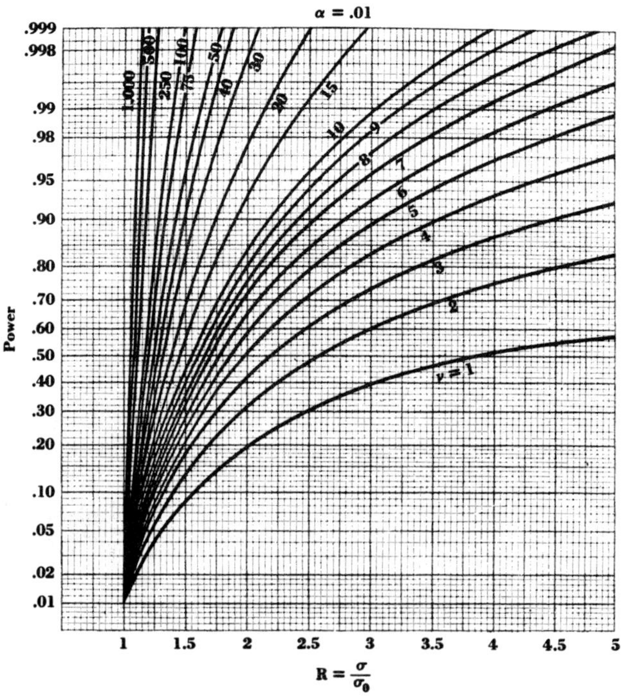

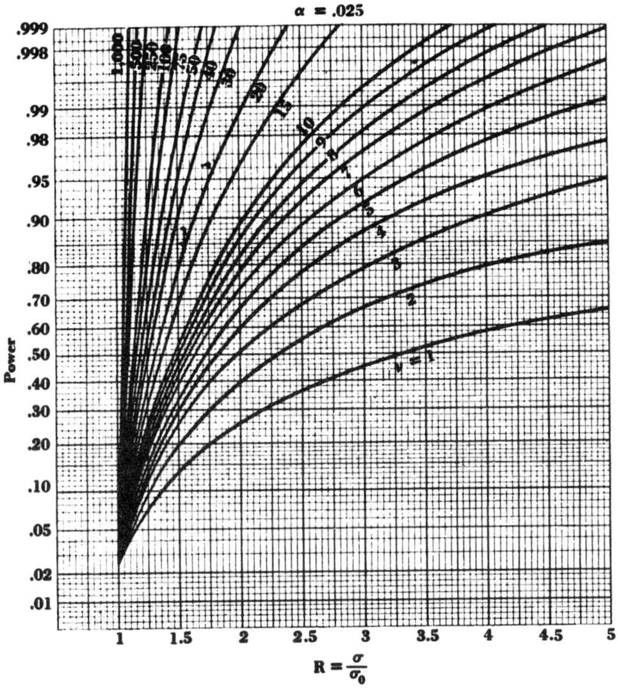
α = .025
Power
R = σ/σ0
ν = 1

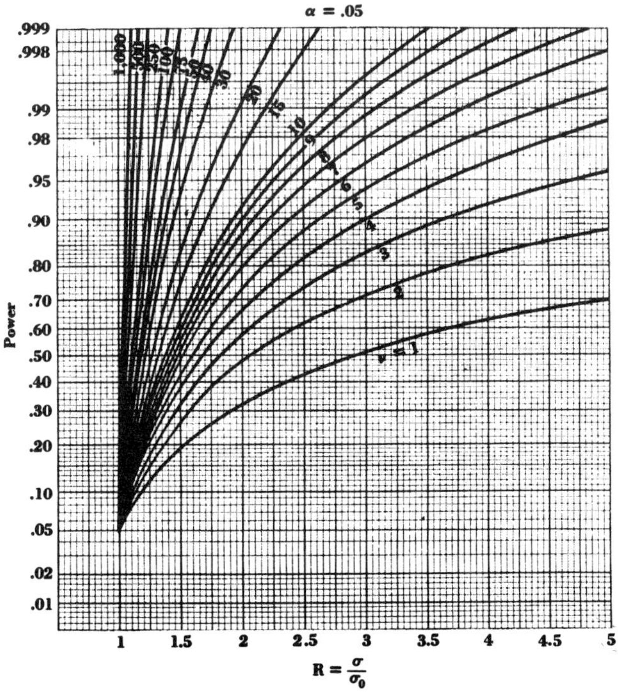
α = .05
Power
R = σ/σ0
ν = 1

Appendix B.3 Confidence Belts for Poisson Means

For the values of α shown, a pair of curves gives a two-sided *CI*. One member of each pair gives a one-sided *CI* for $\alpha/2$.
Enter number of successes, $x.$, on vertical scale, project horizontally to the intersection with the appropriate pair of curves, then project vertically down to horizontal scale. Divide points on horizontal scale by n to obtain the confidence limits.

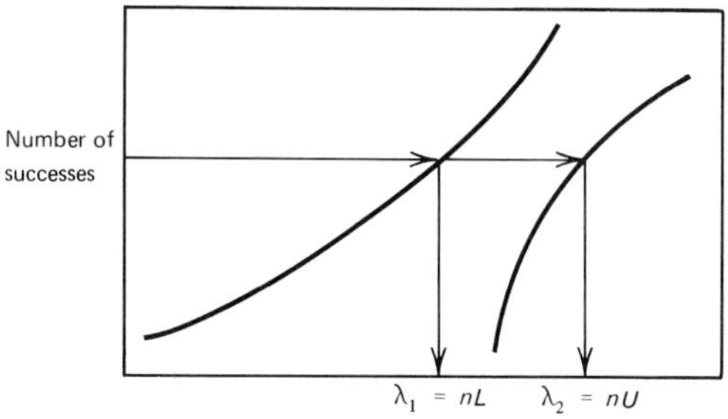

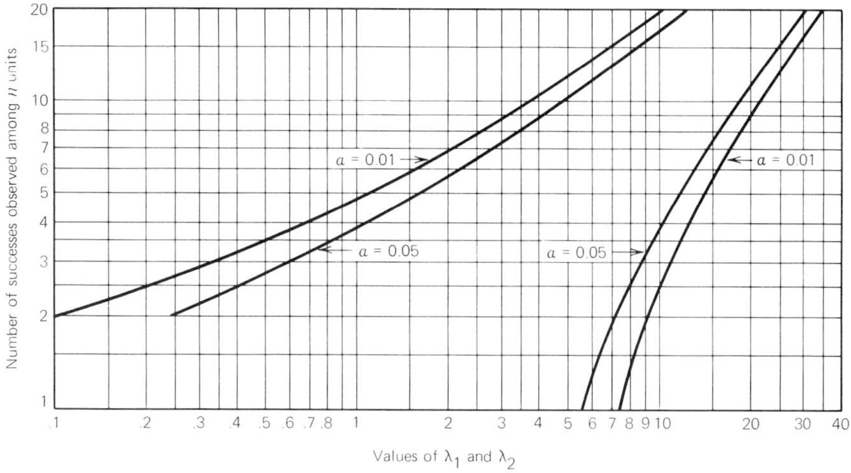
Number of successes observed among n units
Values of λ1 and λ2
α = 0.01
α = 0.05
α = 0.05
α = 0.01
20
15
10
8
7
6
5
4
3
2
1
.1
.2
.3
.4
.5
.6
.7
.8
1
2
3
4
5
6
7
8
9
10
20
30
40

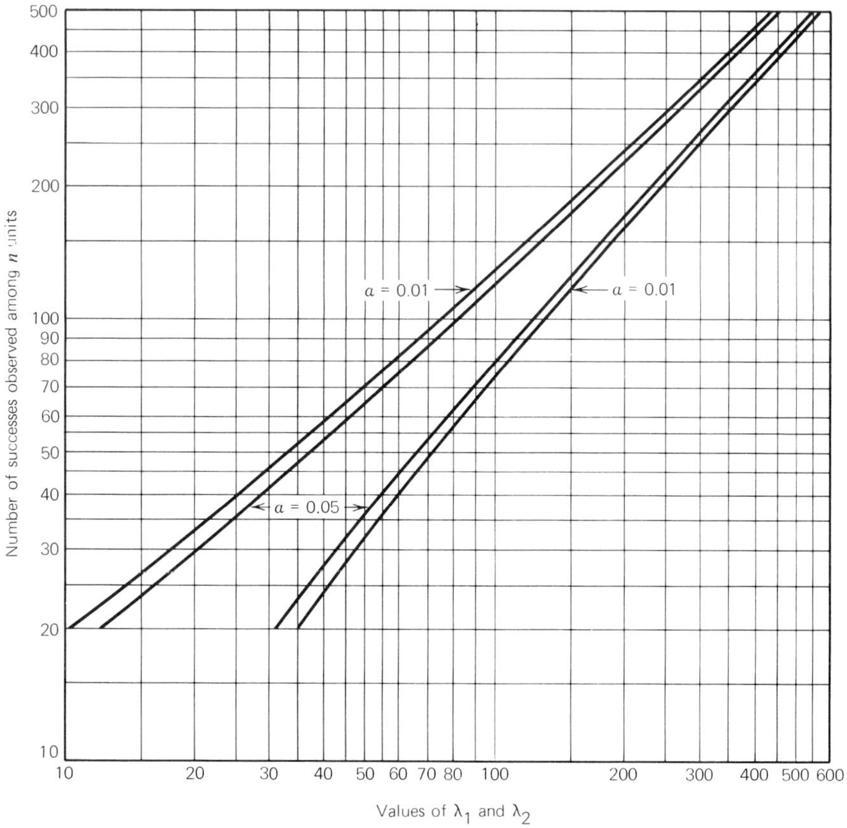
Number of successes observed among n units
Values of λ1 and λ2
α = 0.01
α = 0.01
α = 0.05
500
400
300
200
100
90
80
70
60
50
40
30
20
10
10
20
30
40
50
60
70
80
100
200
300
400
500
600

Appendix B.3 *continued*

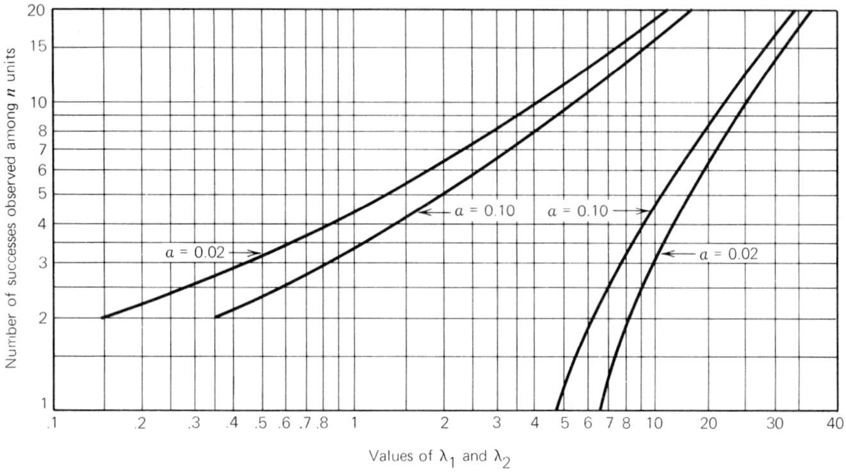

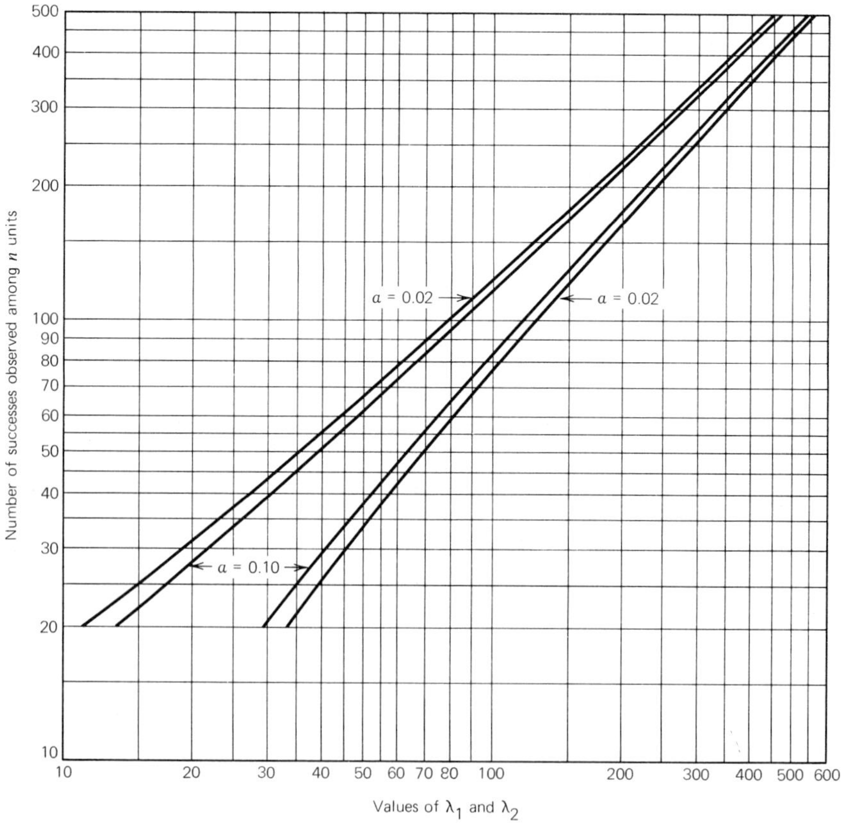

Appendix B.4 Confidence Belts for Binomial Proportions†

For the values of α shown and sample sizes indicated on the curves, a pair of curves gives a two-sided CI. One member of each pair gives a one-sided CI for $\alpha/2$.
Enter sample proportion on horizontal scale, project vertically upward to appropriate curves, and then project horizontally to left scale for confidence limits.

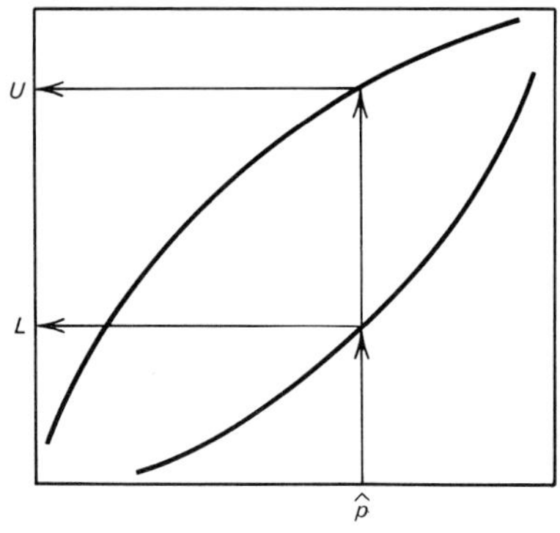

† Reproduced with permission of E. S. Pearson for the Biometrika Trustees from C. J. Clopper and E. S. Pearson, "The Use of Confidence or Fiducial Limits Illustrated in the Case of the Binomial," *Biometrika*, **26**, 1934.

$\alpha = 0.05$

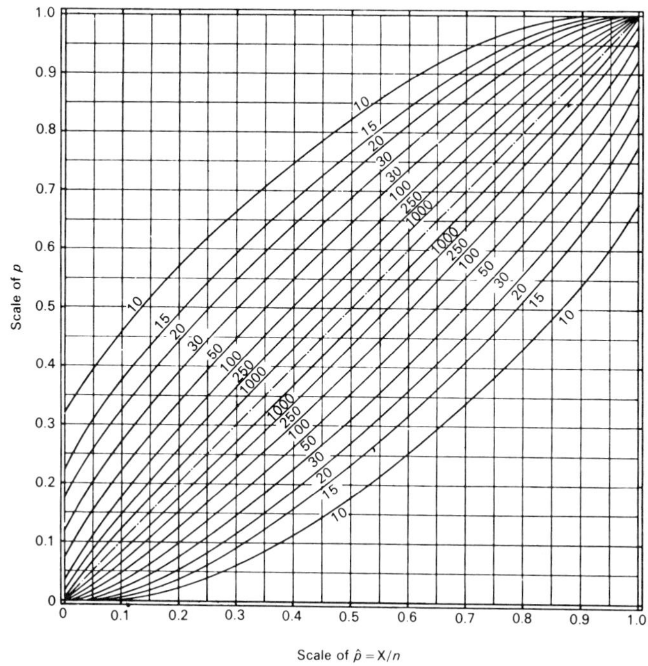

$\alpha = 0.01$

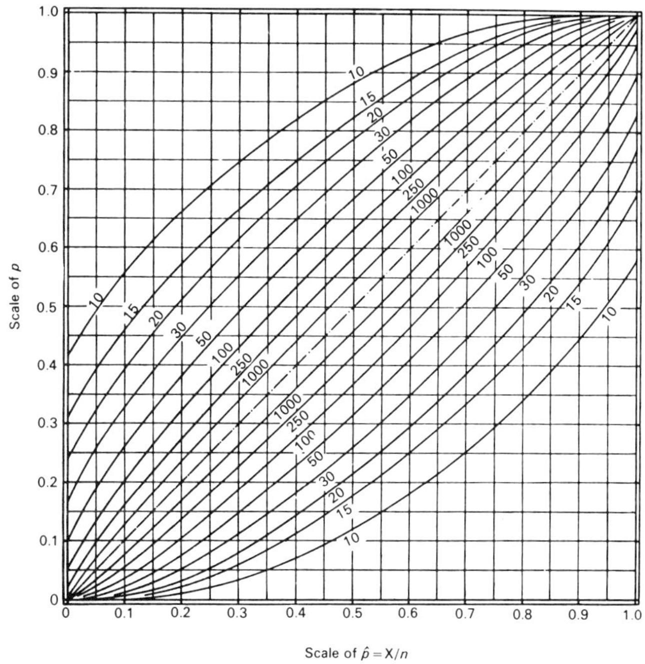

Appendix B.5 Confidence Belts for the Correlation in a Bivariate Normal†

For the values of α shown and sample sizes indicated on the curves, a pair of curves gives a two-sided *CI*. One member of each pair gives a one-sided *CI* for $\alpha/2$.
Enter sample correlation on horizontal scale, project vertically upward to appropriate curves, and then project horizontally to left scale for confidence limits.

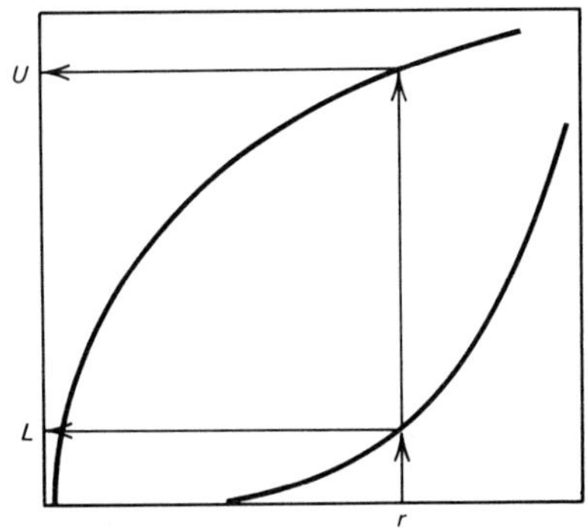

† Reproduced with permission of the author and E. S. Pearson for the Biometrika Trustees from F. N. David, *Tables of the Ordinate and Probability Integral of the Distribution of the Correlation Coefficient in Small Samples*, Biometrika, London, England, 1938.

$\alpha = 0.10$

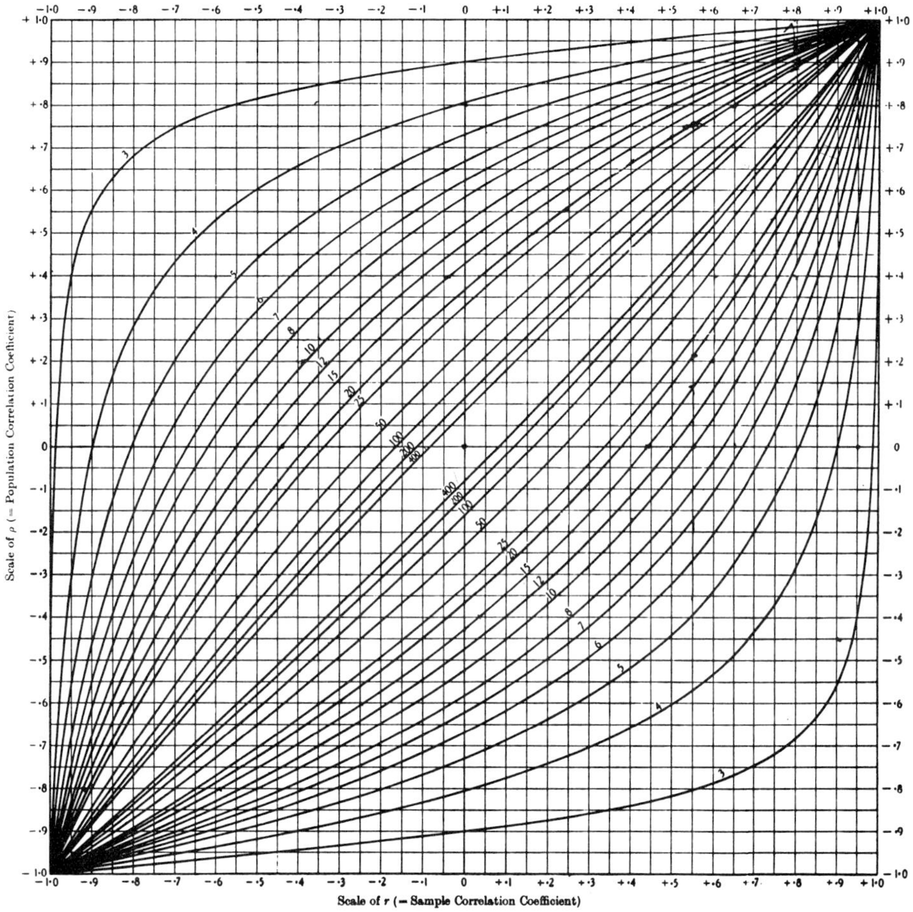

The numbers on the curves indicate sample size

$\alpha = 0.05$

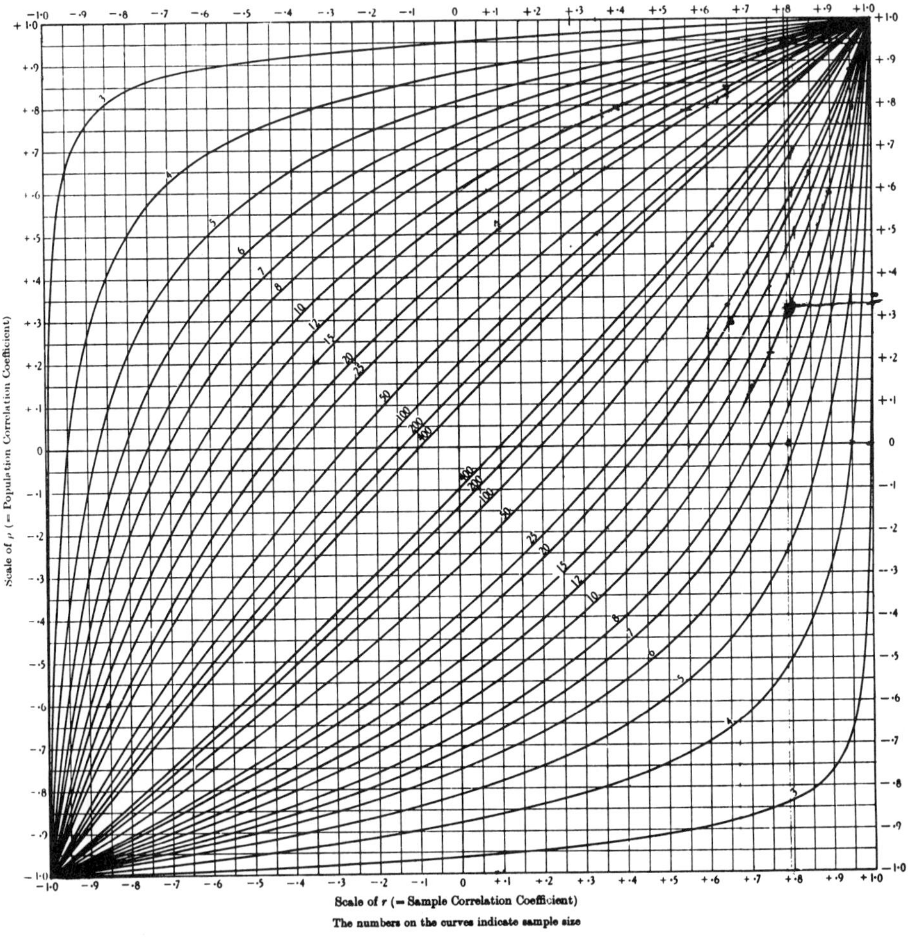
Scale of ρ (= Population Correlation Coefficient)
Scale of r (= Sample Correlation Coefficient)
The numbers on the curves indicate sample size

$\alpha = 0.02$

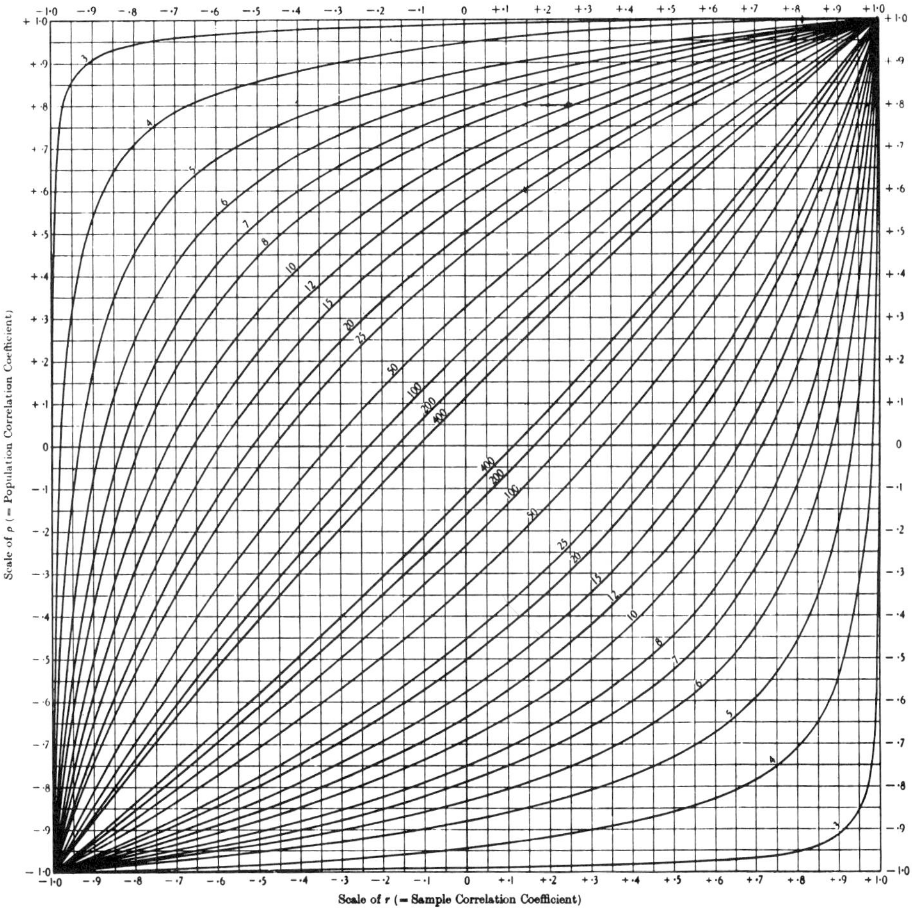

The numbers on the curves indicate sample size

Appendix B.5 *continued*

$\alpha = 0.01$

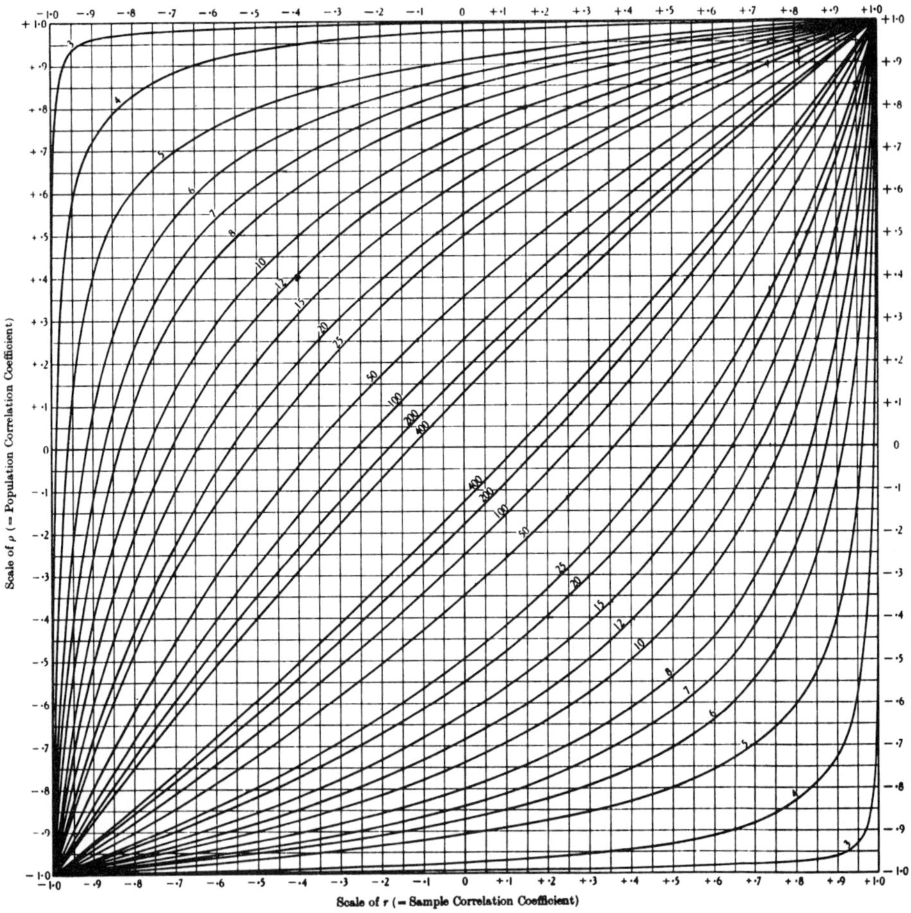

The numbers on the curves indicate sample size

Appendix B.6 Graphs of the Power Function of Tests about Two Normal Variances†

Charts are for *one-sided* tests of size α. Let R_a denote the ratio (less than unity) of the variances under the alternative.

Enter the graphs with R_a on the horizontal scale, project vertically upward to the curve for the appropriate degrees of freedom, and then project horizontally to the left scale for the power.

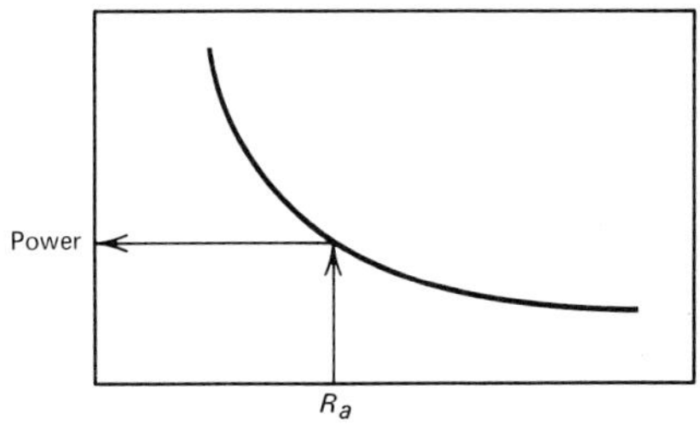

Charts may be used for *two-sided* tests of size 2α; a maximum error of α occurs as R_a approaches unity, and diminishes as the power exceeds 0.30.

† From *Concepts of Statistical Inference*, by W. C. Guenther. Copyright © 1965 by McGraw-Hill, Inc. Used by permission of McGraw-Hill Book Company.

$\alpha = .01$

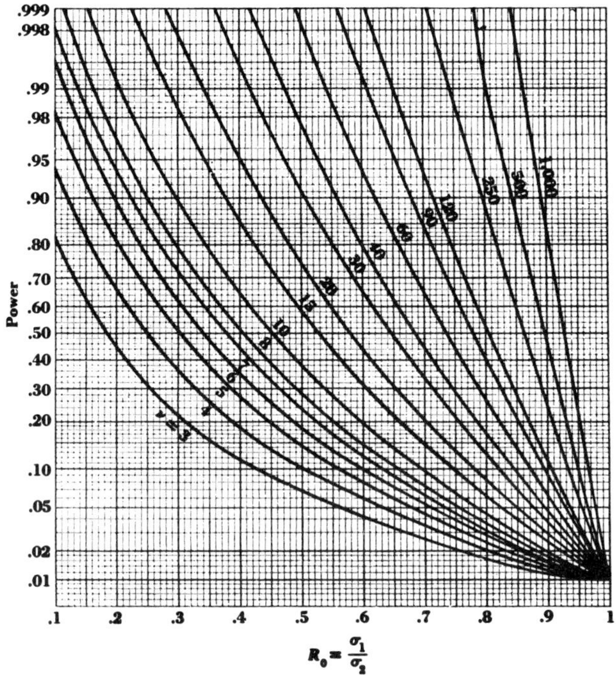

$\alpha = .025$

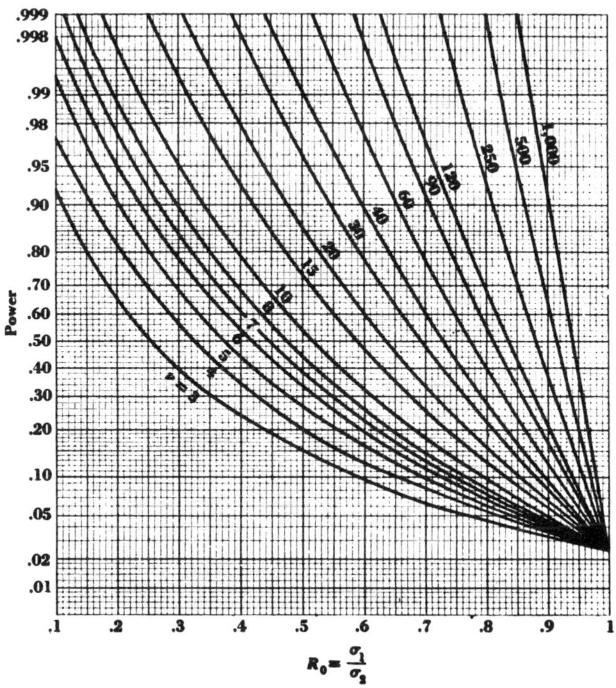

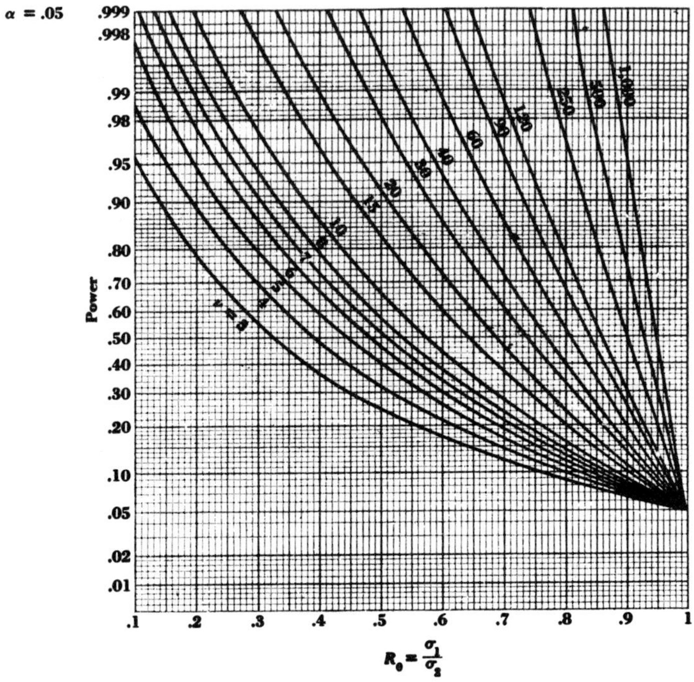

α = .05
Power
.999
.998
.99
.98
.95
.90
.80
.70
.60
.50
.40
.30
.20
.10
.05
.02
.01
.1
.2
.3
.4
.5
.6
.7
.8
.9
1
$R_0 = \frac{\sigma_1}{\sigma_2}$
ν = 3
4
5
6
7
8
10
15
20
30
40
60
90
120
250
500
1,000

Appendix B.7 Power Curves for the Analysis of Variance Test[†]

Curves are for 1 % and 5 % tests for ν_1 treatment degrees of freedom and ν_2 error degrees of freedom.
Enter value of ϕ on horizontal scale, project vertically upward to appropriate curve, then project horizontally to left or right scale for power.

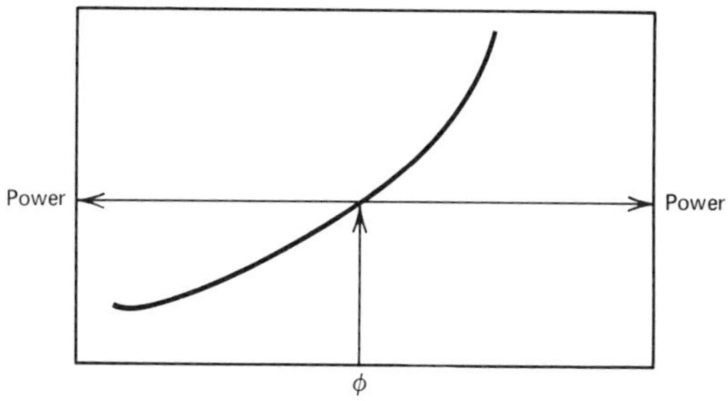

[†] Reproduced with permission of E. S. Pearson for the Biometrika Trustees from E. S. Pearson and H. O. Hartley, "Charts of the Power Function for Analysis of Variance Tests, Derived from the Noncentral F Distribution," *Biometrika*, **38**, 1951. (These eight charts with two additions for $\nu_1 = 12$ and 24 are, it is understood, appearing in *Biometrika Tables for Statisticians*, Vol. 2, now in press.)

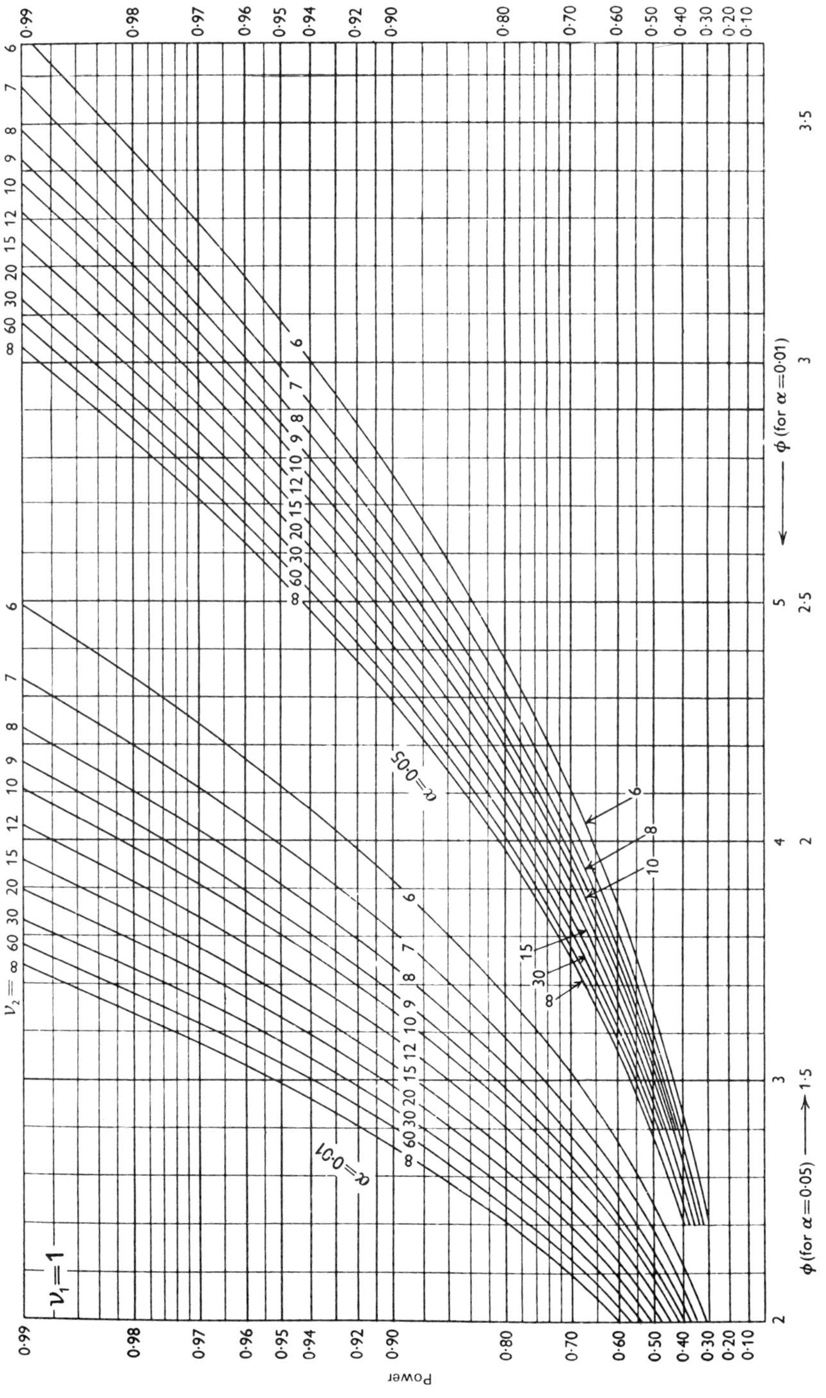

$\nu_1 = 1$
$\nu_2 = \infty$ 60 30 20 15 12 10 9 8 7 6
$\alpha = 0{\cdot}05$
$\alpha = 0{\cdot}01$
Power
0·99 0·98 0·97 0·96 0·95 0·94 0·92 0·90 0·80 0·70 0·60 0·50 0·40 0·30 0·20 0·10
ϕ (for $\alpha = 0{\cdot}01$)
ϕ (for $\alpha = 0{\cdot}05$)
2 3 4 5
1·5 2 2·5 3 3·5

Appendix B.7 *continued*

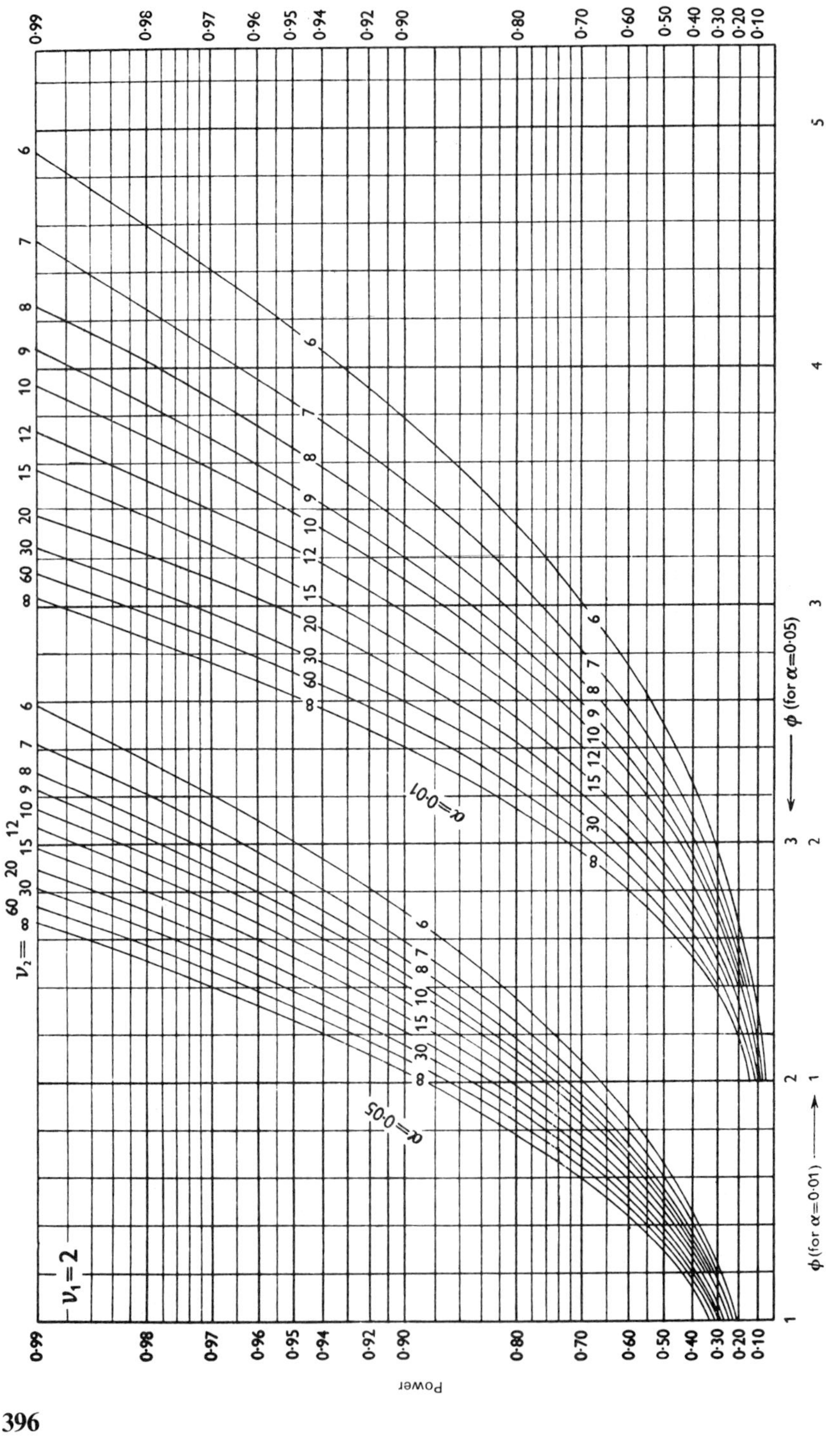

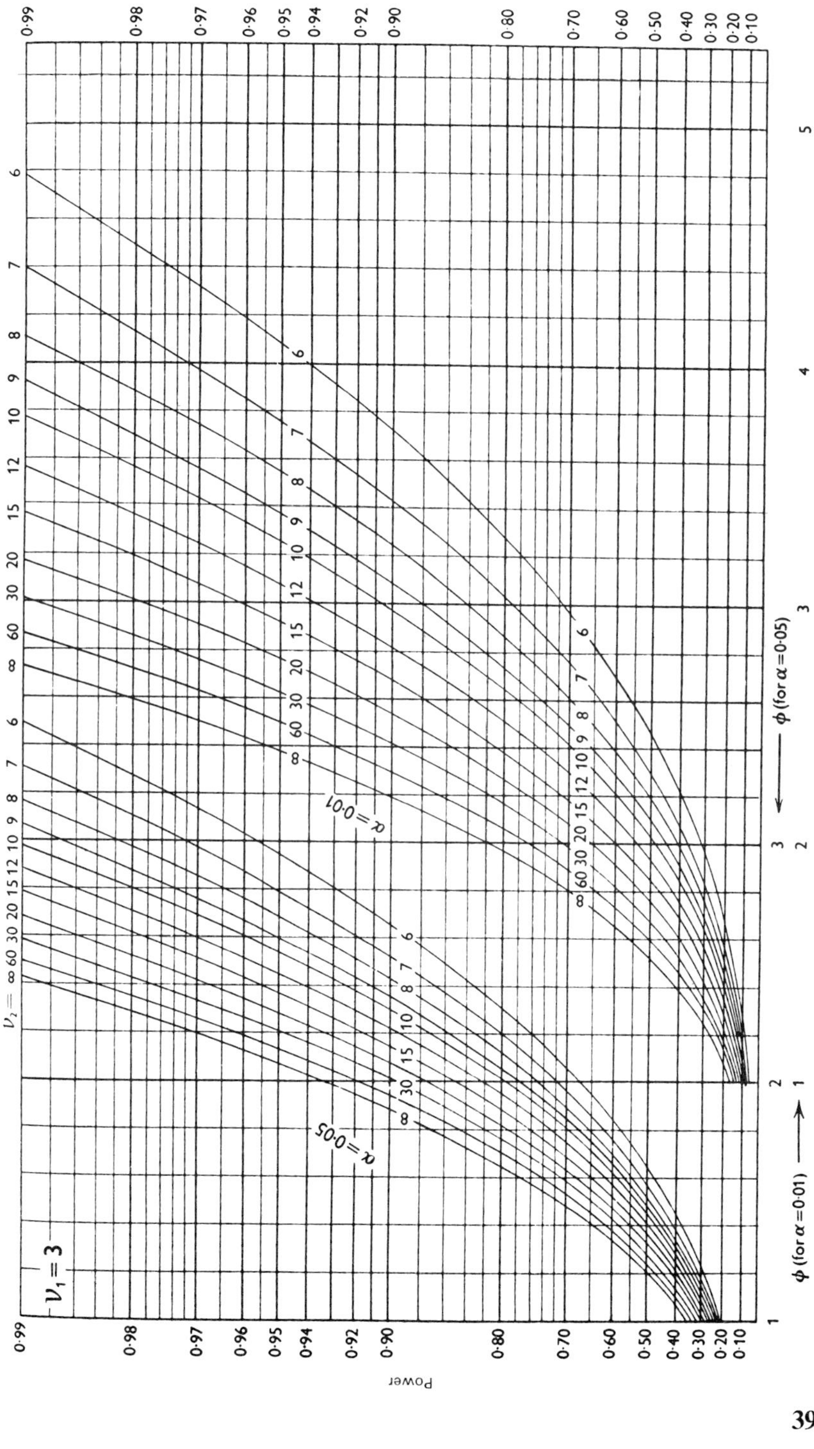
$\nu_1 = 3$
$\nu_2 =$ ∞ 60 30 20 15 12 10 9 8 7 6
Power
φ (for α=0·05)
φ (for α=0·01)
α=0·01
α=0·05

Appendix B.7 *continued*

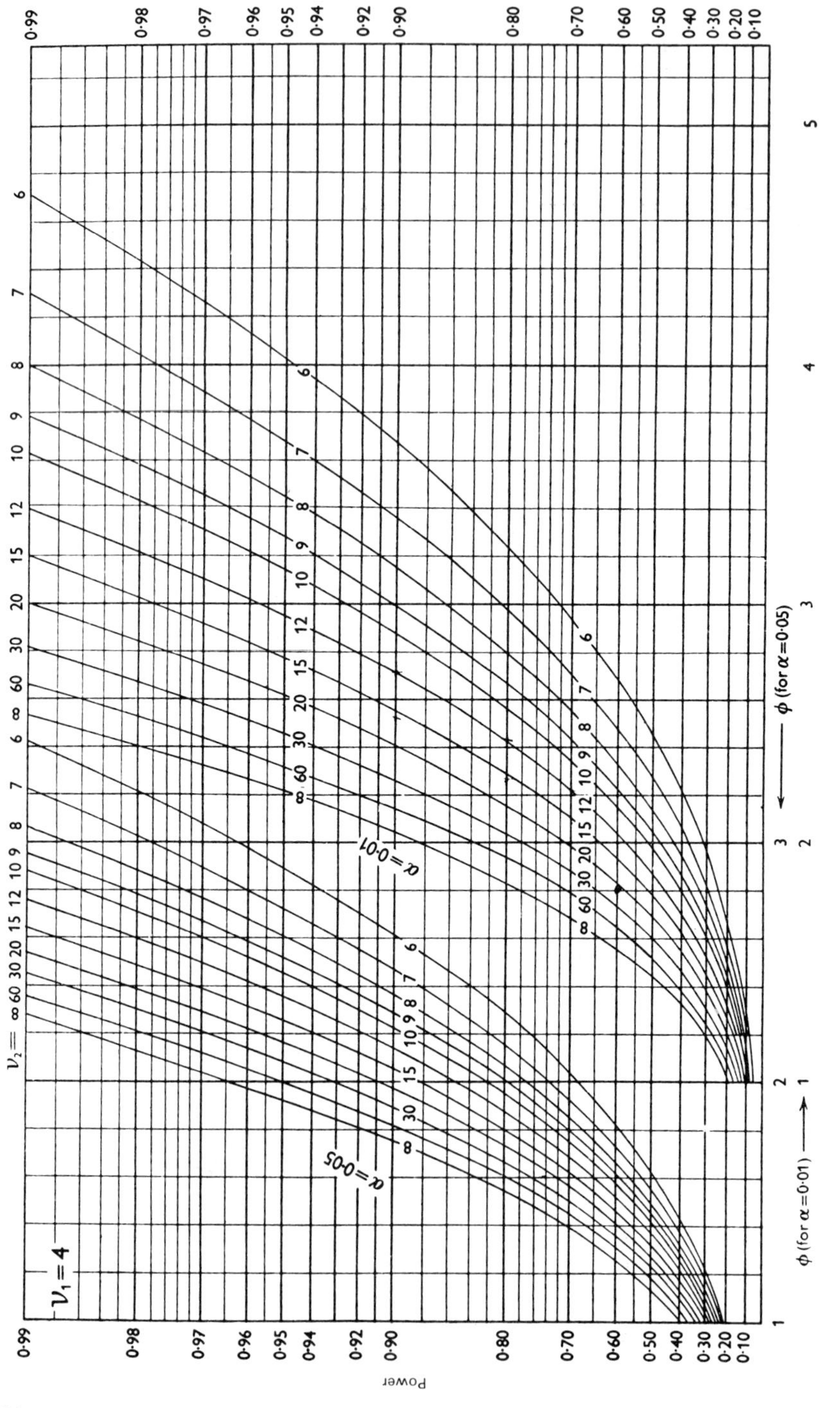

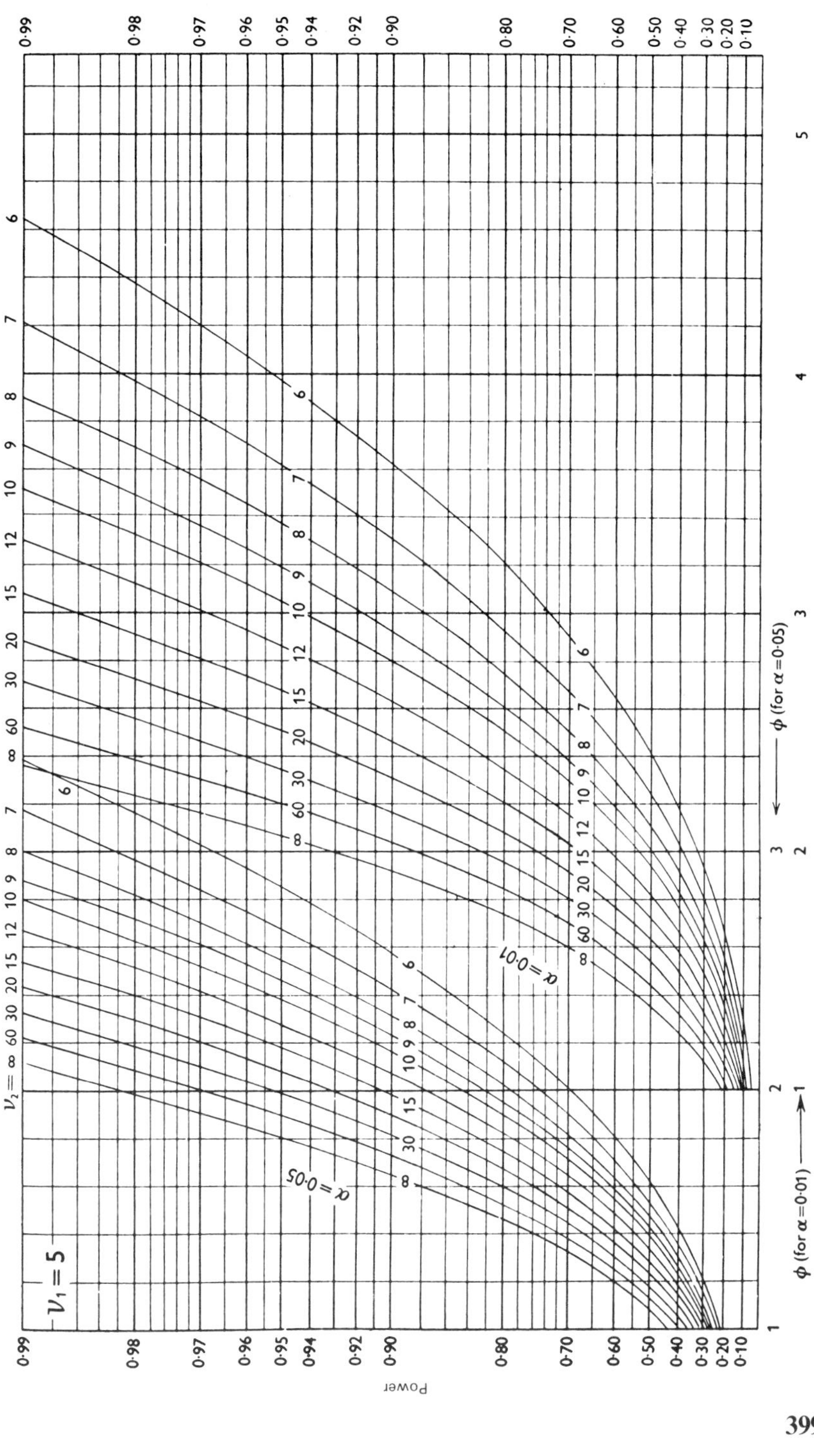

$\nu_1 = 5$
$\nu_2 = \infty$ 60 30 20 15 12 10 9 8 7 6
$\alpha = 0{\cdot}05$
$\alpha = 0{\cdot}01$
Power
ϕ (for $\alpha = 0{\cdot}05$)
ϕ (for $\alpha = 0{\cdot}01$)

Appendix B.7 *continued*

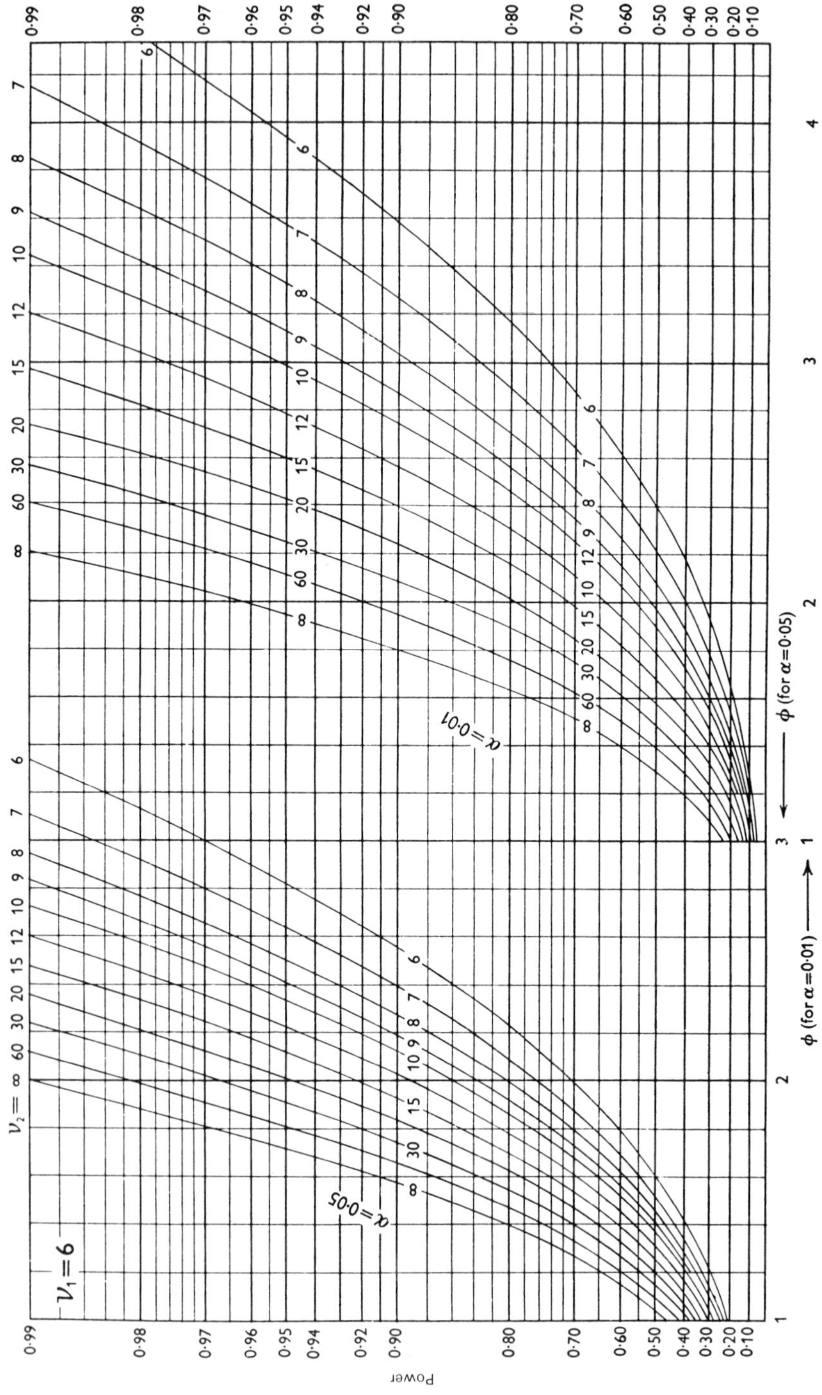

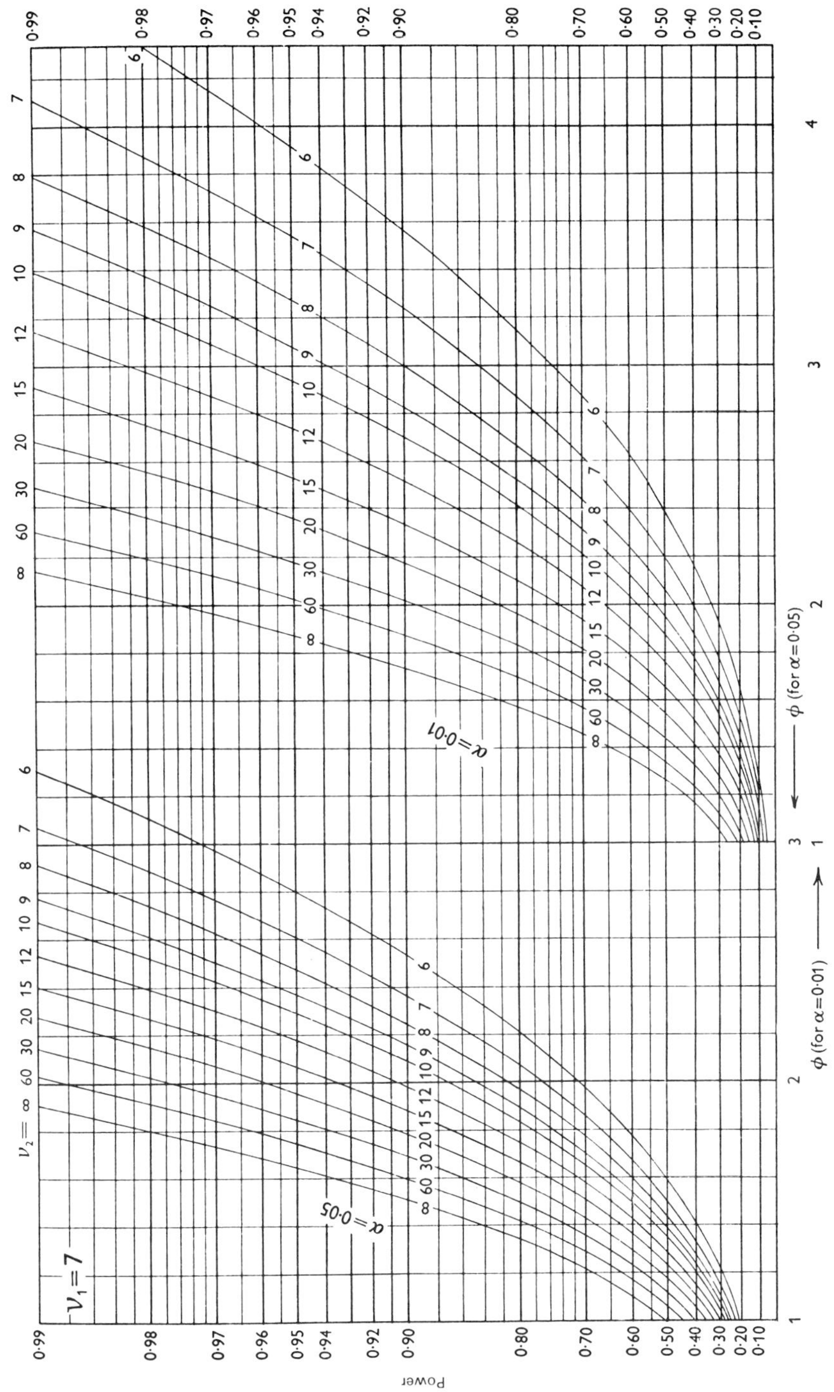

ν1 = 7
ν2 = ∞ 60 30 20 15 12 10 9 8 7 6
α = 0·05
α = 0·01
φ (for α=0·05)
φ (for α=0·01)
Power
0·99 0·98 0·97 0·96 0·95 0·94 0·92 0·90 0·80 0·70 0·60 0·50 0·40 0·30 0·20 0·10

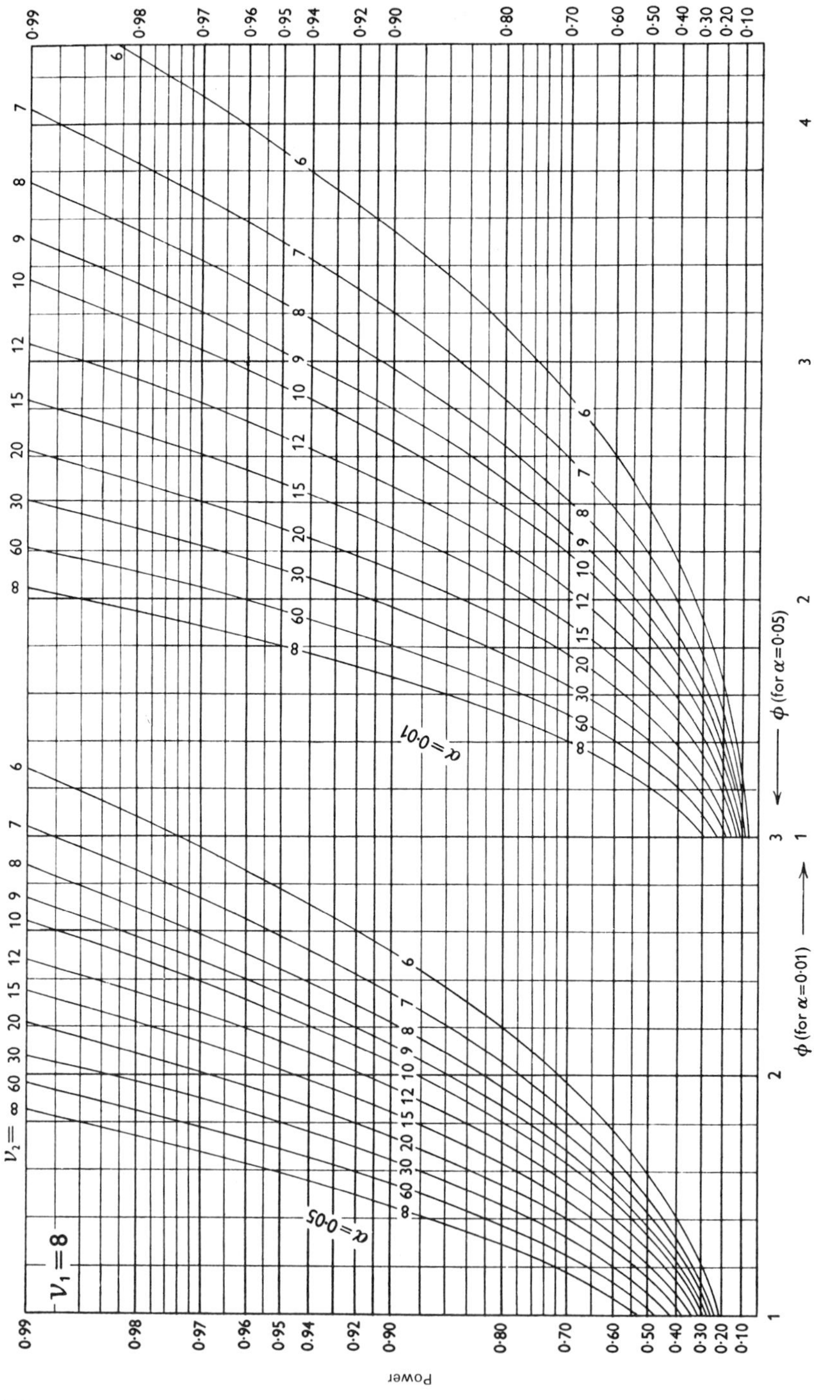
ν₁=8
ν₂=
α=0·05
α=0·01
φ (for α=0·01)
φ (for α=0·05)
Power

Answers to Odd-Numbered Problems

CHAPTER 2

1. (a) {209, 213, 214, 408, 415, 512, 707, 713, 714, 805, 806, 817, 915, 916}
(b) {1, 2, 3, 4, 5, 6, 7, 8, 9, 10, 11, 12, 13, 14, 15}
(c) {General Motors, Chrysler, Ford, American} = {GM, C, F, A}
(d) {0, 1, 2, 3, 4, 5, 6, 7, 8, 9, 10, 11, 12}

3. (a) 14 (b) 15 (c) 4 (d) 13

5. (a) -1 (b) 5 (c) 0 (d) 0 (e) 55

7. (a) 1 (b) 0 (c) $3A - 1$ (d) 0 (e) $3A^2 - 2$ (f) 4

9. (a) $\{-1, 0, 3\}$ (b) $\{0, 1, 2, 3\}$ (c) $-1; 3; -9; \frac{y^2}{4} - y$

11. (a) The domain is given in 1(a).
(b) The range is {6, 7, 8, 10, 11, 12, 13, 14, 15, 16}.
(c) Yes, area codes 209 and 713, 408 and 714, 707 and 806, and 817 and 916.

13. (a) GM; C; F; A
(b) Ignoring order, there are six samples: GM, C; GM, F; GM, A; C, F; C, A; F, A
(c) Ignoring order, there are ten samples: GM, GM; GM, C; GM, F; GM, A; C, C; C, F; C, A; F, F; F, A; A, A

CHAPTER 3

1. (a) {RR, RB, RW, BB, BW, WW}
(b) {RR, BB, WW}
(c) {RB, BB, BW}

3. A: {20, 21, 22, 23, 24, 25, 26, 27, 28, 29}
C: {00, 11, 22, 33, 44, 55, 66, 77, 88, 99}
E: {04, 13, 22, 31, 40}
F: {03, 14, 25, 36, 47, 58, 69, 74, 85, 96, 30, 41, 52, 63}

5. (a) {0, 1, 2, 6, 7} (b) {0, 1, 2, 4, 6} (c) {0, 1, 2, 4, 7}
(d) {1, 4, 7, 8} (e) $\varnothing$ (f) {0, 2}
(g) {0, 1, 2, 4, 6, 7} (h) $\varnothing$ (i) {0, 2}
(j) $\mathscr{S}$ (k) $\varnothing$ (m) {0, 1, 2, 4, 6, 8}
(n) {0, 2, 4, 6, 8} (p) {6, 7, 8}

7.

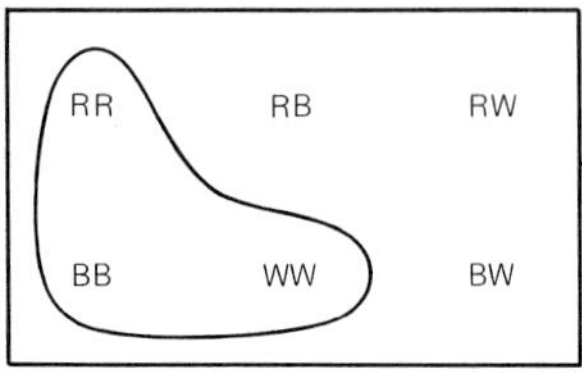

(a)

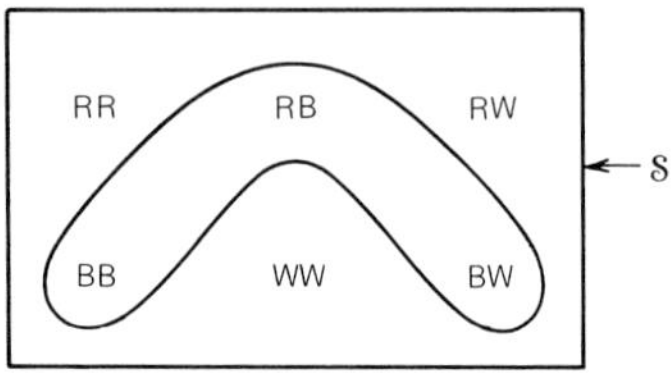

(b)

9. The two events have no outcomes in common and one of the two events must occur.

11. (a) 48 (b) 55 (c) 100 (d) 14

13. (a) 120 (b) 1,001 (c) 364 (d) 12,650 (e) 12,376 (f) 50,388 (g) 286 (h) 3,003

15. (a) C, D (b) B, D (c) $\{1, 3\}$ (d) $\{1, 2, 3, 5, 7, 9\}$

17. (a) $9(26)^2(10)^4$ plates (b) $27(26)^2(10)^4$

19. (a) 0.071 (b) 0.189 (c) 0.530 (d) 0.151 (e) 65,780

21. (a) $\binom{10}{6}$ (b) $6!\binom{10}{6}$ (c) $\binom{10}{2}\binom{8}{4}$

23. (a) 60! (b) 60!/10!

25. 16!/12!

27. 10!/4! 3! 3!

29. (a) $\binom{4}{2}\binom{48}{3}$ (b) $\binom{13}{3}\binom{39}{2}$

(c) $\binom{13}{3}\binom{39}{2}+\binom{13}{4}\binom{39}{1}+\binom{13}{5}\binom{39}{0}$ (d) $\binom{12}{5}$

31. (a) 6 (b) 2

33. (a) 20! (b) 2(6! 5! 9!) (c) 6

35. (a) 960 (b) 600 (c) 500

37. (a) $\binom{100}{5}$ (b) $\binom{10}{5}$ (c) $\binom{20}{5}$

CHAPTER 4

1. (a) 1/4 (b) 1/8 (c) 3/8 (d) 1/2 (e) 1/4 (f) 3/8 (g) 3/4 (h) 0

3. (a) No; $P(A \cup B) < 1$ (b) No; $P(A \cap B) = 0.2 \neq 0$
(c) Yes; $P(A|B) = 0.4 = P(A)$ (d) 0.5 (e) 0.3

5. (a) 15!/5! 5! 5! (b) 10! 5!/15!

7. (a) 1/3 each outcome (b) 1/9 each outcome (c) 1/729 each outcome

9. (a) 10 (b) 1/10

11. (a) $\binom{6}{4}\Big/\binom{13}{4}$ (b) 16/715 (c) 315/715 (d) 135/715

13. (a) 1/140 (b) 3/70 (c) 1/7

15. (a) $\binom{10}{2}\Big/\binom{40}{2}$ (b) $\binom{30}{8}\binom{10}{2}\Big/\binom{40}{10}$

17. (a) $\binom{9}{2}\Big/\binom{12}{2}$ (b) $2\Big/\binom{12}{2}$ (c) $1-\binom{9}{2}\Big/\binom{12}{2}$

19. (a) $\binom{4}{2}\binom{48}{3}\Big/\binom{52}{5}$ (b) $\binom{13}{3}\binom{39}{2}\Big/\binom{52}{5}$

(c) $\left[\binom{13}{3}\binom{39}{2}+\binom{13}{4}\binom{39}{1}+\binom{13}{5}\binom{39}{0}\right]\Big/\binom{52}{5}$

(d) $\binom{12}{5}\Big/\binom{52}{5}$

21. (a) 80/243 (b) 1/243 (c) 32/243 (d) 192/243

23. (a) $\binom{10}{1}\binom{10}{1}\binom{10}{1}\Big/\binom{30}{3}$ (b) $\binom{10}{3}\Big/\binom{30}{3}$

(c) $\binom{10}{3}\Big/\binom{30}{3}$ (d) $\binom{20}{3}\Big/\binom{30}{3}$

CHAPTER 5

1. (a)

Live Weight of Whitefish Caught in the Madison River

Weight	Number Caught
0.75–1.00 pounds	12
1.25–1.50	15
1.75–2.00	5
2.25–2.50	3
2.75–3.00	1

(b)

Live Weight of Whitefish Caught in the Madison River

Weight	Number Caught
less than or equal to 0.625 lb	0
1.125 lb	12
1.625 lb	27
2.125 lb	32
2.625 lb	35
3.125 lb	36

3. (a)

Travel Times to Work

Time	Number of Days
200–299 sec	2
300–399 sec	3
400–499 sec	6
500–599 sec	7
600–699 sec	7
700–799 sec	2
800–899 sec	3

(b)

Travel Times to Work

Time	Number of Days
less than or equal to 199.5 sec	0
299.5 sec	2
399.5 sec	5
499.5 sec	11
599.5 sec	18
699.5 sec	25
799.5 sec	27
899.5 sec	30

5. (a)

Dead Grasshoppers Found on Forty Plots

Number of Hoppers	Number of Plots
0–9	9
10–19	15
20–29	11
30–39	4
40–49	1

(b)

Dead Grasshoppers Found on Forty Plots

Number of Hoppers	Number of Plots
less than or equal to 9.5	9
19.5	24
29.5	35
39.5	39
49.5	40

7. (a) Final Exam Grades in an Elementary Statistics Course

Grades	Number of Students
36–45	2
46–55	5
56–65	6
66–75	5
76–85	23
86–95	19

(b) Final Exam Grades in an Elementary Statistics Course

Grades	Number of Students
less than or equal to 35.5	0
45.5	2
55.5	7
65.5	13
75.5	18
85.5	41
95.5	60

9. (a) 45 (b) 0, 49; 50, 99; etc. (c) −0.5, 49.5, 99.5, etc.
(d) 24.5, 74.5, 124.5, 174.5, 224.5 (e) 0.31, 0.27, 0.20, 0.11, 0.11

CHAPTER 6

1. (a) continuous (b) discrete (c) discrete (d) discrete
(e) continuous (f) discrete

3. (a) real values in some interval, say 0 to 12 inches
(b) 0, 1, 2, . . . , N, where N is the number of reservations originally sold
(c) 0, 1, 2, . . . , N, where N is some integer
(d) 0, 1, 2, . . .
(e) real values in some interval, say 0 to 5 grams
(f) 0, 1, 2, . . .

5. (a)

x	$P(X = x)$
1	3/28
2	15/28
3	10/28

(b) $F(x) = 0$ if $x < 1$
$= 3/28$ if $1 \leqslant x < 2$
$= 18/28$ if $2 \leqslant x < 3$
$= 1$ if $x \geqslant 3$

7. (a) $f(y) = 1/10$ if $y = 7/3, 9/3, 11/3, 12/3, 15/3, 17/3, 18/3, 19/3, 20/3, 22/3$
(b) $g(w) = 6/10$ if $w = 1$
$= 4/10$ if $w = 2$
(c) $F(y) = 0$ if $y < 7/3$
$= 1/10$ if $7/3 \leqslant y < 9/3$
$= 2/10$ if $9/3 \leqslant y < 11/3$
$= 3/10$ if $11/3 \leqslant y < 12/3$
$= 4/10$ if $12/3 \leqslant y < 15/3$
$= 5/10$ if $15/3 \leqslant y < 17/3$
$= 6/10$ if $17/3 \leqslant y < 18/3$
$= 7/10$ if $18/3 \leqslant y < 19/3$
$= 8/10$ if $19/3 \leqslant y < 20/3$
$= 9/10$ if $20/3 \leqslant y < 22/3$
$= 1$ if $y \geqslant 22/3$
(d) $G(w) = 0$ if $w < 1$
$= 6/10$ if $1 \leqslant w < 2$
$= 1$ if $w \geqslant 2$

9. (a)

y_1	$f(y_1)$
-2	1/6
-1	1/4
0	1/3
1	1/6
2	1/12

(b)

y_2	$f(y_2)$
-1	1/6
0	1/6
1	1/4
2	1/12
3	1/6
4	1/12
5	1/12

11. (a) 9/4 (b) 45/112

13.

x_1, x_2	$f(x_1, x_2)$
1,1	1/6
1,2	1/6
1,3	1/6
2,1	1/12
2,2	1/12
2,3	1/12
3,1	1/12
3,2	1/12
3,3	1/12

15. (a) 3 (b) 4.4

17. (a) $F(x) = 0$ if $x < 0$
$= 0.1$ if $0 \leqslant x < 1$
$= 0.3$ if $1 \leqslant x < 2$
$= 0.55$ if $2 \leqslant x < 4$
$= 0.75$ if $4 \leqslant x < 6$
$= 1.0$ if $x \geqslant 6$
(b) 0.75
(c) 0.45
(d) 0

CHAPTER 7

1. (a) 25 (b) 40 (c) 0 (d) 4 (e) 178
(f) 1,600 (g) 18 (h) 18 (i) 2, 3, 3, 3, 4, 4, 4, 5, 5, 7
(j) 4 (k) 4.5

3. (a) 14 (b) 840 (c) 60 (d) 7.75 (e) 2

5. (a) 10.25 (b) 2.93

7. (a) 18.6 (b) 1.5

9. (a) 430.3 sec (b) 11,852 (c) 7.2 (d) 3.3

11. (a) 13, 15, 18, 24, 26, 32, 38, 41, 55, 68, 76, 114 (b) 35

13. (a) 91.7 sec (b) 46.9 (c) 1.25

15. (a) $16,293.75 (b) 150,958 (c) 97.1

17. (a) $13.66 (b) 35.87 (c) $12.03 (d) 3.18
(e) $14.48 (f) $12.445

19. $130.82

21. (a) 8.45 (b) 7.19

23. (a) 101 lb (b) 5.25

25. (a) $2.83 (b) 0.05

27. (a) 32.9 (b) 1.81

29. (a) 95.3 (b) 95.5 (c) 24.5 (d) 4.95

31. $48.62

33. (a) $\prod_{i=1}^{r} (n - i + 1)$ (b) $\prod_{i=1}^{r} (n - i + 1)/i$

35. (a) 3,004; 22,427 (b) 12.6; 188.5
(c) 36,750; 294,000 (d) 93,613; 698,976
(e) 733,213; 5,474,656 (f) 0.26; 2.38

CHAPTER 8

1. (a) binomial; $n = 100$, $p = 1/2$
(b) multinomial; $n = 48$, $p_1 = 0.86$, $p_2 = 0.10$, $p_3 = 0.04$
(c) hypergeometric; $n = 5$, $A = 13$, $B = 39$
(d) perhaps normal but probably some type of Gamma
(e) perhaps normal but probably some type of Gamma
(f) Poisson
(g) extended hypergeometric

3. (a) 0.4096 (b) 0.9933 (c) 0.3277 (d) 4 (e) 4/5

5. (a) 0.1248 (b) 0.0486 (c) 0.7567

7. (a) $\binom{30}{4}\Big/\binom{36}{4}$ (b) $\binom{30}{2}\binom{6}{2}\Big/\binom{36}{4}$ (c) $1 - \binom{30}{4}\Big/\binom{36}{4}$
(d) 4/6 (e) 32/63

9. (a) 0.10565 (b) 0.00009 (c) 0.89426

11. (a) 0.69146 (b) 0.84135 (c) 0.53281

13. (a) $\binom{18}{3}\binom{7}{1}\binom{4}{2}\Big/\binom{35}{6}$ (b) $\binom{18}{3}\binom{7}{1}\binom{10}{2}\Big/\binom{35}{6}$

(c) $$\left[\binom{18}{2}\binom{7}{1}\binom{3}{1}^2\binom{4}{1} + \binom{18}{1}\binom{7}{2}\binom{3}{1}^2\binom{4}{1} + 2\binom{18}{1}\binom{7}{1}\binom{3}{1}\binom{4}{1}\binom{3}{2} + \binom{18}{1}\binom{7}{1}\binom{3}{1}^2\binom{4}{2}\right]\Big/\binom{35}{6}$$

15. (a) 0.74455 (b) 0.69146 (c) 1/2 (d) 0.06658

17. (a) 0.25143 (b) 0.02275 (c) 0.58992 (d) 0

19. (a) 31/56 (b) 16/56 (c) 18/56

21. (a) 0.4407 (b) 0.1366 (c) 0.7869

23. (a) 0.3036 (b) 0.0138 (c) 0.0611

25. (a) 0.1606 (b) 0.9826 (c) 0.0025

27. (a) 0.2059 (b) 0.9978 (c) 0.1841

29. (a) −1.3304 (b) 1.7613 (c) 2.2281 (d) −2.2281
(e) 2.9467 (f) 9 (g) 18

31. (a) 4.10 (b) 0.114 (c) 0.340 (d) 40 (e) 12

33. (a) 15; 30 (b) 27; 54 (c) 48; 96 (d) 127; 254

35. (a) 5/3 (b) 1.2 (c) 5/3
(d) 1.125 (e) 1.002 (f) 3

CHAPTER 9

1.

	(a)				(b)	(d)	(f)
Sample	$\bar{x}$	$\tilde{x}$	s^2	s	$\bar{x} \pm s$	$\tilde{x} \pm s$	$s \pm 1.8$
1, 2, 4	7/3	2	2.33	1.53	0.8; 3.86	0.47; 3.53	−0.27; 3.33
1, 2, 6	9/3	2	7.0	2.65	0.35; 5.65	−0.65; 4.65	0.85; 4.45
1, 2, 12	15/3	2	37.0	6.08	−1.08; 11.08	−4.08; 8.08	4.28; 7.88
1, 4, 6	11/3	4	6.33	2.52	1.15; 6.19	1.48; 6.52	0.72; 4.32
1, 4, 12	17/3	4	32.33	5.69	−0.02; 11.36	−1.69; 9.69	3.89; 7.49
1, 6, 12	19/3	6	30.33	5.51	0.82; 11.84	0.49; 11.51	3.71; 7.31
2, 4, 6	12/3	4	4.0	2.0	2.0; 6.0	2.0; 6.0	0.2; 3.8
2, 4, 12	18/3	4	28.0	5.29	0.71; 11.29	−1.29; 9.29	3.49; 7.09
2, 6, 12	20/3	6	25.33	5.03	1.64; 11.70	0.97; 11.03	3.23; 6.83
4, 6, 12	22/3	6	17.33	4.16	3.17; 11.49	1.84; 10.16	2.36; 5.96

(c) 90% (e) 80% (g) 70%
(h) Yes, the average value of $\bar{X}$ is $5 = \mu$.
(i) No, the average value of $\tilde{X}$ is $4 \neq \mu$.
(j) No, the average value of R is $7.8 \neq \eta = 11$.
(k) No, the average value of S is $4.04 \neq \sigma^* = 4.36$.
(l) Yes, the average value of S^2 is $19 = (\sigma^*)^2$.

3. (a) 1,259.2; 1,268.8 °C (b) 37; 264.1
(c) normality and independent measurements on the melting point

5. (a) 14.3 in. (b) 12.8; 15.8 (c) 9.59 (d) 2.17; 5.16

7. (a) 0.00000122 g (b) 0.00000366 (c) Yes
(d) This balance needs an adjustment.

9. (a) 20.6; 25.4 sec
(b) In general, intervals about μ are longer when the T^* distribution is used instead of the T distribution, since the range is a poorer estimator of variability than is S^2.

11. (a) 0.152; 0.186 g (b) Normality and independent weights
(c) 0.0011; 0.0013 (d) 0.11; 0.13

13. 7,787.7; 7,940.3 g

15. (a) 56.75; 1.48 mg (b) 54.16; 59.34

17. 34.2; 35.2 in.

19. (a) 24.3; 1/2 cm (b) 23; 25.6 (c) 23.1; 25.5
(d) More confidence in (b) results in a wider interval. This is true in general.

21. (a) 89.1; 94.3 sec (b) 5.65; 8.76

23. (a) 4.65; 5.37 lb (b) 0.33; 1.05

CHAPTER 10

Note: Solutions of statistical tests will be given in abbreviated form rather than the standard five-step form.

1. (a) The calculated value of the test statistic $T_{59} = 1.67$ lies between the critical values ± 2.00; therefore, $H_0: \mu = 8.5$ is not rejected.
(b) Normality of paper widths and independence of the widths obtained.

3. (a) The calculated value of the test statistic $T_{24} = 1.94$ exceeds the critical value 1.71; therefore, $H_0: \mu \leqslant 15$ is rejected in favor of $H_1: \mu > 15$.
(b) random sampling of bottles; normality of impurities per bottle
(c) The calculated value of the test statistic $\chi^2_{24} = 34.72$ does not exceed the critical value 39.36; therefore, $H_0: \sigma^2 \leqslant 36$ is not rejected.

5. (a) The calculated value of the test statistic $T_{11} = 0.59$ is not less than the critical value -1.80; therefore, $H_0: \mu \geqslant 180$ is not rejected.
(b) The calculated value of the test statistic $\chi^2_{11} = 8.15$ is not less than the critical value 3.05; therefore, $H_0: \sigma^2 \geqslant 200$ is not rejected.

7. (a) unknown (b) 0.90 (c) 0.10 (d) unknown (e) unknown

9. The calculated value of the test statistic $T_{19} = -2.29$ is less than the critical value -1.73; therefore, $H_0: \mu \geqslant 0$ is rejected.

11. The calculated value of the test statistic $T_{35} = -1.45$ is not less than the critical value -1.69; therefore, $H_0: \mu \geqslant 38$ is not rejected.

13. (a) The calculated value of the test statistic $T_{19} = 2.25$ does not exceed the critical value 2.54; therefore, $H_0: \mu \leqslant 1.25$ is not rejected.
(b) The calculated value of the test statistic $\chi^2_{19} = 24.3$ does not exceed the critical value 30.14; therefore, $H_0: \sigma^2 \leqslant 0.1$ is not rejected.

15. (a) The calculated value of the test statistic $T_{29} = 1.36$ lies between the critical values ± 2.05; therefore, $H_0: \mu = 90$ is not rejected.
(b) The calculated value of the test statistic $\chi^2_{29} = 54.4$ exceeds the critical value 39.09; therefore, $H_0: \sigma^2 \leqslant 25$ is rejected.

17. (a) 0.72 (b) 0.96 (c) 0.9999

19. (a) 0.09 (b) 0.04 (c) 0.30

21. (a) 0.95 (b) 0.70 (c) 0.20

23. (a) Some points on the power curve are:

Δ_a	P_a
$\pm\sigma/4$	0.18
$\pm\sigma/2$	0.42
$\pm\sigma$	0.88

(b) Same power curve as in (a) except that the deviations are now measured relative to 60 instead of 61.

CHAPTER 11

1. (a) The calculated value of the test statistic $F_{11,11} = 1.35$ lies between the critical values 0.288 and 3.47; therefore, $H_0: \sigma_1^2 = \sigma_2^2$ is not rejected.
(b) 2.84; 2.84 (c) 4
(d) The calculated value of the test statistic $T_{22} = -2.9$ is less than the critical value -2.51; therefore, $H_0: \mu_1 \geqslant \mu_2$ is rejected.
(e) Normality and independent random samples in (a); the answers to parts (b) and (c) were calculated under the assumptions of homogeneous population variances and independent random sampling.

3. (a) The calculated value of the test statistic $F_{14,17} = 0.953$ lies between the critical values 0.24 and 3.84; therefore, $H_0: \sigma_1^2 = \sigma_2^2$ is not rejected.
(b) 2
(c) The calculated value of the test statistic $T_{31} = -1.75$ is less than the critical value -1.696; therefore, $H_0: \mu_1 \geqslant \mu_2$ is rejected.

5. (a) 30; 25 (b) 125.3; 100
(c) The calculated value of the test statistic $F_{17,19} = 1.25$ lies between the critical values 0.38 and 2.57; therefore, H_0 is not rejected.
(d) 2.49; 2.37 (e) 5; 3.44 (f) -4.36; 14.36
(g) The calculated value of the test statistic $T_{36} = 1.45$ lies between the critical values ± 2.03; therefore, $H_0: \mu_1 = \mu_2$ is not rejected.
(h) 20.2; 29.8

7. (a) Yes; the mileage with and without the additive is measured on each car.
(b) The calculated value of the test statistic $T_{14} = 2.41$ exceeds the critical value 1.76; therefore, $H_0: \mu_D \leqslant 0$ is rejected in favor of $H_1: \mu_D > 0$.

9. (a) The calculated value of the test statistic $F_{16,15} = 2.05$ lies between the critical values 0.36 and 2.84; therefore, $H_0: \sigma_1^2 = \sigma_2^2$ is not rejected.
(b) 6.78; 0.706
(c) The calculated value of the test statistic $T_{31} = 2.51$ exceeds the critical value 1.70; therefore, $H_0: \mu_1 - \mu_2 \leqslant 5$ is rejected.

11. (a) The calculated value of the test statistic $F_{29,29} = 1.33$ lies between the (interpolated) critical values 0.38 and 2.64; therefore, $H_0: \sigma_1^2 = \sigma_2^2$ is not rejected.
(b) The calculated value of the test statistic $T_{58} = 4.2$ exceeds the critical value 1.67; therefore, $H_0: \mu_1 \leqslant \mu_2$ is rejected.

13. The calculated value of the test statistic $T_{14} = -0.32$ is not less than the critical value -2.62; therefore, $H_0: \mu_D \geqslant 12$ is not rejected.

15. The calculated value of the test statistic $T_{182} = 3.40$ exceeds the critical value 2.35; therefore, $H_0: \mu_2 \leqslant \mu_1$ is rejected.

17. The calculated value of the test statistic $T_{16}^* = 0.116$ exceeds the critical value 0.083; therefore, $H_0: \mu_1 - \mu_2 \leqslant 5$ is rejected.

19. Some points on the power curve are

Δ_a	P_a
$\sigma/4$	0.13
$\sigma/2$	0.28
σ	0.69
$3\sigma/2$	0.94

21. (a) 0.09 (b) 0.15 (c) 0.20 (d) 0.34

23. Some points on the power curve are

R_a	Power
0.83	0.08
0.67	0.24
0.50	0.62
0.33	0.94

25. (a) 0.19 (b) 0.91 (c) 0.49 (d) 0.91

CHAPTER 12

1. (a) $\hat{Y} = 3X - 1$ (b) 20 (c) 0.133 (d) 0.828

3. (a) $\hat{Y} = 2X - 5$ (b) 10 (c) 0.5 (d) 0.5

5. (b) $\hat{Y} = 2.24 + 0.444X$ (c) 0.063 (d) 0.94

7. (b) $\hat{Y} = 202.9 + 1.09X$ (c) 411.75 (d) 0.949
(e) The calculated value of the test statistic $T_8 = 12.26$ exceeds the critical value 1.86; therefore, $H_0: \beta_1 \leqslant 0$ is rejected.

9. (a) $\hat{Y} = 0.257 - 0.00128X$ (b) 0.992
(c) The calculated value of the test statistic $T_9 = -33.7$ lies outside the critical values ± 3.25; therefore, $H_0: \beta_1 = 0$ is rejected.
(d) 0.2545; 0.2595 0.1263; 0.1317
(e) 0.231; 0.011

11. (a) $\hat{Y} = 5.11X - 9{,}941.87$ (b) 10.74 (c) 0.972
(d) This would involve extrapolating; that is, assuming that the same relationship held after 1962.
(e) 89.1

13. (a) Adj. $y_i = y_i - 5.11(x_i - 1957)$

Year	Adj. y_i
1951	58.2
1952	57.1
1953	56.6
1954	60.7
1955	58.4
1956	60.4
1957	66.7
1958	61.7
1959	60.6
1960	59.1
1961	58.6
1962	54.1

(b) No

15. (a) 0.12 (b) 0.36 (c) 0.67

17. (a) 0.68 (b) 0.93 (c) 0.993

CHAPTER 13

1. (a) 0.67 (b) 0.40; 0.82
(c) The sample correlation coefficient $r = 0.67$ exceeds the critical value 0.3061; therefore, $H_0 : \rho = 0$ is rejected.

3. (a) 0.81
(b) The value 0.81 indicates that the linear relationship between X and Y is moderately strong.

5. -0.98

7. (a) $r = 0$
(b) The sample correlation coefficient measures linear not quadratic or curvilinear relationships.

9. (a) No, since ice cream sales are expected to be low during cold months and high during hot months, while cottage cheese usage would tend to be more uniform throughout the year.
(b) 0.607

11. (a) -0.885 (b) -0.96; -0.67
(c) The sample correlation coefficient $r = -0.885$ lies outside the critical values ± 0.6411; therefore, $H_0 : \rho = 0$ is rejected.

13. (a) 0.628
(b) Yes, one would not expect a large correlation since gas usage tends to fluctuate throughout the year, while electrical usage is fairly uniform.

15. (a) 0.974
(b) The calculated value of the test statistic $Z \doteq 2.47$ exceeds the critical value 2.33; therefore, $H_0: \rho \leqslant 0.9$ is rejected.

17. (a) 0.03 (b) 0.145 (c) 0.71 (d) 0.955

CHAPTER 14

1. Using Appendix B.3, the approximate interval is (0.18, 0.95).

3. The calculated value of the test statistic, $Z \doteq -2.25$, is not less than the critical value, -2.33; therefore, $H_0; p \geqslant 0.8$ is not rejected.

5. (a) 6.25; 9.50
(b) The calculated value of the test statistic, $Z \doteq 3.29$, exceeds the critical value, 1.645; therefore, $H_0: \lambda \leqslant 6$ is rejected.

7. (a) 0.05419; 0.56750
(b) The calculated value of the test statistic, $Z \doteq -1.67$, is less than the critical value, -1.645; therefore, $H_0: p \geqslant 0.6$ is rejected.

9. (a) The calculated value of the test statistic, $\chi_4^2 \doteq 4.19$, does not exceed the critical value, 9.49; therefore, $H_0: p_1 = 0.1, p_2 = p_4 = 0.15, p_3 = 0.48, p_5 = 0.12$ is not rejected.
(b) 0.443; 0.0287

11. The calculated value of the test statistic, $\chi_2^2 \doteq 39.12$, exceeds the critical value, 5.99; therefore, H_0: "Length of mammology papers and number of authors are independent factors" is rejected.

13. The calculated value of the test statistic, $Z \doteq -1.87$, is less than the critical value, -1.645; therefore, $H_0: p \geqslant 0.5$ is rejected.

15. (a) The calculated value of the test statistic, $\chi_2^2 \doteq 5.76$, does not exceed the critical value, 5.99; therefore, $H_0: p_1 = p_3 = 1/6, p_2 = 4/6$ is not rejected.
(b) 0.117; 0.022 0.742; 0.03 0.141; 0.024

17. The calculated value of the test statistic, $Z \doteq 1.88$, exceeds the critical value, 1.645; therefore, $H_0: \lambda \leqslant 4$ is rejected.

19. The calculated value of the test statistic, $Z \doteq -1.26$, is not less than the critical value, -2.58; therefore, $H_0: \lambda \geqslant 8$ is not rejected.

21. (a) 0.98341 (b) 0.86864 (c) 0.07966

23. (a) 12.15 (b) 0.78 (c) 10.25; 14.25
(d) The calculated value of the test statistic, $Z \doteq 3.04$, exceeds the critical value, 1.645; therefore, $H_0: \lambda = 10$ is rejected.

25. (a) 0.212 (b) 0.000034 (c) 0.20; 0.224

27. The individual chi squares are:

fall	$\chi_2^2 \doteq 2.78$	winter	$\chi_2^2 \doteq 7.90$
spring	$\chi_2^2 \doteq 2.67$	summer	$\chi_2^2 \doteq 2.72$

Reject $H_0: p_1 = p_2 = p_3 = 1/3$ for winter quarter.
For fall, spring, and summer the pooled chi square is $\chi_2^2 \doteq 2.62$.

29. (a) 0.55962 (b) 0.99036 (c) 0.93189

31. (a) 0.56750 (b) 0.83646 (c) 0.95907

33. (a) 0.97558 (b) 0.91309 (c) 0.77637

CHAPTER 15

1. (a)

ANOVA Table

Source	*df*	*SS*	*MS*	*EMS*
Treatments	5	510	102	$\sigma_\varepsilon^2 + 2\sum_i \tau_i^2$
Residual	54	540	10	σ_ε^2
Total	59	1,050		

(b) 2

(c) The calculated value of the test statistic $F_{5,54} = 10.2$ exceeds the (interpolated) critical value 3.40; therefore, equality of treatment means is rejected.

3. (a)

ANOVA Table

Source	*df*	*SS*	*MS*	*EMS*
Treatments	5	16	3.2	$\sigma_\varepsilon^2 + 5\sigma_\tau^2$
Residual	24	24	1	σ_ε^2
Total	29	40	XXX	

(b) The calculated value of the test statistic $F_{5,24} = 3.2$ exceeds the critical value 2.62; therefore, $H_0: \sigma_\tau^2 = 0$ is rejected.

(c) $\hat{\sigma}_\varepsilon^2 = 1$ and $\hat{\sigma}_\tau^2 = 0.44$.

5. (a)

ANOVA Table

Source	*df*	*SS*	*MS*	*EMS*
Treatments	3	152	50.7	$\sigma_\varepsilon^2 + \sum_i r_i\tau_i^2/3$
Residual	28	150	5.4	σ_ε^2
Total	31	302	XXX	

(b) The calculated value of the test statistic $F_{3,28} = 9.46$ exceeds the (interpolated) critical value 4.6; therefore, equality of treatment means is rejected.

(c) -0.4; 6.4.

(d)

i	i^*	$\bar{x}_{i\cdot} - \bar{x}_{i^*\cdot}$	$s_{\bar{x}_{i\cdot} - \bar{x}_{i^*\cdot}}$
1	2	−3	1.24
1	3	−5	1.16
1	4	0	1.16
2	3	−2	1.16
2	4	3	1.16
3	4	5	1.09

7. (a)

ANOVA Table

Source	*df*	*SS*	*MS*
Treatments	3	220	73.3
Residual	21	420	20
Total	24	640	xxx

(b) The calculated value of the test statistic $F_{3,21} = 3.66$ exceeds the critical value 3.07; therefore, equality of treatment means is rejected.

(c) Independent observations; measurements for each treatment are normally distributed; variance of measurements is the same for all treatments; linear model is appropriate.

(d) −15.4; −0.6

9. (a) $x_{ij} = \mu + \tau_i + \varepsilon_{ij};\ i = 1, 2, 3, 4$

$$j = 1, 2, \ldots, 8$$

x_{ij} represents the j-th observation of the i-th filtration method.

μ represents the overall mean.

τ_i represents the increase or decrease in impurities with the i-th filtration method.

ε_{ij} represents the extraneous, unexplained components of x_{ij}.

(b)

ANOVA Table

Source	*df*	*SS*	*MS*	*EMS*
Treatments	3	7.30	2.43	$\sigma_\varepsilon^2 + 8\sum_i \tau_i^2/3$
Residual	28	5.24	0.19	σ_ε^2
Total	31	12.54	xxx	

(c) The calculated value of the test statistic $F_{3,28} = 12.8$ exceeds the (interpolated) critical value 4.6; therefore, equality of mean impurities for the four methods is rejected.

(d) For each method, the measurements are independent, have the same variance, and are normally distributed.

11. (a) $x_{ij} = \mu + \tau_i + \varepsilon_{ij};\ i = 1, 2, 3$

$j = 1, 2, \ldots, 10$

x_{ij} represents the j-th weight gain with the i-th percentage of potassium.

μ represents the overall mean.

τ_i represents the increase or decrease in weight gain from the i-th percentage of potassium.

ε_{ij} represents the extraneous, unexplained components of x_{ij}.

(b)

ANOVA Table

Source	*df*	*SS*	*MS*	*EMS*
Percentages	2	1,674.5	837.25	$\sigma_\varepsilon^2 + 5\sum \tau_i^2$
Residual	27	392.5	14.54	σ_ε^2
Total	29	2,067	XXX	

(c) The calculated value of the test statistic $F_{2,27} = 57.6$ exceeds the (interpolated) critical value 5.50; therefore, reject the hypothesis of equal mean gains under the three percentages.

(d) The calculated value of the test statistic $T_{27} = 5.32$ lies outside the critical values ± 2.05; reject $H_0: \mu_2 = \mu_3$.

13. (a) 0.58 (b) 0.92 (c) 0.93

15. (a) 0.74 (b) 0.12

17. (a) 0.90 (b) 0.05 (c) 0.01

INDEX

TABLE OF SQUARES AND SQUARE ROOTS

N	N^2	$\sqrt{N}$	$\sqrt{10N}$
1	1	1	3.162
2	4	1.414	4.472
3	9	1.732	5.477
4	16	2	6.325
5	25	2.236	7.071
6	36	2.449	7.746
7	49	2.646	8.367
8	64	2.828	8.944
9	81	3	9.487
10	100	3.162	10
11	121	3.317	10.488
12	144	3.464	10.954
13	169	3.606	11.402
14	196	3.742	11.832
15	225	3.873	12.247
16	256	4	12.649
17	289	4.123	13.038
18	324	4.243	13.416
19	361	4.359	13.784
20	400	4.472	14.142
21	441	4.583	14.491
22	484	4.69	14.832
23	529	4.796	15.166
24	576	4.899	15.492
25	625	5	15.811
26	676	5.099	16.125
27	729	5.196	16.432
28	784	5.292	16.733
29	841	5.385	17.029
30	900	5.477	17.321
31	961	5.568	17.607
32	1024	5.657	17.889
33	1089	5.745	18.166
34	1156	5.831	18.439
35	1225	5.916	18.708
36	1296	6	18.974
37	1369	6.083	19.235
38	1444	6.164	19.494
39	1521	6.245	19.748
40	1600	6.325	20
41	1681	6.403	20.248
42	1764	6.481	20.494
43	1849	6.557	20.736
44	1936	6.633	20.976
45	2025	6.708	21.213
46	2116	6.782	21.448
47	2209	6.856	21.679
48	2304	6.928	21.909
49	2401	7	22.136
50	2500	7.071	22.361